THE CHEMISTRY OF
NONAQUEOUS SOLVENTS

Volume VA
PRINCIPLES AND BASIC SOLVENTS

Contributors

STEN AHRLAND

DENISE BAUER

ROBERT L. BENOIT

PHILIPPE GAILLOCHET

MARCEL JOZEFOWICZ

CHRISTIAN LOUIS

JEAN-MAXIME NIGRETTO

E. PUNGOR

K. TÓTH

STANLEY WAWZONEK

PETR ZUMAN

THE CHEMISTRY OF
NONAQUEOUS SOLVENTS

Edited by J. J. LAGOWSKI

DEPARTMENT OF CHEMISTRY
THE UNIVERSITY OF TEXAS AT AUSTIN
AUSTIN, TEXAS

Volume VA

PRINCIPLES AND BASIC SOLVENTS

1978

ACADEMIC PRESS   New York   San Francisco   London
A Subsidiary of Harcourt Brace Jovanovich, Publishers

ACADEMIC PRESS, INC.
111 Fifth Avenue, New York, New York 10003

United Kingdom Edition published by
ACADEMIC PRESS, INC. (LONDON) LTD.
24/28 Oval Road, London NW1 7DX

Library of Congress Cataloging in Publication Data

Lagowski, J J   ed.
    The chemistry of nonaqueous solvents.

    Includes bibliographies.
    CONTENTS: v. 1. Principles and techniques. --
v. 2. Acidic and basic solvents. --v. 3. Inert,
aprotic, and acidic solvents. [etc.]
    1. Solvents. I. Title.
TP247.5.L3   660.2'2'482   66-16441
ISBN 0-12-433805-4 (v. 5A)

Contents

1. Solvation and Complex Formation in Protic and Aprotic Solvents
STEN AHRLAND

2. Solvent Basicity
ROBERT L. BENOIT AND CHRISTIAN LOUIS

List of Contributors

Numbers in parentheses indicate the pages on which the authors' contributions begin.

STEN AHRLAND, Inorganic Chemistry 1, Chemical Center, University of Lund, Lund, Sweden (1)

DENISE BAUER, Laboratoire de Chimie Analytique, École Supérieure de Physique et de Chimie de Paris, Paris, France (251)

ROBERT L. BENOIT, Département de Chimie, Université de Montréal, Montréal, Quebec, Canada (63)

PHILIPPE GAILLOCHET, Laboratoire de Chimie Analytique, École Supérieure de Physique et de Chimie de Paris, Paris, France (251)

MARCEL JOZEFOWICZ, Department of Chemistry, University of Paris, Villetaneuse, France (179)

CHRISTIAN LOUIS, Département de Chimie, Université de Montréal, Montréal, Quebec, Canada (63)

JEAN-MAXIME NIGRETTO, Department of Chemistry, University of Paris, Villetaneuse, France (179)

E. PUNGOR, Institute for General and Analytical Chemistry, Technical University, Budapest, Hungary (145)

K. TÓTH, Institute for General and Analytical Chemistry, Technical University, Budapest, Hungary (145)

STANLEY WAWZONEK, Department of Chemistry, University of Iowa, Iowa City, Iowa (121)

PETR ZUMAN, Department of Chemistry, Clarkson College of Technology, Potsdam, New York (121)

Preface

Volume V of this treatise completes the original plan established in 1965 to attempt a critical analysis of the subject from two points of view, viz., (1) a discussion of the theoretical aspects of nonaqueous solution chemistry independent of solvent and (2) a consideration of individual solvents or solvent types for which reasonably comprehensive information has been gathered to this point in time. Chapters 1–4 in this volume contribute to the first point of view whereas the remaining chapters, for the most part, contain information on individual solvent systems; there is, of course, some discussion of special aspects of theory in the latter class of chapters.

Taken as a whole, the 38 chapters in the 5 volumes of this treatise would have been organized according to the following outline, if the Editor had been able to overcome the logistic problems associated with the personal work schedules of the contributors.

The Chemistry of Nonaqueous Solvents

Part I.   Practical Aspects

Part II.   Theoretical Aspects

The division of some chapters between Parts I and II in the outline is somewhat arbitrary since many chapters in Part I are organized about well-established or evolving principles.

I should like to acknowledge the help of Ms. R. Schall who assisted in numerous ways during the preparation of Volume V. The cooperation of the staff of Academic Press in many ways since the inception of this treatise has been outstanding. Finally, the contributions of the numerous authors, both in terms of the manuscripts they produced and their numerous suggestions are gratefully appreciated. It is apparent that a very large number of persons contributed materially to the success of this effort since its inception in 1965.

J. J. LAGOWSKI

Contents of Other Volumes

VOLUME I   PRINCIPLES AND TECHNIQUES

VOLUME II   ACIDIC AND BASIC SOLVENTS

~ 1 ~

Solvation and Complex Formation in Protic and Aprotic Solvents

∽

STEN AHRLAND

Inorganic Chemistry 1, Chemical Center
University of Lund, Lund, Sweden

I. Solvation, Solubility, and Complex Stability

A. Lattice versus Solvation Energies

Due to strong electrostatic interactions, the formation of ionic crystals from gaseous ions is always a strongly exothermic reaction, i.e., the lattice enthalpies ΔH_{lat}^0 are always large and negative. On the other hand, crystallization always implies an increase of order so the lattice entropies ΔS_{lat}^0 are always negative, thus counteracting the reaction. At normal temperatures, and even more at low temperatures, however, ΔH_{lat}^0 strongly predominates over $T\Delta S_{lat}^0$ so the lattice free energy

$$\Delta G_{lat}^0 = \Delta H_{lat}^0 - T\Delta S_{lat}^0 \tag{1}$$

is virtually determined by the value of ΔH_{lat}^0. For 1 : 1 electrolytes at the standard temperature of 25°C, the value of $T\Delta S_{lat}^0$ hardly exceeds 10% of ΔH_{lat}^0 (Table I). The same certainly applies to ionic compounds of other types.

In order that a substance be soluble in a certain solvent, ΔG_{lat}^0 must be compensated by the solvation free energy, ΔG_{sv}^0. In the simple case of a fully ionized 1 : 1 electrolyte, ML, ΔG_{sv}^0 is the sum of the solvation free energies of the cation and the anion

$$\Delta G_{sv}^0 = \Delta G_{sv}^0(M) + \Delta G_{sv}^0(L) \tag{2}$$

The difference

$$\Delta G_s^0 = \Delta G_{sv}^0(M) + \Delta G_{sv}^0(L) - \Delta G_{lat}^0 \tag{3}$$

is the solution free energy which is directly related to the solubility product K_s:

$$\Delta G_s^0 = -RT \ln K_s \tag{4}$$

A value of $K_s = 10^{-4}\ M^2$ which means a perceptible though still low solubility corresponds to a value of $\Delta G_s^0 \approx 23$ kJ at 25°C; more soluble salts have values of ΔG_s^0 less than that.

Also, in the case of ΔG_{sv}^0, the enthalpy term ΔH_{sv}^0 greatly predominates over the entropy term $T\Delta S_{sv}^0$ (see Table I). Because the solvation of gaseous ions implies an increase of order, ΔS_{sv}^0 is always negative and counteracts the reaction, just as ΔS_{lat}^0.

Energetically, the extensive neutralization of electric charges is the main feature of the formation of both crystals and solvates from gaseous ions. If the compound is at all soluble, ΔH_{lat}^0 and ΔH_{sv}^0, constituting the main parts of ΔG_{lat}^0 and ΔG_{sv}^0, respectively, therefore have to be of much the same

TABLE I

ENERGY TERMS CONNECTED WITH THE SOLUBILITY OF ALKALI HALIDES IN WATER AND PROPYLENE CARBONATE[a]

Compound	Water							PC[i]				
	$-\Delta H_{lat}^0$ [b]	$-T\Delta S_{lat}^0$ [c]	$-\Delta H_{sv}^0$ [d]	$-T\Delta S_{sv}^0$ [e]	ΔG_s^0 [f]	ΔH_s^0 [g]	ΔS_s^0 [h]	$-\Delta H_{sv}^0$	$-T\Delta S_{sv}^0$	ΔG_s^0	ΔH_s^0	ΔS_s^0
LiF	1040	78	1036	88	14.2	4.6	-32	—	—	96.2	-7.1	—
LiCl	861	73	899	70	-40.6	-37.2	11	869	102	22.1	-28.5	-98
LiBr	819	72	869	66	-55.6	-49.4	21	848	99	-5.4	-28.5	-77
LiI	762	69	825	57	-75.3	-63.2	41	825	101	-31.4	-63.0	-106
NaF	923	78	923	82	4.2	0.4	-13	—	—	761	—	—
NaCl	787	73	783	60	-8.8	3.8	42	761	83	43.9	25.5	-62
NaBr	752	72	753	56	-16.3	-0.8	52	741	89	28.5	11.7	-56
NaI	702	70	710	49	-28.9	-7.9	70	723	95	4.2	-21.1	-85
KF	822	75	839	71	-21.8	-17.6	14	—	—	41.4	—	—
KCl	717	72	700	51	-4.6	17.2	73	694	88	39.3	23.4	-53
KBr	689	72	669	46	-5.9	20.1	87	673	85	29.7	16.3	-45
KI	646	68	626	38	-9.6	20.5	102	651	87	13.8	-4.6	-62
RbF	781	68	807	56	-38.5	-26.4	41	—	—	15.5	—	—
RbCl	687	64	671	39	-8.4	16.7	84	668	71	26.4	20.1	-21
RbBr	661	64	640	34	-7.9	21.8	100	647	67	18.4	14.6	-13
RbI	624	62	598	27	-8.4	25.9	115	604	50	7.9	-1.9	-33
CsF	731	56	769	35	-58.6	-37.7	70	—	—	-13.8	—	—
CsCl	659	53	641	26	-8.8	18.0	90	640	51	16.7	19.2	8
CsBr	636	49	610	20	-2.9	25.9	97	619	47	14.6	16.7	7
CsI	602	48	569	15	-0.4	33.1	112	597	51	6.7	4.2	-8

[a] Data given in kJ at 25°C; for ΔS_s^0 in JK^{-1}.

[b] M. Salomon, J. Phys. Chem. 74, 2519 (1970); much the same values (except for the cesium halides) in g.

[c] From Eq. 1, ΔG_{lat}^0 from ref. b.

[d] From Eq. 5.

[e] $T\Delta S_{sv}^0 = \Delta H_{sv}^0 - \Delta G_{sv}^0$; ΔG_{sv}^0 from ref. b.

[f] From Eqs. 2 and 3.

[g] D. F. C. Morris, Struct. Bonding (Berlin) 4, 63 (1968).

[h] $\Delta S_s^0 = (\Delta H_s^0 - \Delta G_s^0)/T$.

[i] Calculated from ΔH_{sv}^0, ΔG_{sv}^0, etc., given in ref. b.

order of magnitude. For the same reason, ΔH^0_{sv}, and consequently ΔG^0_{sv}, for a certain salt cannot change very much between solvents that really do dissolve it, even if the solvents differ greatly in other ways. This is amply illustrated by the values of ΔH^0_{sv} for the alkali halides in the protic solvent water and the aprotic solvent propylene carbonate (PC) (see Table I).

Consequently, the solution free energy ΔG^0_s, Eq. 3, is small relative to ΔG^0_{lat} and ΔG^0_{sv}, and the solution enthalpy

$$\Delta H^0_s = \Delta H^0_{sv} - \Delta H^0_{lat} = \Delta H^0_{sv}(M) + \Delta H^0_{sv}(L) - \Delta H^0_{lat} \qquad (5)$$

is small relative to ΔH^0_{lat} and ΔH^0_{sv}. Furthermore, the energies of transfer between various solvents, ΔG^0_{tr} and ΔH^0_{tr}, are also small relative to ΔG^0_{sv} and ΔH^0_{sv}.

Thermodynamically, only the overall functions ΔG^0_{sv}, ΔH^0_{sv}, and ΔS^0_{sv} referring to a certain electrolyte can be calculated. The contributions from the individual ions making up these sums (Eqs. 2 and 5) cannot be determined except by the introduction of some extrathermodynamic assumption. The fundamental reason for this is that any part of the solution extending for more than a few interionic distances has to remain electrically neutral. Several assumptions that would allow the calculation of quantities referring to individual ions have been proposed, however, and some of them yield very concordant results. This strongly indicates that reasonably correct values are in fact obtained. These will be fully discussed below.

For gaseous nonionic compounds, crystallization and solvation does not involve the neutralization of net charges. The lattice and solvation enthalpies and the free energies are therefore generally much lower than for electrolytes. For polar molecules they may nevertheless be quite sizable and also of a mainly electrostatic character. The values of ΔH_s and ΔG_s might be as large for neutral as for ionic compounds which means that they are in the former instance generally a much larger fraction of the lattice or solvation enthalpies and free energies, respectively.

For neutral molecules and complexes the thermodynamic functions can be calculated without any extrathermodynamic assumptions. Furthermore, as the energy terms involved are both fewer and often also better known, the accuracy is often considerably higher than can be achieved for ionic compounds.

Even if, as pointed out, the values of ΔG^0_s are always small relative to ΔG^0_{lat} and ΔG^0_{sv}, they nevertheless determine the solubilities. The energy differences characterizing the dissolution of soluble and slightly soluble compounds are in fact not very large, as is immediately evident from Eq. 4. Between values of K_s of 10^{-2} M^2 and 10^{-12} M^2, representing fairly soluble and fairly insoluble compounds, respectively, the difference in ΔG^0_s does not amount to more than 57 kJ at 25°C. Evidently, the various values of ΔG^0_s

listed in Table I refer to solubilities of very different order of magnitude. Moreover, a preferential solvation in one of the solvents of one or the other of the ions involved may bring about a large increase of solubility even if the corresponding increase of $-\Delta H_{sv}$ amounts to only a few percent.

For the 1 : 1 electrolytes listed in Table I, ΔH_s^0 is positive about as often as it is negative. Definite though fairly intricate trends are discernable and are further discussed in Section III.

The values of ΔS_s^0 are, on the other hand, almost always positive in water but negative in PC. Exceptions are LiF and NaF in water and (where the fluorides are too insoluble to be measured) CsCl and CsBr in PC. Evidently, dissolving crystalline alkali halides in water generally means a decrease of order, in PC generally an increase of order. As is more fully discussed in Section II,C this certainly partly depends upon the fact that water is a more structured solvent than PC.

The formation of ionic solvates in water therefore implies a net increase of order only in those two cases where the very strongly structuring F^- acts together with cations which are also fairly strongly solvated, namely, Li^+ and Na^+. In the less structured PC, on the other hand, the formation of solvates generally implies a net increase of order, except in cases where the exceptionally weakly solvated Cs^+ is present.

The net values of ΔS_s^0 are also much influenced by differences between water and PC in the solvation of the individual ions, as is further discussed in Section III.

B. Solvates versus Complexes

So far, solvation has only been discussed for electrolytes of the simple type exemplified by the alkali halides in Table I. For these solvation is strong enough, at least in water, to prevent the formation of complexes between the constituent ions. Also, in other strongly solvating solvents of high dielectric constant like PC (cf. Sections II and III), this is most probably the case, at least in not too concentrated solutions, though the evidence on this point is still scarce and somewhat conflicting.[1]

For other metal ions, however, the tendency to coordinate anions is often so strong that these enter the inner coordination sphere of the metal ion as ligands, displacing solvent molecules in the process. The solvent molecules might also be displaced by other neutral molecules of stronger affinity to the metal ion. Such substitution reactions are generally termed complex formation, more specifically in the inner sphere. To what extent these substitutions take place evidently depends very much upon the strength of the solvation of all species involved. The stabilities of the complexes formed are, to a very high degree, a function of the stabilities of the solvates to be broken up. As the latter is likely to vary significantly between different solvents, and moreover in quite different ways for different species, the complex stabilities must also vary quite considerably between different solvents.[2]

It should be stressed, however, that not only the metal ions, but also the ligands, are solvated. For neutral ligands, especially of low polarity, the solvation is of course weak. Anionic ligands are, on the other hand, solvated at least as strongly as metal ions of the same ratio of charge to radius, as is further discussed in Section III,A. Though this ratio is generally higher and the solvation therefore stronger for metal ions, the solvation of especially anionic ligands is by no means negligible. The formation of a complex in solution is therefore not only a competition for the metal ion between the ligand and the solvent, but also a competition for the ligand between the metal ion and the solvent.

In the simplest case, the formation of a complex does not involve any major change of the coordination figure of the metal ion. Solvent molecules are just replaced by ligands. Quite often, however, a complete disruption of the structure occurs. A typical case in point occurs when mercury(II) coordinates halide ions.[3–4a,74,78] Octahedral solvates are then first transformed into linear halide complexes, HgL_2, and finally, at high ligand concentrations, into tetrahedral complexes, HgL_4^{2-}. Changes from octahedral to tetrahedral coordination take place in the zinc and cadmium halide systems, as is further discussed in Section IV,B.

The solubilities of neutral complexes like HgL_2 again depend upon the balance between the lattice and the solvation free energies. For $HgCl_2$ in aqueous solution, ΔG_{sv} is large enough to bring about a fairly high solubility in spite of the fact that the solvating water molecules are now relegated to the outer coordination sphere of the metal ion. For HgI_2, on the other hand, ΔG_{sv}^0 is so small relative to ΔG_{lat}^0 that the complex is only very slightly soluble in water. In a strongly dipolar aprotic solvent such as dimethyl sulfoxide (DMSO), where the solvation of metal ions and metal complexes tends to be stronger than in water (cf. Section III, Tables III–V), not only $HgCl_2$ but also HgI_2 is very soluble.[4] Similar increases of solubilities in dipolar aprotic solvents relative to protic ones are observed for many other compounds, e.g., for the neutral copper(I) complexes[5,6] CuCl, CuBr, CuI, and CuSCN. Aprotic solvents therefore allow a straightforward investigation of many systems which present great difficulties in aqueous solution because of their very low solubilities.

Solvents of different types might also prefer different oxidation states of an element. Consequently, oxidation states which are unstable, e.g., in water, are sometimes quite stable in aprotic solvents. A case in point is the enormous stabilization of copper(I)[5] and even more of gold(I)[7] in DMSO relative to water. At least for copper(I), the stabilization is even more marked in acetonitrile[6] (AN), also an aprotic solvent but solvating via nitrogen which means that special affinities come into play.

C. Solvolysis

The metal ion solvates formed in protic solvents generally act as more or less strong Brønsted acids, able to transfer protons to the solvent molecules or to other bases present. This is obviously not possible for the solvates formed in aprotic solvents. Consequently, solvolytic reactions do not occur in such media, which is a most important difference from the conditions prevailing in protic solvents. Many reactions and investigations are therefore feasible in aprotic solvents which cannot be realized, e.g., in water where hydrolytic reactions completely predominate the chemistry of many metal acceptors.

D. Conclusions on the Influence of the Solvent

Evidently vast differences exist between the solvating properties of protic and aprotic solvents, and also between solvents coordinating via oxygen and solvents coordinating via more selective donor atoms, such as nitrogen. The latter two categories are, in the following sections, referred to as oxic and anoxic solvents, respectively. The differences in solvating properties in turn cause drastic variations between different solvents in the solubilities and complex formation of various electrolytes, as well as in the stabilities of various oxidation states. In order to elucidate these influences, the behavior of a number of suitably chosen ions and complexes should be compared in solvents representative of the various categories discerned.

II. SELECTION AND PROPERTIES OF SOLVENTS TO BE COMPARED

A. Solvent Bonding Modes

In a good solvent for electrolytes, the neutralization of ionic charges by solvation has, as was stressed in Section I, to be about as effective as in ionic crystals. In order to achieve this, the solvent has to be strongly polar. The ion-dipole interactions may be further strengthened if the solvent molecules are capable of specific interactions with the ions. One mode of such selective bonding takes place via the hydrogen atoms of protic solvents to ligand atoms of high negative charge density. Very different kinds of selective bonds, of a markedly covalent character, are formed between soft metal ions and solvents coordinating via more or less soft ligand atoms, such as N, or, more typically, S. In certain solvents, which might be called "multifunctional," several of the factors mentioned act simultaneously as, e.g., in the protic as well as highly dipolar formamide which may coordinate via N as well as via O. In such complicated cases, the effects of the various influences is indeed difficult to estimate, especially as these vary considerably between different solutes. Multifunctional solvents are, as should be expected, very versatile, dissolving solutes of very different characteristics. They are therefore potentially useful from a practical point of view though in the case of formamide the usefulness is restricted by the inherent instability of the solvent molecule.[8]

Strongly hydrogen-bonding protic solvents should be particularly efficient in dissolving salts of anions prone to participate in such bonding. Among these are the fluoride ion and ions coordinating via oxygen atoms (sulfate, carboxylates, etc.) and also, though less marked, the chloride and bromide ions. From the values of ΔG_s^0 quoted in Table I it is evident that this is really the case. The fluorides especially are indeed much more soluble in the protic solvent water than in the aprotic solvent PC, while the difference is smallest for the iodides.

In oxygen-containing protic solvents like water and alcohols, strong hydrogen bonds are also formed between the solvent molecules, which very much increases the structural order. Primarily this is reflected in dielectric constants ε which are very high relative to the dipole moments in the gas phase μ (Table II). Especially in the case of water, where hydrogen bonding is exceptionally strong, the melting and boiling points are also abnormally high, as is well known.

Strong hydrogen bonds between the solvent molecules might influence the solvation, and hence the solubility, complex formation, etc., in several different ways. First, this bonding involves strong competition with the cations present for the coordinating oxygen atoms of the solvent which might well result in a weaker solvation than in, e.g., a dipolar aprotic solvent. As will be shown in Section III,A this is in fact the case for most cations. For the same reason, those anions which do not form hydrogen bonds tend to be more weakly solvated in protic than in dipolar aprotic solvents. Contrary to this, those anions which do form hydrogen bonds of fair strength are, as has been stressed above, more strongly solvated in protic than in aprotic solvents.

Though the hydrogen bonding thus counteracts the solvate formation by making ΔH_{sv}^0 less favorable, it will on the other hand promote it by making

ΔS_{sv}^0 more favorable which will be a consequence of the high degree of structural order imposed already on the bulk solvent.

Finally, the high dielectric constant ensuing as the result of increased structural order decreases the electrostatic interactions between the ions present. This should bring about a higher solubility and also a decreasing tendency for complex formation.

It should be pointed out, however, that the various effects mentioned are indeed closely interwoven. For one thing, the dielectric properties in the spheres of action of the ionic charges are certainly very different from those of the bulk solvent. This may be expressed by introducing effective dielectric constants varying with the solute, and also with the distance between the charges, but always considerably lower than the macroscopic constant valid for the bulk solvent.[9] Just because of the structuring influence of the ionic charges, solutions of electrolytes cannot be regarded as dielectric continua.

B. Solvents Selected

For the solvents selected, enough information has to be available at the standard temperature of 25°C to allow extensive comparisons of solvation and complex formation. All protic solvents fulfilling this condition are oxic, though some also contain further coordinating atoms as, e.g., formamide (cf. Section II,A). Among the anoxic protic solvents, ammonia has been very extensively investigated. Quite naturally, most of the data refer to low temperatures in the neighborhood of the boiling point, −33.4°C, and an immediate comparison with other solvents is therefore not possible. Ammonia has therefore been only briefly discussed, especially as exhaustive recent reviews of its solvent properties are available.[10]

The choice among the protic solvents available should in the first hand be governed by the consideration that very different hydrogen bonding capacities ought to be represented. As a protic solvent forming exceptionally strong hydrogen bonds, water is the obvious choice, as it is both very typical and very important. Methanol, also fairly extensively investigated, will be chosen as a protic solvent forming much weaker hydrogen bonds than water.

Among the strongly polar aprotic solvents, the oxic propylene carbonate and dimethylsulfoxide and the anoxic acetonitrile have been selected. For all these, considerable amounts of pertinent data exist. This is obviously much due to their convenient liquid ranges (see Table II, part A) which allow reasonably easy handling at atmospheric pressure and standard temperature, 25°C. No aprotic solvent coordinating via donor atoms other than O or N has as yet been so thoroughly investigated as to warrant any extensive

TABLE II

PHYSICAL PROPERTIES[a] OF THE SOLVENTS DISCUSSED, INCLUDING ESTIMATES OF THE LONDON, ΔH_{vL}^0, AND DIPOLE, ΔH_{vD}^0, CONTRIBUTIONS TO THE VAPORIZATION ENTHALPY $\Delta H_v^{0\ b}$

Solvent	MW	C_s (M)	Mp (°C)	Bp (°C)	μ (D)	ε	ΔH_v^0 (kJmole^{-1})	ΔH_{vL}^0 (kJmole^{-1})	ΔH_{vD}^0 (kJmole^{-1})	ΔS_v^0 (JK^{-1}mole^{-1})
A. Solvents selected for comparison of solvation and complex formation (see Tables III–XV)										
H_2O	18.02	55.3	0.0	100	1.85	78.5	44.0[h]	5	39	119[h]
CH_3OH	32.04	24.6	−97.7	64.7	1.70	32.6	37.4[h]	11	26	111[h]
PC[c]	102.09	11.7	−49.2[d]	241.7[d]	4.98[e]	64.4[d]	59.6[h]	29	31	120[o]
$(CH_3)_2SO$	78.13	14.0	18.5[f]	189.0[f]	3.96[f]	46.4[f]	52.9[i]	24	29	118[o]
CH_3CN	41.05	18.9	−43.8	81.6	3.92	36.0[g]	34.7[h]	14	21	99[h]
B. Solvents selected for further discussion of connections between physical properties and chemical structure										
H_3N	17.0	—	−77.7	−33.4	1.47	16.9	19.9[j]	6	14	86[m]
C_2H_5OH	46.1	—	−114.1	78.3	1.69	24.3	42.3[h]	16	26	121[h]
$(CH_3)_2O$	46.1	—	−138.5	−23	1.30	5.02	22.6[k]	16	7	78[p]
$(C_2H_5)_2O$	74.1	—	−116.3	34.6	1.15	4.25	29.1[k]	27	2	95[q]
C_5H_{10}	70.1	—	−93.9	49.3	0	1.96	28.6[l]	—	—	89[l]
C_6H_{12}	84.2	—	6.5	80.7	0	2.02	33.1[l]	—	—	94[l]
CCl_4	153.8	—	−23.0	76.8	0	2.23	32.6[h]	—	—	95[h]
BCl_3	117.2	—	−107.3	12.5	0	≈2	23.0[h]	—	—	81[h]
BBr_3	250.5	—	−46.0	91.3	0	2.51	34.3[h]	—	—	95[h]

[a] From J. A. Riddick and W. A. Bunger, *in* "Technique of Chemistry" (A. Weissberger, ed.), 3rd ed. Vol. 2 Wiley (Interscience), New York, 1970, or (primarily for values of μ and ε) from R. C. Weast, ed., "Handbook of Chemistry and Physics," 53rd ed. Chem. Rubber Publ. Co, Cleveland, Ohio, 1972–1973, if not stated otherwise.

[b] Values of μ, ε, ΔH_v^0, and ΔS_v^0 refer to the standard state, 25°C, 1 atm, if not otherwise stated.

[c] Propylene carbonate, $H_3C \cdot CH-O-CO-O-CH_2$.

[d] T. Fujinaga and K. Izutsu, *J. Pure Appl. Chem.* **27**, 273 (1971).

[e] C. M. Criss and M. Salomon, *in* "Physical Chemistry of Organic Solvent Systems," (A. K. Covington and T. Dickinson, eds.). p. 293. Plenum, New York, 1973.

[f] W. L. Reynolds, *Prog. Inorg. Chem.* **12**, 1 (1970).

[g] M. Spiro, *in* "Physical Chemistry of Organic Solvent Systems" (A. K. Covington and T. Dickinson, eds.), p. 638. Plenum, New York, 1973.

[h] "Selected Values of Physical and Thermodynamic Properties," Nat. Bur. Stand., Circ. No. 500. U.S. Govt. Printing Office, Washington D.C., 1952.

[i] T. B. Douglas, *J. Am. Chem. Soc.* **70**, 2001 (1948).

[j] At 25° and saturation pressure, 9.90 atm, "Handbook," p. E21 (see note a). At the boiling point, $-33.4°$C and 1 atm, $\Delta H_v^0 = 23.4$ kJ mole^{-1}, J. J. Lagowski and G. A. Moczygemba, *in* "The Chemistry of Nonaqueous Solvents" (J. J. Lagowski, ed.), Vol. 2, Chapter 7. Academic Press, New York, 1967.

[k] "Handbook," p. D153 (see note a).

[l] "Selected Values of Physical and Thermodynamic Properties of Hydrocarbons and Related Compounds," Am. Pet. Inst. Res. Proj. No. 44. Carnegie Press, Pittsburgh, Pennsylvania, 1953.

[m] For 25°C, 1 atm, by combination of ΔH_v^0 and $\Delta G_v^0 = -RT \ln p$, on the assumption that the activity of the saturated vapor is approximately equal to its pressure, and that ΔH_v^0 does not change appreciably within the pressure range applied. At the normal boiling point, $\Delta S_v^0 = 97.4$ JK^{-1} mole^{-1} (see note j).

[n] G. Olofsson and I. Olofsson, *J. Am. Chem. Soc.* **95**, 7231 (1973).

[o] Calculated as described in note m, with $p = 0.053$ torr for PC and 0.599 torr for DMSO at 25°C, determined by S. Y. Lam and R. L. Benoit, *Can. J. Chem.* **52**, 718 (1974). The latter value agrees with that earlier found by Douglas, namely, $p = 0.600$ torr (see note i).

[p] R. M. Kennedy, M. Sagenkahn, and J. G. Aston, *J. Am. Chem. Soc.* **63**, 2267 (1941).

[q] Calculated as described in note m, with $p = 533.8$ torr at 25°C, determined by R. S. Taylor and L. B. Smith, *J. Am. Chem. Soc.* **44**, 2450 (1922).

comparison. Some interesting measurements[11] have been performed with
N,N-dimethylthioformamide, however, which amply confirm the strong and
selective solvation of soft metal ions like Cu^+ and Ag^+ by solvents coordinating via soft donor atoms, like S.

C. Properties Connected with the Solvent Structure

As to the solvent properties, those directly connected with the dipole
character and the structural order of the solvents are of prime importance.
These are the dipole moment in the gas phase, μ, the dielectric constant, ε,
the vaporization enthalpy, ΔH_v^0, and the vaporization entropy, ΔS_v^0, which
have all been listed in Table II.

The high degree of structural order due to hydrogen bonding in water is,
as pointed out above, reflected in an exceptionally high ratio ε/μ. In
methanol where the hydrogen bonding is much weaker, ε is much lower in
spite of the fact that μ is almost as high as for water. Also for methanol,
however, the ratio ε/μ is considerably higher than for any of the aprotic
solvents. Among these the two oxic solvents have a somewhat higher ratio
than the anoxic acetonitrile (see Table II, part A).

The forces acting between the solvent molecules are directly measured by
the values of ΔH_v^0 which is the energy required to tear apart the molecules
of the liquid state. The resultant interaction is partly due to the London
forces, originating from dispersion interactions between all atoms of solvent
molecules in contact, and partly to the dipole forces, also including, in the
case of protic solvents, the hydrogen bonding. For the internal order of the
solvent the latter forces are mainly responsible. It is therefore of primary
interest to devise a procedure allowing an estimation of the part of ΔH_v^0 due
to these forces, ΔH_{vD}^0. This has been done by subtracting from ΔH_v^0 the
part due to the London forces, ΔH_{vL}, which in turn have been evaluated
empirically in the following way.

In nonpolar solvents, the London forces are the only ones acting between
the molecules. By comparing the values of ΔH_v^0 for suitably chosen nonpolar solvents it is therefore possible to estimate the London contributions due
to particular atoms or groups of atoms. In order to make the values assigned
consistent, a few more polar solvents have also been considered. All solvents
employed are listed in Table II, part B.

From the values of ΔH_v of C_5H_{10} and C_6H_{12}, each CH_2 group is found to
give a London contribution of ≈ 5.5 kJ. Much the same increase of ΔH_v,
namely, ≈ 5 kJ, is also found between CH_3OH and C_2H_5OH. This is to be
expected as these solvents have practically the same value of μ and ought
moreover to have about the same hydrogen bonding capacity. The difference

in ΔH_v^0 should therefore mainly be caused by the dispersion interaction due to another CH_2 group. The increase per CH_2 group between $(CH_3)_2O$ and $(C_2H_5)_2O$ is smaller, but this might essentially be due to the perceptibly lower μ of the latter molecule. The London contribution in H_2O should be much the same as for CH_2, presumably somewhat smaller, ≈ 5 kJ, on account of the lower polarizability of O relative to C. For NH_3, ≈ 6 kJ seems to be a reasonable estimate. For OH and O, contributions ≈ 4 kJ cannot be far off the mark. A higher value is expected for the much larger and hence more easily distorted Cl. From the values of ΔH_v for the tetrahedral CCl_4 and the trigonally planar BCl_3, both certainly only with the Cl atoms in contact with neighbor molecules, the contribution from Cl is found to be ≈ 8 kJ. Approximately the same value should apply to S. On the even more plausible assumption that in the likewise trigonally planar BBr_3 only the Br atoms are in contact with other molecules, a reasonable London contribution of ≈ 11.5 kJ is calculated for Br.

The sums of these contributions yield the values of ΔH_{vL} listed in Table II. By subtraction from ΔH_v^0, the values of the wanted quantity ΔH_{vD} listed in the same Table are finally found.

By far the highest value of ΔH_{vD}^0, 39 kJ, is found for water, implying that this solvent indeed possesses the highest degree of structural order among all those discussed. The strong hydrogen bonding in water thus more than compensates the much higher dipole moments of the oxic aprotic solvents PC and DMSO which have, however, the next largest values of ΔH_{vD}^0, namely, 31 and 29 kJ, respectively. These values are somewhat larger than those found for the hydrogen-bonding methanol and ethanol. Evidently, the hydrogen bonds between the alcohol molecules, considerably weaker than those formed in water, cannot fully compensate the much higher dipole moments of PC and DMSO. For the anoxic AN, the value of ΔH_{vD}^0 seems to be lower, ≈ 21 kJ, but is nevertheless still high, as is to be expected from the high value of μ. Ammonia is an especially interesting case, being a protic as well as an anoxic solvent. Its value of $\Delta H_{vD} = 14$ kJ is certainly higher than would be expected for a solvent of this low value of μ forming no hydrogen bonds. This is evident from the values of ΔH_{vD}^0 calculated for the aprotic dimethyl and diethyl ethers which are, as expected, very low. On the other hand, the hydrogen bonding is evidently weaker than in the other protic solvents, especially much weaker than in water. This of course confirms the well-known fact that much stronger hydrogen bonds are generally formed to oxygen than to nitrogen. Nevertheless, ΔH_{vD}^0 constitutes the larger part, 70%, of ΔH_v^0 for ammonia. This is about the same as for methanol and exceeded only by the figure found for water, 89%—which is indeed high. For the aprotic but strongly polar PC, DMSO, and AN, the figures are 52, 55, and 60%, respectively. For the polyatomic PC and

DMSO, the contributions from the London forces are so large, however, that the total ΔH_v^0 become the highest ones quoted.

For H_2O and NH_3, the contributions due to dipole and dispersion forces have also been calculated by a theoretical approach.[12] The dipole parts of ΔH_v^0 are found to be 81% and 50%, respectively. Especially in the case of H_2O, these results are fairly compatible with those of the present estimations. It might therefore be inferred that the ΔH_{vD}^0 listed in Table II do give a fairly correct measure of the dipole interaction, and hence of the structural order of the solvents. For solvents containing a fair number of atoms, the method presently employed seems moreover to be the only one available for a rough but still useful estimate of the various contributions to ΔH_v^0.

Strong intermolecular forces in the solvents are also reflected in high boiling points. It should be remembered, however, that while the values of ΔH_v^0 listed reflect the forces acting at the standard temperature, 25°C, this does not apply to the boiling points which is one reason why the two quantities cannot be expected to run quite parallel.

Another reason is that the vaporization entropies, ΔS_v^0, differ between different solvents. A favorable, i.e., strongly positive value of ΔS_v^0 might compensate a high value of ΔH_v^0 so that a vapor pressure of 1 atm, i.e., the boiling point, is reached already at a relatively low temperature. A case in point is ethanol which, in spite of the much higher value of ΔH_v^0, has about the same boiling point as CCl_4 or C_6H_{12}. This is evidently due to the more positive ΔS_v^0 of ethanol.

The formal concentration, C_s, of the solvent naturally decreases with increasing size of the solvent molecule (see Table II) which means that the value of C_s is much higher for water, and also considerably higher for methanol, than for any of the three aprotic solvents discussed. The activity of the solvent does not primarily depend upon C_s, however, but rather on those forces between the solvent molecules which restrict their freedom of orientation and coordination. As pointed out, ΔH_{vD}^0 provides a good measure of these restrictive forces. Among the especially selected solvents in Table II, part A, the difference between the highest and lowest values of ΔH_{vD}^0, for water and AN, respectively, is ≈ 18 kJ. The ratio of solvent concentration between these solvents is 2.9, corresponding to a free energy contribution RT ln 2.9 = 2.6 kJ. Even between the solvents of the highest and lowest concentrations, namely, water and PC, this ratio is not higher than 4.7, corresponding to an energy contribution of 3.8 kJ. It is evident that the free energy contributions due to these concentration differences are generally much smaller than those arising from differences in the dipole forces, including hydrogen bonding. In fact, it might be rather safely concluded that, in spite of its high concentration, water has the lowest solvent activity by far of all the solvents in Table II where values of ΔH_{vD}^0 have been stated. Among the solvents selected for further comparison (see Table II, part A), AN certainly has the highest solvent activity, with methanol and DMSO coming next. Their activity should still be considerably higher than that of water. Even for PC, with $\Delta H_{vD}^0 \approx 31$ kJ, the solvent activity should be somewhat higher than that of water, as the larger structural freedom can hardly be fully compensated by the much lower value of C_s. This larger freedom is, as pointed out in Section I,A, also reflected in the very negative values of ΔS_s^0 observed when alkali halides are dissolved in PC (cf. Table I). In a solvent as little structured as PC, virtually all ions evidently become structure makers.

III. SOLVATION ENTHALPIES OF INDIVIDUAL IONS AND NEUTRAL MOLECULES IN SOME REPRESENTATIVE SOLVENTS

A. Solvation Enthalpies of Individual Ions

As stressed above, the contributions of the individual ions to the solvation enthalpies of salts cannot be found by purely thermodynamic methods. As these quantities reflect important properties of both the ions and the solvents, however, great effort has been made to effect a reasonable partition of the total ΔH_{sv}^0 between the ions involved by means of various extrathermodynamic assumptions. The greatest interest has naturally been devoted to aqueous solutions. The focal point of this discussion has been how to calculate the "best" value of the solvation enthalpy of the proton, $\Delta H_h^0(H^+)$. The number of studies referring to this quantity is truly impressive.[13-16] It is comforting to find that several of the fairly different approaches used end up with a value of $\Delta H_h^0(H^+)$ around 1100 kJ. In this discussion, the value proposed by Morris,[15] $\Delta H_h^0(H^+) = 1103$ kJ, will be adopted. This value is calculated according to the method of Halliwell and Nyburg,[14] with the use of a seemingly more realistic set of ionic radii than that used by the original authors. The difference from the original value,[14] $\Delta H_h^0 = 1091$ kJ, is not larger than the magnitude of the estimated uncertainty, however. It is also interesting to note that these values agree, within the errors stated, with the very first estimation of $\Delta H_h^0(H^+) = 1096$ kJ, due to Fajans[17] in 1919. Once the value of $\Delta H_h^0(H^+)$ has been settled, values of ΔH_h^0 for all other ions can in principle be calculated. In Table IIIA, values of ΔH_h^0 for some representative cations and anions are listed.

As stressed in Section I,A, the solvation enthalpies of ions differ relatively little between different solvents. The differences, called the enthalpies of transfer, ΔH_{tr}^0, nevertheless represent important differences in the ion-solvent interactions. Values of ΔH_{tr}^0 for salts are readily determined as

TABLE IIIA

HYDRATION ENTHALPIES,[a] ΔH_h^0, AND ENTHALPIES OF TRANSFER, $\Delta H_{tr}^0(H_2O \rightarrow S)$, BETWEEN WATER AND SOLVENTS OF DIFFERENT SOLVATING PROPERTIES FOR SOME REPRESENTATIVE CATIONS AND ANIONS; INTERNALLY CONSISTENT CRYSTALLOGRAPHIC IONIC RADII,[b] r_c, COMPARED FOR SOME OF THE LARGEST IONS WITH THEIR STOKES RADII IN THE POORLY SOLVATING SOLVENT NITROMETHANE,[c] $r_s(NM)$

| Ions | r_c | $r_s(NM)$ | $-\Delta H_h^0$ | $\Delta H_{tr}^0(H_2O \rightarrow S)$ | | |
				Methanol[d]	PC[e]	DMSO[e]
H^+	—	—	1103	—	43.9^v	-25.5^v
Li^+	0.93	—	533	-18.9	3.1	-26.4
Na^+	1.17	—	417	-20.4	-10.2	-27.7
K^+	1.49	—	333	-20.0	-21.9	-34.9
Rb^+	1.64	—	307	-19.7	-24.6	-33.5
Cs^+	1.83	—	289	-15.9	-26.8	-30.0
Ag^+	1.12^f	—	483^n	-21.3^u	-16.7^u	-54.0^u
Tl^+	1.48^g	—	341^n	—	—	—
Zn^{2+}	0.75^h	—	2063^p	—	—	-60^p
Cd^{2+}	0.95^h	—	1831^p	—	—	-67^p
Hg^{2+}	1.02^h	—	1845^p	—	—	-76^p
H_4N^+	1.61^i	—	302^r	—	—	—
Me_4N^+	2.6^j	2.4	157^s	-0.2	-16.3	-15.3
Et_4N^+	2.7^j	2.7	127^s	5.9	0.7	-2.8
Pr_4N^+	3.1^j	3.3	121^s	—	11.7	6.5
Ph_4P^+	4.15^k	—	46^t	—	-13.4	-9.3
Ph_4As^+	4.3^k	—	42^t	3.3^q	-14.6^q	-11.9^q
F^-	1.16	—	502	—	—	—
Cl^-	1.64	—	366	7.9	26.4	18.8
Br^-	1.80	—	335	3.8	13.6	3.5
I^-	2.04	—	294	-1.6	-3.3	-12.8
CF_3COO^-	—	—	—	10^x	32.6	—
ClO_4^-	2.3^l	—	—	-4.6	-16.4	-19.2
BPh_4^-	4.05^m	4.1	47^t	3.3^q	-14.6^q	-11.9^q

[a] Hydration enthalpies calculated on the basis of $\Delta H_h^0(H^+) = 1103$ kJ, proposed by D. F. C. Morris, *Struct. Bonding* (Berlin) **4**, 63 (1968). Also, the values of ΔH_h^0 for the alkali metal and halide ions are from this paper (Table 7, Method 1). For other ions, see notes n–t. Enthalpy data in kJ mole^{-1}; radii in Å; at 25°C.

[b] The values for the alkali metal and halide ions are those proposed by D. F. C. Morris (see note a), on the basis of electron density determinations. Hence the values for the other ions have been estimated as described in notes f–m.

[c] J. F. Coetzee and G. P. Cunningham, *J. Am. Chem. Soc.* **87**, 2529 (1965).

[d] G. Choux and R. L. Benoit, *J. Am. Chem. Soc.* **91**, 6221 (1969) (except for Ag^+ and CF_3COO^-, see notes u and x). The value for Li^+ has been corrected so as to be consistent with the values of $\Delta H_{tr}^0(H_2O \rightarrow M)$ given for LiCl and Cl^- (-11.0 and $+7.9$ kJ, respectively).

[e] C. V. Krishnan and H. L. Friedman, *J. Phys. Chem.* **73**, 3934 (1969) (except for H^+, Ag^+, Zn^{2+}, Cd^{2+}, and Hg^{2+}—see notes v, u, and p).

[f] From the interionic distances in the face-centered, cubic AgCl (2.78 Å) and AgBr (2.89 Å); *Struct. Rep.* **8**, 249 (for 1940–1941).

[g] From the interionic distances in the face-centered, cubic TlCl (3.15 Å) and TlBr (3.29 Å); from TlI (3.47 Å) the slightly lower $r_c(Tl^+) = 1.43$ Å is obtained; L. G. Schulz, *J. Chem. Phys.* **18**, 996 (1950).

[h] R. D. Shannon and C. T. Prewitt, *Acta Crystallogr., Sect. B* **25**, 925 (1969). These authors apply the Goldschmidt approach but their values are based on many more experimental data than the original Goldschmidt set. For Ag^+ and Tl^+, the values of r_c are close to those deduced above (notes f and g). It therefore seems reasonable to adopt the values of r_c found for Zn^{2+}, Cd^{2+}, and Hg^{2+} in this way in order to arrive at a consistent set. Recent structure determinations of DMSO solvates have also confirmed that $r_c(Hg^{2+}) - r_c(Cd^{2+}) = 0.07$ Å; S. Ahrland, I. Persson, and M. Sandström, unpublished data.

[i] From the distances N—Cl (3.27 Å), N—Br (3.45 Å), and N—I (3.60 Å) in the face-centered cubic NH_4Cl, NH_4Br, and NH_4I; G. Bartlett and I. Langmuir, *J. Am. Chem. Soc.* **43**, 84 (1921).

[j] For Me_4N^+ from the distances N—Cl (4.35 Å), N—Br (4.39 Å), and N—I (4.57 Å) in Me_4NCl, Me_4NBr, and Me_4NI, see L. G. Hepler, J. M. Stokes, and R. H. Stokes, *Trans. Faraday Soc.* **61**, 20 (1965) and further from the distance N—Cl (4.86 Å) in Me_4NClO_4, see J. D. Cullough, *Acta Crystallogr.* **17**, 1067 (1964). For Et_4N^+ from the distance N—I (4.76 Å) in Et_4NI, see E. Wait and H. M. Powell, *J. Chem. Soc.* p. 1872 (1958). For Pr_4N^+ from the distance N—Br (4.94 Å) in n-Pr_4NBr, see A. Zalkin, *Acta Crystallogr.* **10**, 537 (1957). The resulting radii, consistent with the set of structurally determined halide radii employed here, are lower than the widely quoted values due to R. A. Robinson and R. H. Stokes, "Electrolyte Solutions," 1st ed. Butterworth, London, 1955. It is believed, however, that the present values are physically more significant as they are founded on direct experimental evidence and moreover agree quite well with the Stokes' radii in poorly solvating solvents (see text).

[k] For Ph_4P^+ from the distance P—I (6.19 Å) in Ph_4PI, see T. L. Khocjanova and Yu. Stručkov, *Kristallografiya* **1**, 669 (1956) [as cited in *Struct. Rep.* **20**, 588 (for 1956)]; cf. also E. Grunwald, G. Baughman, and G. Kohnstam, *J. Am. Chem. Soc.* **82**, 5801 (1960). For Ph_4As^+ from the distance As—I (6.35 Å) in Ph_4AsI, see R. C. L. Mooney, *J. Am. Chem. Soc.* **62**, 2955 (1940). The difference 0.16 Å between the distances As—I and P—I in these two isostructural compounds is, as expected, practically the same as between the As—C bonds (1.95 Å) and P—C bonds (1.80 Å).

[l] Various estimates of $r_c(ClO_4^-)$ differ widely. The present value is founded on that quoted by H. F. Halliwell and S. C. Nyburg, *J. Chem. Soc.* p. 4603 (1960). The original value of 2.45 Å has been modified to the present 2.3 Å, however, to account for that general decrease of anionic radii which results from the new approach for their evaluation from electron density measurements (see note b).

[m] The bond B—C in BPh_4^- is $\simeq 1.7$ Å [I. Bertini, P. Dapporto, G. Fallini, and L. Sacconi, *Inorg. Chem.* **10**, 1703 (1971); T. L. Blundell and H. M. Powell, *Acta Crystallogr., Sect. B* **27**, 2304 (1971)], i.e., 0.1 Å and 0.25 Å shorter than P—C in Ph_4P^+ and As—C in Ph_4As^+, respectively (see note k). Hence $r_c(BPh_4^-)$ should be correspondingly shorter than $r_c(Ph_4P^+)$ and $r_c(Ph_4As^+)$.

[n] ΔH_h^0 for the silver and thallium halides from M. Salomon, *J. Phys. Chem.* **74**, 2519 (1970). Hence $\Delta H_h^0(Ag^+)$ and $\Delta H_h^0(Tl^+)$ by subtraction of ΔH_h^0 for the halide ions.

[p] S. Ahrland, L. Kullberg, and R. Portanova, *Acta Chem. Scand., Ser. A* **32**, (1978). In press.

[q] Pairs of values set equal according to the Grunwald assumption are italicized.

[r] $\Delta H_h^0 = \Delta H_{lat}^0 + \Delta H_s^0$ for the ammonium halides from H. F. Halliwell and S. C. Nyburg, *Trans. Faraday Soc.* **59**, 1126 (1963), with the values of ΔH_{lat}^0 recalculated employing the values of r_c in this table.

differences between the values of ΔH_s^0 in the solvents concerned. Just as for ΔH_{sv}^0, however, the values for the individual ions cannot be calculated except by the introduction of an extrathermodynamic assumption. Again, several such assumptions have been proposed. Of these, the Grunwald assumption has become most widely used, as being both very reasonable and easy to apply. The basic postulate is that large monovalent ions of approximately the same size that also show the same face to the outer world interact in the same way with a given solvent molecule. For an electrolyte composed of one cation and one anion of the properties specified, this implies, among other things, that the solvation enthalpies, and also the enthalpies of transfer between different solvents, should be the same for the two constituent ions, namely, half of the total for each. The pair of ions originally suggested by Grunwald et al.[18] were the tetraphenylphosphonium cation, Ph_4P^+, and the tetraphenylborate anion, BPh_4^-. Later, the tetraphenylarsonium ion, Ph_4As^+, has most often been recommended, and chosen, as the cation of the pair.[19-22] The values of ΔH_{tr}^0 listed in Table IIIA have all been calculated on this latter assumption. When other reasonable assumptions are applied, much the same picture emerges, however.[21,22]

A strong indication that the Grunwald assumption is indeed close to the truth is provided by the limiting molar conductivities of Ph_4As^+ and BPh_4^- in AN, determined by measurements of transference numbers.[23] They are very nearly the same, 56.4 and 57.8 cm^2 ohm^{-1} $mole^{-1}$ at 25°C, respectively, which certainly proves that the forces acting between the ions and the solvent molecules must be of much the same strength for both ions.

It might also be investigated whether the Grunwald assumption is consistent with the set of hydration enthalpies given in Table IIIA for the alkali and halide ions. Values of ΔH_s^0 have been measured for several salts $MBPh_4$, Ph_4PL, and Ph_4AsL where M^+ is an alkali and L^- is a halide ion.[16,20,24,25] The values of ΔH_{lat}^0 of these can be calculated by means of an empirical formula given by Halliwell and Nyburg.[26] This formula has been found to reproduce fairly well such values of ΔH_{lat}^0 which can be determined experimentally, i.e., primarily those pertaining to the metal halides where the

---

[s] $\Delta H_h^0 = \Delta H_{lat}^0 + \Delta H_s^0$ for Me_4NCl, Me_4NBr, Me_4NI, Et_4NI, and Pr_4NBr from D. A. Johnson and J. F. Martin, J. Chem. Soc., Dalton Trans. p. 1585 (1973).

[t] Values of ΔH_s^0 for Ph_4PCl, Ph_4AsCl, and Ph_4AsBr from C. V. Krishnan and H. L. Friedman (note e), for Ph_4AsI from E. M. Arnett and D. R. McKelvey, J. Am. Chem. Soc. **88**, 2598 (1966); for $LiBPh_4$ and $NaBPh_4$ from M. Salomon (note n). Values of ΔH_{lat}^0 calculated by means of the formula of H. F. Halliwell and S. C. Nyburg (note r).

[u] B. G. Cox and A. J. Parker, J. Am. Chem. Soc. **95**, 402 (1973).

[v] R. Domain, M. Rinfret, and R. L. Benoit, Can. J. Chem. **54**, 2101 (1976).

[x] Calculated from the values of ΔH_s^0 for trifluoroacetates in water and methanol measured by C. V. Krishnan and H. L. Friedman [J. Phys. Chem. **74**, 2356 (1970); **75**, 388 (1971)], and the values of $\Delta H_{tr}(H_2O \rightarrow M)$ for the alkali ions quoted in this table.

ionization potentials and electron affinities involved in the calculations are sufficiently well known.[15,27] Once ΔH_s^0 and ΔH_{lat}^0 are known, ΔH_{sv}^0 can be calculated from Eq. 5.

The ionic radii r_c used in the calculation of ΔH_{lat}^0 according to Halliwell and Nyburg are listed in Table IIIA. The set adopted is founded on recent determinations of the electron density in alkali halides.[15] From the radii of the alkali metal and halide ions thus found, those of the other ions quoted in Table IIIA have been derived from various structural data, for details see the notes of the table.

Especially for the largest ions discussed, the calculation of r_c is somewhat uncertain, and an independent check of the values arrived at is desirable. Just for these ions, however, such a check can be performed by the application of Stokes' law for the movement of spherical bodies in a fluid medium.[28] Furthermore, as the ions are of low charge and are not only bulky but show very unattractive alkyl and phenyl groups to the outer world, they must be very poorly solvated in most media. The Stokes radii, r_s, found should therefore be about the same as those derived from structural data. In fact the values found in nitromethane[29] for R_4N^+ and BPh_4^-, $r_s(NM)$, are virtually identical with the crystallographic radii (Table IIIA). Only slightly higher values (at most 0.2 Å) are found for R_4N^+ in acetonitrile and nitrobenzene; for BPh_4^- the values of r_s are practically the same in the three solvents.[29] The values of r chosen here seem therefore to be physically significant. This applies not only to those large ions where values of r_s are available, but also to Ph_4P^+ and Ph_4As^+ as the differences in bond distance between the bonds B—C, P—C, and As—C are well known.

Though the values of r thus seem to be fairly reliable, allowing a good estimate of ΔH_{lat}^0 and hence of the total ΔH_h^0 for the various electrolytes employed, the values of ΔH_h^0 for BPh_4^-, Ph_4P^+, and Ph_4As^+ found can nevertheless not be expected to be very accurate. They are so much smaller than the known ΔH_h^0 of all counter ions employed that the relative errors of the differences calculated are fairly large. In view of this, the values actually found, namely, 47, 46, and 42 kJ (Table IIIA), should be considered as confirming fairly well the Grunwald assumption, that these ions have virtually the same ΔH_h^0. Moreover, the small differences observed are in the direction expected from the small but significant differences in ionic radius between the ions. To sum up, the indications are that the Grunwald assumption is valid within about $\pm 5\%$. Consequently, the errors due to this source in the values of $\Delta H_{tr}^0(H_2O \rightarrow S)$ for the individual ions, as listed in Table IIIA, should not exceed 1 or, at most, 2 kJ. In a given solvent, they will, moreover, be the same for all ions of the same charge.

Several of the important differences foreseen between protic and aprotic solvents stand out very clearly in Table IIIA (cf. Section II,A). For the metal ions listed, and also for H^+, the values of $\Delta H_{tr}^0(H_2O \to DMSO)$ are all rather exothermic, indicating a preference for DMSO. An essential reason for this is certainly the sequestering of the water molecules by strong inter-molecular hydrogen bonds which have to be broken on the formation of a solvate. In DMSO, the forces acting between the dipoles are considerably weaker, as discussed in Section II,B. Other factors are of course also at work, however. This is especially evident from the value of $\Delta H_{tr}^0(H_2O \to DMSO)$ for Ag^+ which is very high relative to ΔH_h^0, and also relative to the ionic charge. The values of $\Delta H_{tr}^0(H_2O \to PC)$ are less negative for all these ions, i.e., the values of $\Delta H_{tr}^0(DMSO \to PC)$ are endothermic, indicating a stronger solvation by DMSO than by PC. The difference is especially large for H^+, Li^+, and Ag^+, with $\Delta H_{tr}^0(DMSO \to PC) = 69.4, 29.5,$ and 37.3 kJ, respectively. For the first two ions, $\Delta H_{tr}^0(H_2O \to PC)$ even turns endothermic, and for H^+ very strongly so. In these cases hydration is thus preferred before solvation by PC, in spite of the strong competition for the water molecules offered by the strong intermolecular bonding in aqueous solution. Evidently, these ions have a fairly low affinity for PC.

For ions of the types represented by R_4N^+, Ph_4P^+, Ph_4As^+, and BPh_4^- the values of $\Delta H_{tr}^0(H_2O \to DMSO)$ and $\Delta H_{tr}^0(H_2O \to PC)$ are, on the other hand, not very different. These weakly solvated ions do not strongly prefer one or the other of the two aprotic solvents. Except for Et_4N^+ and Pr_4N^+, they markedly prefer these solvents to water, however. The smaller the tetraalkylammonium ions are, the more strongly they are favored by the aprotic solvents.

Remarkably, ClO_4^- also strongly prefers the aprotic solvents, although this ion certainly shows a very different face to the outer world. The values of $\Delta H_{tr}^0(H_2O \to S)$ prove unequivocally, however, that if the oxygens of this ion do form hydrogen bonds with water, these bonds must in any case be weak. The other oxygen donor listed in Table IIIA, CF_3COO^-, behaves very differently indeed. This ion strongly prefers the protic water to the aprotic PC. Water is also preferred to methanol though $\Delta H_{tr}^0(H_2O \to M)$ is much smaller than $\Delta H_{tr}^0(H_2O \to PC)$. Evidently CF_3COO^- is a strongly hydrogen bonding ligand.

The affinities of the halide ions to the various solvents display especially interesting and revealing patterns. Only I^- is more strongly solvated in DMSO and in PC than in water while the opposite is true for Br^- and, to a much higher degree, for Cl^-. The hydrogen bonds formed in water by the latter ligands are evidently strong enough to make this protic solvent more attractive than either of the two aprotic ones. Of these, DMSO turns out to be more strongly solvating than PC, just as for the metal ions and H^+. The value of $\Delta H_{tr}^0(DMSO \to PC)$ is of the same magnitude, $\simeq 10$ kJ, for Cl^-, Br^-, and I^-. In methanol, the hydrogen bonding is considerably weaker than in water (see Section II,C). The solvation pattern for the halide ions should therefore be intermediate to those found in water, and in the aprotic solvents. As is evident from Table IIIA, this is in fact also the case.

For the fluoride ion, no values of $\Delta H_{tr}^0(H_2O \rightarrow DMSO)$ or $\Delta H_{tr}^0(H_2O \rightarrow PC)$ can be determined, on account of the slight solubility of fluorides in the aprotic solvents (see Table I, Section II,A and Kenttämaa[30]). The very marked decrease in solubility relative to water strongly indicates, however, that the solvation of F^- must be much weaker in the aprotic solvents, just as expected for an ion exceptionally prone to hydrogen bonding. As already pointed out, the solvation of the alkali cations is generally stronger in the aprotic solvents (Table IIIA). This term cannot, therefore, be responsible for the decrease of the solubility. Furthermore, in aprotic solvents the values of ΔS_s^0, the sum of the solution entropies for the two ions involved, do not seem to differ very much between the various halides of a given cation (see Table I). The extremely low solubility of the fluorides relative to the other halides must therefore be due to an exceptionally unfavorable value of $\Delta H_{sv}^0(F^-)$ in the aprotic solvents, i.e., to very positive values of $\Delta H_{tr}^0(H_2O \rightarrow DMSO)$ and $\Delta H_{tr}^0(H_2O \rightarrow PC)$ for the fluoride ion. This lowers $\Delta G_{sv}^0(F^-)$ so that the total solvation free energies cannot any longer match the lattice free energy. Hence $\Delta G_s^0 \gg 0$ (Eq. 3), i.e., the solubility becomes very low.

Hydrogen bonding to the halide ions must also be the main cause of the much higher values of $-\Delta H_h^0$ found for F^-, Cl^-, and Br^- relative to alkali metal ions of approximately the same radii, namely, Na^+, Rb^+, and Cs^+, respectively. For methanol, a similar effect is also found, although it is smaller, as is to be expected due to weaker hydrogen bonding. In Table IIIB, the differences $\Delta H_{sv}^0(M) - \Delta H_{sv}^0(L)$ have been listed for all the solvents discussed, insofar as pertinent data have been available. As is to be expected, the differences found for the two protic solvents decrease in the sequence $F^- > Cl^- > Br^-$, i.e., as the halide ions become less prone to form hydrogen bonds. It should also be noted that $r_c(Cs^+) = 1.83$ Å is in fact somewhat larger than $r_c(Br^-) = 1.80$ Å. If it were possible to compare Br^- with an alkali metal ion of $r_c = 1.80$ Å, the differences found in the last row of Table IIIB would be even smaller, by $\simeq 2$ kJ. The decrease along the sequence should therefore be somewhat more marked. As important, the differences found for the two aprotic solvents are all quite small, and even smaller than stated in Table IIIB if the same reasonable correction as above, 2 kJ, is deducted from the values in the last row. The remaining differences are in fact inside the limits of error set by the accuracy of the individual values of ΔH_{sv}^0. Where no hydrogen bonding is involved, the values of ΔH_{sv}^0 are indeed much the same for halide and alkali ions of the same size.

This comparison of the solvation enthalpies of halide and alkali metal

TABLE IIIB

DIFFERENCES BETWEEN SOLVATION ENTHALPIES OF ALKALI METAL AND
HALIDE IONS OF APPROXIMATELY THE SAME RADIUS IN PROTIC AND APROTIC
SOLVENTS[a]

Enthalpy	Water	Methanol	PC	DMSO
$\Delta H_{sv}^0(Na^+) - \Delta H_{sv}^0(F^-)$	85	—	—	—
$\Delta H_{sv}^0(Rb^+) - \Delta H_{sv}(Cl^-)$	59	31	8	6
$\Delta H_{sv}^0(Cs^+) - \Delta H_{sv}^0(Br^-)$	46	26	5	12

[a] Data from Table IIIA, in kJ.

ions presupposes, of course, that correct partitions have been effected between anions and cations, i.e., that the extrathermodynamic assumptions
introduced are valid. The fact that such a coherent picture emerges provides
another strong indication that this is indeed so.

The different modes of change found for ΔH_{sv}^0 of the halide ions between
protic and aprotic solvents means that their relative competitive power for
metal ions changes between the two types of solvents. Complexes of
hydrogen bonding ligands should be more stable relative to complexes of
ligands not able to form hydrogen bonds as the systems are transferred from
protic to aprotic solvents. The effects should be considerable as seen from
the values of $\Delta H_{tr}^0(H_2O \rightarrow PC)$ and $\Delta H_{tr}^0(H_2O \rightarrow DMSO)$ (see Table IIIA).
Thus between Cl^- and I^-, which have very different ability to form hydrogen
bonds, the values of $\Delta H_{tr}^0(H_2O \rightarrow S)$ differ by 29.7 and 31.6 kJ for PC
and DMSO, respectively. These changes correspond to an increase in the
stabilities of chloride relative to iodide complexes amounting to 5.2 and 5.5
powers of ten in the equilibrium constants, respectively. The differences are
evidently independent of the absolute values ascribed to the ion solvation
enthalpies and the conclusions are therefore valid independent of the extrathermodynamic assumptions introduced in order to find these.

The solvation data so far discussed do not allow, on the other hand, any
prediction as to whether the halide complexes will be absolutely more stable
in aprotic than in protic solvents. One relation important for the absolute
stability is the relative strength of solvation of the free metal ions and the
complexes formed. The stronger solvation found for the free metal ions in
aprotic solvents will of course counteract the complex formation but this
influence will be partly compensated by that stronger solvation of the complexes which is also likely to occur. The data available for this latter effect
are, as might perhaps not be too surprising, even more scarce than for the

free metal ions. The results of some recent investigation, all pertaining to neutral complexes, is discussed in the next section.

B. Solvation Enthalpies of Neutral Molecules and Complexes

The calculation of solvation enthalpies for neutral entities is in principle simple, as they are found as the differences between the heat of solution, ΔH_s^0, and the heat of vaporization, ΔH_v^0, or sublimation, ΔH_{sub}^0. In so far as these latter quantities are accurately known which is fairly often the case, accurate values of ΔH_{sv}^0 can therefore be found by a simple determination of ΔH_s^0, providing of course that the compound does not dissociate in solution. If so, the enthalpies for the further reactions also have to be taken into account.

For comparison, neutral entities of two essentially different categories have been considered, i.e., simple protic molecules and metal complexes of net zero charge. For the former group data are available in the literature though surprisingly few concerning the nonaqueous solvents (see Table V). For the neutral complexes the calculations are mostly based on unpublished studies of the halides of the zinc group elements.[31] The results of these investigations have been summarized in Table IV.

The zinc halides are so extensively dissociated in aqueous solution that no values of ΔH_{sv}^0 pertaining to the neutral complexes can be calculated. In DMSO, on the other hand, this is possible as the complexes are much more stable in this solvent. The corrections that nevertheless have to be introduced on account of dissociation and association reactions have been determined in separate measurements of the thermodynamics of the systems in solution (see Section V). For the more stable cadmium halides, such a procedure can be applied not only to the measurements in DMSO but also to those performed in aqueous solution. The neutral mercury halide complexes are so stable in all the solvents considered, including in this case methanol,[32] that no corrections for dissociation or association have to be applied.

Once the values of $\Delta H_{tr}^0(H_2O \rightarrow DMSO)(L)$ of the halide ions have been settled according to the procedure described in Section III,A, the data now available for the zinc, cadmium, and mercury halides also allow the calculation of $\Delta H_{tr}^0(H_2O \rightarrow DMSO)(M)$ of Zn^{2+}, Cd^{2+}, and Hg^{2+}, from the equation

$$\Delta H_{tr}^0(H_2O \rightarrow DMSO)(M) = \Delta H_s^0(DMSO)(ML_2) - \Delta H_s^0(H_2O)(ML_2) -$$

$$(\Delta H_{\beta 2}^0(DMSO) - \Delta H_{\beta 2}^0(H_2O)) - 2\Delta H_{tr}^0(H_2O \rightarrow DMSO)(L) \quad (6)$$

where $\Delta H_{\beta 2}^0$ denotes the enthalpy changes of the association reactions $M^{2+} + 2L^- \rightarrow ML_2$ in the solvents indicated. The enthalpies of transfer

TABLE IV

Solvation Enthalpies for Neutral Complexes ML_2 in the Protic Solvents Water and Methanol and the Aprotic Solvent Dimethylsulfoxide[a]

ML_2	Water			Methanol		DMSO	
	ΔH^0_{sub}	ΔH^0_s	ΔH^0_{sv}	$\Delta H^{0\ i}_s$	ΔH^0_{sv}	$\Delta H^{0\ e}_s$	ΔH^0_{sv}
$ZnCl_2$	149.0[b]	—	—	—	—	−70.0	−219.0
$ZnBr_2$	145.2[b]	—	—	—	—	−75.9	−221.1
ZnI_2	140.6[b]	—	—	—	—	−84.1	−224.2
$CdCl_2$	180[c]	−13.7[e]	−194	—	—	−41.2	−221
$CdBr_2$	163[c]	−7.3[e]	−170	—	—	−46.2	−209
CdI_2	146[c]	4.9[e]	−141	—	—	−45.2	−191
$HgCl_2$	83.1[d]	14.0[f]	−69.1	−3.0	−86.1	−21.2	−104.3
$HgBr_2$	83.7[c]	20[g]	−64	−1.7	−85.4	−17.0	−100.7
HgI_2	91.2[c]	28.9[h]	−62.3	7.7	−83.5	−4.3	−95.6

[a] $\Delta H^0_{sv}(ML_2)$ for the reactions $ML_2(g) \to ML_2(sv)$, calculated from the dissolution and sublimation enthalpies of the solid compounds according to $\Delta H^0_{sv} = \Delta H^0_s - \Delta H^0_{sub}$. Values in kJ, 25°C.

[b] D. W. Rice and N. W. Gregory, J. Phys. Chem. **72**, 3361 (1968).

[c] L. Brewer, G. R. Somayajulu, and E. Brackett, Chem. Rev. **63**, 111 (1963).

[d] D. Cubicciotti, H. Eding, and J. W. Johnson, J. Phys. Chem. **70**, 2989 (1966).

[e] S. Ahrland, L. Kullberg, and R. Portanova, unpublished data (1975). Media used: 1 M $NaClO_4$(water) and 1 M NH_4ClO_4(DMSO).

[f] P. K. Gallagher and E. L. King, J. Am. Chem. Soc. **82**, 3510 (1960); in 1 M $HClO_4$.

[g] Calculated from the solubility data of H. J. V. Tyrrel and J. Richards, J. Chem. Soc. p. 3812 (1953).

[h] J. J. Christensen, R. M. Izatt, L. D. Hansen, and J. D. Hale, Inorg. Chem. **3**, 130 (1964); at an ionic strength $I = 0.5\ M$ ($NaClO_4$, $HClO_4$).

[i] K. Hartley, H. Pritchard, and H. Skinner, Trans. Faraday Soc. **47**, 254 (1951); $I \to 0$.

from water to DMSO become smoothly more exothermic in the sequence $Zn^{2+} < Cd^{2+} < Hg^{2+}$ (Table IIIA).

Also in DMSO, these ions are all solvated via oxygen, and moreover, by six solvent molecules in a regular octahedral arrangement. Originally, this was deduced from infrared spectra[3] but later on, in the case of Hg^{2+} and Cd^{2+}, also confirmed by x-ray diffraction studies on the solvate solutions.[74] As to Hg^{2+}, this result is rather remarkable as this acceptor generally much prefers sulfur to oxygen and moreover does not often coordinate six equivalent ligands. It should be noted that also Ag^+ is seemingly solvated by DMSO via oxygen.[3] The coordinating properties of sulfur in sulfoxides are evidently quite weak, especially compared with those displayed by sulfur in sulfides.[32a]

Both for the cadmium and the mercury halides, $-\Delta H_{sv}^0(ML_2)$ is indeed considerably higher in DMSO than in water, as could certainly be foreseen from the corresponding changes of $\Delta H_{sv}^0(M)$, expressed in the exothermic values of $\Delta H_{tr}^0(H_2O \rightarrow DMSO)$ for Zn^{2+}, Cd^{2+}, and Hg^{2+} listed in Table IIIA. The changes are relatively much larger for the complexes than for the free ions, however. This is evident from the ratio $\Delta H_{tr}^0(H_2O \rightarrow DMSO)/\Delta H_h^0$ which for both Cd^{2+} and Hg^{2+} is ≈ 0.04, much lower than for any of the complexes (Table V). For the complexes HgL_2, the ratio even reaches values as high as ≈ 0.5. The solvents thus interact much more specifically with the complexes ML_2 than with the free ions M^{2+}. This is reasonable as the effects due to a simple neutralization of charges must be very much larger for the free ions.

In fact the absolute values of $\Delta H_{tr}(H_2O \rightarrow DMSO)$ for the complexes are not so much smaller than for the free ions (see Tables IIIA and V). In the most extreme case, CdI_2, the value is as high as 75 % of that of the free ion and, even if this fraction is smaller for the other complexes, it is nevertheless considerable in all cases. This means that the stronger solvation in DMSO relative to water is compensated to a very considerable extent by a corre-

TABLE V

Hydration Enthalpies and Enthalpies of Transfer between Water and Selected Solvents for Some Neutral Molecules and Complexes[a]

| | $-\Delta H_h^0$ | $\Delta H_{tr}^0(H_2O \rightarrow S)$ | | | $\Delta H_{tr}^0(H_2O \rightarrow DMSO)/\Delta H_h^0$ |
		Methanol	PC	DMSO	
H_2O	44.0[b]	—	8.1[c]	-5.4^c	0.12
CH_3OH	44.7[b]	7.32[b]	—	5.9[d]	-0.13
H_3N	34.6[b]	—	13.8[e]	11.9[e]	-0.34
H_2S	19.2[b]	—	—	—	
$CdCl_2$	194	—	—	-27	0.14
$CdBr_2$	170	—	—	-39	0.23
CdI_2	141	—	—	-50	0.35
$HgCl_2$	69.1	-17.0	—	-35.2	0.51
$HgBr_2$	64	-21	—	-37	0.58
HgI_2	62.3	-21.2	—	-33.3	0.53

[a] For the metal complexes data from Table IV; for the other molecules, see notes b–e.

[b] "Selected Values of Physical and Thermodynamic Properties," Nat. Bur. Stand., Circ. No. 500. US Govt. Printing Office, Washington, D.C., 1952.

[c] R. L. Benoit and S. Y. Lam, *J. Am. Chem. Soc.* **96**, 7385 (1974). Much the same result has been found for PC by Y.-C. Wu and H. L. Friedman, *J. Phys. Chem.* **70**, 501 (1966), and for DMSO by E. M. Arnett and D. R. McKelvey, *J. Am. Chem. Soc.* **88**, 2598 (1966).

[d] E. M. Arnett and D. R. McKelvey, see note c.

[e] R. L. Benoit, personal communication (1976).

sponding strengthening of the solvation of the complexes. As the forces acting between the ions forming the complexes are moreover stronger in DMSO, on account of its lower ε (cf. Section II,A), the complexes are indeed likely to be at least as strong in DMSO as in water, or for ligands like Cl^- and Br^- which are less strongly solvated in DMSO, even considerably more stable in this solvent.

Though the values of $\Delta H_{sv}^0(ML_2)$ are small relative to $\Delta H_{sv}^0(M)$, they nevertheless represent quite substantial amounts of energy, of much the same order of magnitude as ΔH_{sv}^0 for most of the monovalent ions listed in Table IIIA. The value of $-\Delta H_h(H^+)$ is of course much higher; the values of the very bulky ions Ph_4P^+, Ph_4As^+, and BPh_4^- are much lower, on the other hand. This of course reflects the marked partition of positive and negative charges within the formally neutral ML_2 complexes while the net charge of the large ions of tetrahedral symmetry is fairly evenly distributed and of low density.

For both water and DMSO, the values of $-\Delta H_{sv}^0(ML_2)$ are much higher for the zinc and cadmium than for the mercury halides (see Table IV). The reason is certainly that the solvent molecules enter the inner coordination sphere in the case of ZnL_2 and CdL_2, but not in the case of HgL_2. The second zinc and cadmium complexes are either octahedral or tetrahedral in solution (see Section IV) while the mercury complexes are linear.[4,4a,33] In the latter case, the solvent molecules have to stay in the outer coordination sphere where their interaction with the central ion is much weaker.

Another very striking fact is that, while for HgL_2 and CdL_2 the values of ΔH_{sv}^0 both in water and DMSO distinctly decrease in the order $Cl^- > Br^- > I^-$, the values of ΔH_{sv}^0 for ZnL_2 which are, as mentioned, available only in water, increase is this order. This is probably due to a combination of the following two effects.

First, in the ZnL_2 complexes, where the forces between L^- and the hard acceptor Zn^{2+} are of a mainly electrostatic character,[33] the effective charge on the metal ion will be higher the lower the charge density on the ligand; the higher the effective charge, the stronger the solvation. As the charge density decreases in the order stated above, the effective charge will increase and the solvation consequently becomes stronger in the same order. On the other hand, in complexes like CdL_2 and HgL_2 where the metal ion acceptors are soft, the interactions between the ligands and the central ions are of a rather covalent character.[33] The effective charge on the metal ion will in such cases be lower the more extensive the orbital overlap with the ligand, i.e., the softer the ligand. As the softness increases in the order $Cl^- < Br^- < I^-$, the effective change will decrease for these soft metal ions, and the solvation consequently becomes weaker, in this same order.

Second, a change of the coordination number around the central ion certainly takes place not only at the formation of the mercury halides but also in the other complex systems discussed, though for the zinc and cadmium halides the change is from octahedral to tetrahedral coordination. As is more fully demonstrated in Section V, this change occurs preferentially at one step, characterized by abnormally positive values of the stepwise enthalpy and entropy changes, ΔH_j^0 and ΔS_j^0. For all the cadmium halides, this particular step is the third one in aqueous solution, but the second one in DMSO. For the zinc halides, really reliable results can only be obtained in DMSO, due to the low stability of the complexes in water. In the chloride and bromide systems the change mainly occurs at the first step and seems to be practically complete with the second, while in the iodide system the change mainly occurs at the second step and certainly extends beyond that step. This means that the solvated $ZnCl_2$ and $ZnBr_2$ complexes are exclusively tetrahedral, while for ZnI_2 an equilibrium exists between tetrahedral and octahedral complexes, a circumstance which is apt to result in a relatively higher value of ΔH_{sv}^0 for ZnI_2 than for $ZnCl_2$ or $ZnBr_2$.

The two effects discussed are most likely interwoven as the high effective charge left on zinc in the iodide complexes permits the coordination of more solvent molecules at a later stage of the complex formation than in the case of the chloride and bromide systems where the effective charge is lower.

These differences in the patterns of coordination and bonding must also be responsible for another striking difference between the CdL_2 and HgL_2 complexes, i.e., that the values of $\Delta H_{tr}^0(H_2O \rightarrow DMSO)$ and consequently the ratios $\Delta H_{tr}^0(H_2O \rightarrow DMSO)/\Delta H_h^0(ML_2)$ also increase strongly in the order $Cl^- < Br^- < I^-$ for CdL_2 while they stay approximately constant for HgL_2. For the latter complexes, the ratios are also considerably higher than for the complexes CdL_2.

The neutral protic molecules H_2O, CH_3OH, H_3N, and H_2S entered in Table V for comparison all have lower values of $-\Delta H_h^0$ than any of the complexes. The values for HgL_2, coordinating in the outer sphere, are not so much higher, however, and the difference is certainly mainly due to the much stronger London forces exerted by the heavier atoms (cf. Section II,C). The values for CdL_2, coordinating in the inner sphere, are so much higher, however, that bonds of a different character are clearly indicated, as has also been postulated above.

Among the molecules H_2S, H_3N, and H_2O, which exert modest London forces of much the same magnitude, the values of $-\Delta H_h^0$ clearly increase with the capacity for hydrogen bonding. Also the increase of μ (0.97, 1.47, and 1.85 D, respectively) evidently acts in the same direction. In the case of CH_3OH, the larger London forces make up for the lower hydrogen bonding capacity relative to water, resulting in practically the same value of ΔH_h^0 for both compounds.

Among the few values of $\Delta H^0_{tr}(H_2O \rightarrow S)$ determined for these simple protic molecules, five are endothermic and one exothermic. The exothermic value of $\Delta H^0_{tr}(H_2O \rightarrow DMSO)$ for water is rather surprising as one would be inclined to think that the hydrogen bonds would make a water molecule prefer its peers, as in fact it does relative to PC. Also $\Delta H^0_{tr}(H_2O \rightarrow DMSO)$ for methanol is, as expected, endothermic. More data are obviously needed before a coherent pattern can be discerned here. The values of $\Delta H^0_{tr}(H_2O \rightarrow S)$ for these protic molecules are further much smaller than those found for the neutral complexes (see Table V). The latter evidently interact with the solvents in a more specific manner. Especially for the complexes involving the soft acceptor Hg(II) the relative differences of ΔH^0_{sv} between the protic water and the aprotic DMSO are indeed large compared with those exhibited by the simple molecules.

IV. Stabilities of Complexes in the Solvents Selected

A. Brønsted Acids

Brønsted acids are complexes where the proton acts as acceptor. As is evident from the foregoing discussion, the stability of such a complex, i.e., its strength as an acid, very much depends upon the solvating and dielectric properties of the solvent. The acid becomes stronger, the more strongly the proton and the corresponding base are solvated relative to the undissociated acid. A further important factor is of course the intrinsic basicity of the corresponding base. The formation of the complex implies a competition for the proton between this base and the solvent, acting as another base. Thus if the corresponding base has a strong proton affinity, it may compete successfully even with a strongly solvated proton, i.e., a strongly basic solvent. If, on the other hand, its proton affinity is weak, the acid may be extensively dissociated even in a solvent of low basicity. Finally, the lower the dielectric constant, the more the electrostatic interactions favor the formation of species which are uncharged, or at least of low charge density.

The relative importance of the various factors involved differs considerably between acids of different types. Consequently, the influence of the solvent on their dissociation equilibria will also vary tremendously.

Among the aprotic solvents discussed here, DMSO is more basic than water.[34] The stronger affinity for H^+ is reflected in the negative value of $\Delta H^0_{tr}(H_2O \rightarrow DMSO) = -25.5$ kJ (Table IIIA). Contrary to that, PC is much less basic than water, $\Delta H^0_{tr}(H_2O \rightarrow PC) = 43.9$ kJ. This applies even more to AN where $\Delta H_{tr}(H_2O \rightarrow AN) = 56.0$ kJ.[35] Acids should thus tend to be stronger in DMSO and weaker in PC and AN than in water.

As for the bases, the hydrogen bonding ones are more strongly solvated by water than by aprotic solvents. In addition to the values listed in Table IIIA for Cl^-, Br^-, and, especially, CF_3COO^-, this is also borne out by the value of $\Delta H_{tr}^0(H_2O \rightarrow AN) = 23.4$ kJ found[35] for Cl^-. On the other hand, bases not able to form hydrogen bonds are more strongly solvated by aprotic solvents (Table IIIA). Acids corresponding to hydrogen bonding bases should therefore tend to be stronger in water than in aprotic solvents, while the opposite should be true for acids corresponding to bases forming no hydrogen bonds.

The concept of a protic solvent implies that the solvent possesses protons prone to chemical interactions. Such protons might also dissociate relatively easy. The protic solvents therefore act as acids of measurable strength. Contrary to this, the aprotic solvents are so weakly acidic that the most drastic means have to be employed in order to remove a proton.[36] This means that dissociation equilibria involving even extremely weak acids can be measured in aprotic solvents as the corresponding strong bases do not protolyze the solvent.

The differences in this respect between the protic and aprotic solvents discussed here, are indeed tremendous. The ion products of the autoprotolytic reactions are for water and methanol 10^{-14} and $10^{-16.6}$ M^2, respectively (King,[37] Table 3.3.4). For DMSO[34,38] and AN[39] these products are so small that their determination becomes exceedingly difficult, but in any case they are $< 10^{-32}$ M^2. Consequently, in DMSO it has been possible to measure[38] stability constants K_1 of association reactions $H^+ + L^{n-} \rightleftharpoons HL^{1-n}$ as high as $K_1 \approx 10^{31}$ M^{-1}, i.e., to determine the acidities of extremely weak acids or the basicities of extremely strong bases, much stronger than can ever exist in an aqueous solution.

For perchloric, trifluoromethane sulfonic, and fluorosulfonic acids, the corresponding bases ClO_4^-, $CF_3SO_3^-$, and FSO_3^- have such a low proton affinity that the acids are completely dissociated in both water and DMSO.[34,36,40,41] At least for perchloric acid, this is the case even in the weakly basic AN.[42] For these acids the solvent influence is overshadowed by the low basicities of the anions.

Among such acids whose dissociation equilibria can be measured in water as well as in DMSO, two groups can be discerned: those which are much stronger in water than in DMSO and those which are not (see Table VI). Among the former are acetic, benzoic, hydrochloric, and hydrobromic acids; among the latter are picric and nitric acids, as well as the ammonium, anilinium and pyridinium ions. Both groups comprise strong as well as weak acids.

The preference of the proton for DMSO over water, as reflected in $\Delta H_{tr}^0(H_2O \rightarrow DMSO) = -25.5$ kJ (see Table IIIA) would mean that the

STEN AHRLAND

TABLE VI

STABILITIES OF BRØNSTED ACIDS IN PROTIC AND APROTIC SOLVENTS[a]

Acid	Water	Methanol	DMSO	AN
CH_3COOH	4.76^b	9.7^e	12.6^i	22.3^i
C_6H_5COOH	4.20^b	9.4^e	11.1^i	20.7^i
CF_3COOH	-0.3^j	—	3.5^j	—
CH_3SO_2OH	—	—	1.6^k	10.0^n
$C_6H_2(NO_2)_3OH$	0.3^b	—	-0.3^j	11.0^n
HNO_3	-1.4^c	—	1.4^k	8.9^m
HCl	-7^c	1.2^f	2.1^k	8.9^m
HBr	-9^c	0.4^g	1.0^l	5.5^m
NH_4^+	9.25^c	10.8^e	10.5^i	16.5^i
$C_6H_5NH_3^+$	4.61^d	—	3.6^i	10.6^i
$C_5H_5NH^+$	5.22^d	5.27^h	3.4^i	12.3^i

[a] Expressed by log K_1 (M^{-1}). $I = 0$, 25°C.

[b] G. Kortüm, W. Vogel, and K. Andrussow, *Pure Appl. Chem.* **1**, 187 (1961).

[c] D. D. Perrin, *Pure Appl. Chem.* **20**, 133 (1969).

[d] D. D. Perrin, "Dissociation Constants of Organic Bases in Aqueous Solution." Butterworth, London, 1965 (Supplement to *Pure Appl. Chem.*).

[e] C. D. Ritchie, *J. Am. Chem. Soc.* **91**, 6749 (1969).

[f] T. Shedlovsky and R. L. Kay, *J. Phys. Chem.* **60**, 151 (1956).

[g] E. Schreiner, *Z. Phys. Chem.* **111**, 419 (1924).

[h] C. H. Rochester, *J. Chem. Soc. B*, p. 33 (1967).

[i] I. M. Kolthoff, M. K. Chantooni, Jr., and S. Bhowmik, *J. Am. Chem. Soc.* **90**, 23 (1968).

[j] R. L. Benoit and C. Buisson, *Electrochim. Acta* **18**, 105 (1973).

[k] C. McCallum and A. D. Pethybridge, *Electrochim. Acta* **20**, 815 (1975).

[l] J. A. Bolzan and A. J. Arvia, *Electrochim. Acta* **16**, 531 (1971).

[m] I. M. Kolthoff, S. Bruckenstein, and M. K. Chantooni, Jr., *J. Am. Chem. Soc.* **83**, 3927 (1961). In AN, equilibria $HL + L^- \rightleftharpoons HL_2^-$ also exist, with log K_2: HCl, 2.2; HBr, 2.4; HNO_3, 2.3.

[n] I. M. Kolthoff and M. K. Chantooni, Jr., *J. Am. Chem. Soc.* **87**, 4428 (1965).

values of log K_1 would be ≈ 4 units lower in DMSO than in water. The total difference of basicity[34] only seems to correspond to ≈ 2 units, however, which would mean that the entropy term works in the other way. This is indeed most probable as will be further discussed in Section V.

On the other hand, the lower value of ε in DMSO would tend to increase log K_1 relative to water. This would also occur if the solvation of the undissociated acid would be stronger in DMSO. Finally, for acids corresponding to strongly hydrogen bonding bases, a very severe decrease of the acid strength should occur in DMSO where the solvation of the base is much weaker than in water.

In fact all those acids that are much stronger in water than in DMSO correspond to strongly hydrogen bonding bases while, on the other hand, those acids that have about the same strength in both solvents do not. Evidently, the strong solvation of the base due to hydrogen bonding keeps acids of the first group much more extensively dissociated in water. When this factor is absent, as is the case for acids of the second group, the changes in the other energy terms between water and DMSO balance each other fairly nicely.

In methanol, which is protic but less strongly hydrogen bonding than water, the acids of the first group are, as expected, weaker than in water but stronger than in DMSO (Table VI). Since the value of ε is fairly low in methanol it is understandable that the values of log K_1 are generally closer to those found in DMSO than to those found in water. For acids of the second group, the values of log K_1 in methanol are, also as expected, much the same as those found in the two other solvents.

In AN, all the acids listed in Table VI are much weaker than in the other solvents. The main reason is no doubt the poor solvation of H^+ in AN. The difference $\Delta H_{tr}(DMSO \rightarrow AN) = 81.5$ kJ corresponds to a difference Δ log $K_1 = 14.3$ between the two solvents. The actual differences are clustered around eight units (see Table VI). Other energy terms evidently counteract, reducing the differences to about half of that expected from the considerable change of proton solvation.

B. Metal Halide and Pseudohalide Complexes

1. SELECTION OF THE METAL IONS

In water, two opposite sequences are observed for the stabilities of complexes formed between metal and halide ions, namely (a) $F^- \gg Cl^- > Br^- > I^-$ and (b) $F^- \ll Cl^- < Br^- < I^-$. For those metal ions where (a) sequences are found, the complex formation mainly depends on electrostatic interaction. On the other hand, for those metal ions where (b) sequences are found, the bonding is markedly covalent, involving a true electron acceptor–donor interaction.[9,33,43 45] Metal ions displaying (a) sequences in aqueous solution have been termed hard, those displaying (b) sequences soft.[44,45] The same terms have also been applied to the ligands preferred by each category so that F^- is the hardest and I^- the softest of the halide ions. The aptitude for hydrogen bonding decreases with increasing softness, as is of course to be expected from the essentially electrostatic character of the hydrogen bonds.

In aprotic solvents, where complexes of hydrogen-bonding ligands should be more stable than in water relative to complexes of ligands not prone to such bonding (see Section III,A), the affinity sequences will evidently be modified, and moreover in the direction from (b) to (a). Acceptors having (a) sequences in water will have even more marked (a) sequences in aprotic solvents. Acceptors with mild (b) sequences in water may well turn to (a) sequences in aprotic solvents. Acceptors with very marked (b) sequences in water are likely to keep (b) sequences also in aprotic solvents but the differences in stabilities between the various halides will be much less marked than in water.

In order to substantiate these predictions, halide systems of metal ions displaying different affinity sequences in aqueous solution should evidently be investigated. To facilitate the comparison, the acceptors should preferably have the same charge and outer electron configuration. The divalent d^{10} ions of the zinc group fulfill these requirements,[2] ranging from the (a) acceptor Zn^{2+}, via Cd^{2+}, of mild (b) character, to the very marked (b) acceptor Hg^{2+}. For these acceptors, extensive data on the stability of the halide complexes in aqueous solution are also available.[2]

It is natural to extend the comparison to the monovalent d^{10} acceptors Cu^+, Ag^+, and Au^+, in order to investigate the influence of the ionic charge. Also the halide complexes of these acceptors have been fairly extensively investigated. This applies especially to Ag^+ where numerous data exist also for aprotic solvents. For copper(I), and even more for gold(I), the extensive disproportionation has very much restricted their investigation in aqueous solution. In aprotic solvents, the monovalent oxidation states are much more stable.[5,46] Thus for copper the disproportionation constant, at 25°C, for the reaction $2\,Cu^+ \rightleftharpoons Cu(s) + Cu^{2+}$ is in DMSO[5] $K_d = 2\ M^{-1}$ as against $10^6\ M^{-1}$ in aqueous solution.[47] Such a stabilization of copper(I) is in fact to be expected from those values of $\Delta H_{tr}^0(H_2O \rightarrow DMSO)$ for Cu^+ and Cu^{2+} which may reasonably be deduced from values measured for analogous ions. For Cu^+, the value ought not to be very different from that found for Ag^+, while for Cu^{2+} the value should be close to those found for Zn^{2+} and Cd^{2+} (see Table IIIA). This would mean that the equilibrium above would be displaced to the left by ≈ 40 kJ corresponding to $\approx 10^7$ in K_2, a figure of the magnitude actually observed. For gold(I), the stabilization from water to DMSO is even more spectacular. Gold(III) is in fact not stable in DMSO; attempts to produce it by anodic oxidation results instead in an oxidation of the solvent.[7]

Except for d^{10} ions, data also exist for other transition metal ions, primarily Cu^{2+} and Ni^{2+}. These are worthwhile to consider in order to find how the stabilities vary between acceptors of the same charge but with different electron configuration.

In the following tables, the stabilities are expressed by the stability constants K_j for the consecutive formation equilibria $ML_{j-1} + L \rightleftharpoons ML_j$.

2. Zinc(II), Cadmium(II), and Mercury(II)

In water, the zinc halide complexes are extremely weak but a marked (a) sequence can nevertheless be discerned (Table VII). In methanol, only the chloride has been investigated. The complexes are much more stable in this solvent, due to a stronger electrostatic interaction between the ions and also, presumably, to a weaker solvation, though $\Delta H^0(H_2O \rightarrow Met)$ is not known for Zn^{2+}. In DMSO, all the systems have again been investigated. The complexes formed are more stable throughout than in water but the stability increase is by far largest for the chloride system (Table VII). For iodide

TABLE VII

Stabilities of Zinc Halide and Pseudohalide Complexes in Protic and Aprotic Solvents at 25°C

Water, NaClO$_4$:	3 M^a			1 M^b	3 M^c
	Cl$^-$	Br$^-$	I$^-$	SCN$^-$	CN$^-$
log K_1	−0.19	−0.57	−1.5	0.71	5.34
log K_2	−0.40	−0.8	—	0.34	5.68
log K_3	0.75	0.5	—	0.15	5.65
log K_4	—	—	—	0.32	4.90
K_1/K_2	1.6	2	—	2.3	0.46
K_2/K_3	0.07	0.05	—	1.6	1.1
K_3/K_4	—	—	—	0.7	5.6
Methanol,^{d} $I \rightarrow 0$	Cl$^-$				
log K_1	3.9	—	—	—	—
log K_2	4.3	—	—	—	—
K_1/K_2	0.4	—	—	—	—
DMSO,^{e} NH$_4$ClO$_4$:		1 M			
	Cl$^-$	Br$^-$	I$^-$	SCN$^-$	
log K_1	1.94	0.84	−0.70	1.38	
log K_2	3.89	2.89	1.41	1.41	
log K_3	2.26	1.34	0.15	2.40	
log K_4	—	—	—	1.65	
K_1/K_2	0.011	0.009	0.008	0.9	
K_2/K_3	43	35	18	0.10	
K_3/K_4	—	—	—	5.6	

[a] P. Gerding, *Acta Chem. Scand.* **23**, 1695 (1968).

[b] S. Ahrland and L. Kullberg, *Acta Chem. Scand.* **25**, 3692 (1971).

[c] H. Persson, *Acta Chem. Scand.* **25**, 543 (1971).

[d] H. Hoffmann, G. Platz, and M. Franke, *Proc. Int. Conf. Coord. Chem.*, *16th*, 1974 Paper 3.35. Interpolated between the values for 20°C and 30°C given by the authors.

[e] S. Ahrland and N.-O. Björk, *Acta Chem. Scand., Ser. A* **30**, 270 (1976).

the increase is very modest while bromide is, as expected, intermediate. The (a) sequence found in water is thus much more marked in DMSO, as could be foreseen from the nonexistence of hydrogen bonding in the latter solvent.

A very striking feature of all the zinc halide systems in DMSO is the very narrow range of existence of the first complex, expressed by abnormally low values of the ratio K_1/K_2 (Table VII). Most often the ratios K_j/K_{j+1} for systems of consecutive complexes of monodentate ligands are between 1 and 20. With these ratios, the maximum share of the complex ML_j of the total metal ion concentration C_M which is reached at the ligand number $\bar{n} = j$ will be approximately 33 and 70%, respectively.[48] The low ratio of $K_1/K_2 \approx$ 0.01, representative for the present systems, means on the other hand that the maximum share of the ZnL^+ complex, reached at $\bar{n} = 1$, will only be $\approx 5\%$. Constants of complexes present in such low relative concentrations are generally difficult to determine precisely. In this case, however, the reproducibility of the emf's measured is good enough to allow a satisfactory precision.[49]

Extremely narrow as well as extremely wide ranges of existence of intermediate complexes are generally connected with changes of the coordination figure.[2,32a,49] Such changes provoke extremes of stability by involving extraordinary enthalpy and entropy changes. For a safe interpretation of the processes taking place, a determination of all these thermodynamic functions is therefore necessary, in addition to the determination of K_j, which provides the free energy changes according to $\Delta G_j^0 = -RT \ln K_j$. Further discussion will therefore be postponed until Section V where such measurements are reported.

The zinc thiocyanate complexes show a modest increase of stability from water to DMSO, similar to that found for the iodide complexes (see Table VII). This is to be expected as SCN^- forms only weak if any hydrogen bonds in water.

The cadmium halides display a mild (b) sequence in water (Table VIII). Much the same picture emerges in 1 M and 3 M sodium perchlorate media, though the complexes are weaker throughout in the 1 M medium where a stability minimum occurs, as is generally the case for systems of this type.[50] In methanol, the complexes are considerably stronger, obviously for the same reasons that operate in the case of zinc chloride. The affinity sequence in the protic solvent methanol is the same as in water. As for zinc, all the complexes are more stable in DMSO than in water, and again the increase is largest for the chloride, smaller for the bromide, and very modest for the iodide complexes. As a consequence, the mild (b) sequence found in water turns into an (a) sequence in DMSO. That switch of the stability sequence between protic and aprotic solvents, which was foreseen as a possibility, is thus realized.

TABLE VIII

STABILITIES OF CADMIUM HALIDE AND PSEUDOHALIDE COMPLEXES IN PROTIC AND APROTIC SOLVENTS AT 25°C

Water, NaClO$_4$:	1 M^a				3 M^b				
	Cl$^-$	Br$^-$	I$^-$	SCN$^-$	Cl$^-$	Br$^-$	I$^-$	SCN$^-$	CN$^-$
log K_1	1.35	1.56	1.88	1.32	1.59	1.76	2.08	1.39	5.62
log K_2	0.43	0.46	0.78	0.67	0.64	0.59	0.70	0.59	5.20
log K_3	−0.37	0.23	1.69	0.04	0.18	0.98	2.14	0.60	4.90
log K_4	—	0.41	1.28	−0.15	—	0.38	1.60	—	3.48
K_1/K_2	8.3	12	13	4.4	8.8	15	24	6.3	2.6
K_2/K_3	6	1.2	0.12	4.3	2.9	0.41	0.04	1.0	2.0
K_3/K_4	—	0.7	2.6	1.6	—	4.0	3.4	—	27

Methanol, LiNO$_3$:	2 M^c		$I \to 0^d$
	Cl$^-$	Br$^-$	Cl$^-$
log K_1	2.1	4.0	5.95
log K_2	1.5	2.0	2.74
log K_3	1.5	0.9	—
log K_4	1.0	1.3	—

DMSO, NH$_4$ClO$_4$:	1 M^e				0.1 M^f			0.1 M^g
	Cl$^-$	Br$^-$	I$^-$	SCN$^-$	Cl$^-$	Br$^-$	I$^-$	CN$^-$
log K_1	3.23	2.92	2.18	1.81	4.32	3.71	2.59	—
log K_2	1.98	1.91	1.40	0.91	2.90	2.52	1.68	—
log K_3	2.57	2.75	2.93	0.20	3.42	3.25	2.91	—
log K_4	1.75	1.68	1.17	—	2.29	1.80	1.08	—
K_1/K_2	18	10	6	7.9	26	15	8	—
K_2/K_3	0.26	0.14	0.03	5	0.30	0.19	0.06	—
K_3/K_4	7	12	58	—	13	28	68	—

[a] Water, 1 M medium; Cl$^-$: C. E. Vanderzee and H. J. Dawson, *J. Am. Chem. Soc.* **75**, 5659 (1953); Br$^-$: P. Kivalo and P. Ekari, *Suom. Kemistil.* **B 30**, 116 (1957); I$^-$, SCN$^-$: P. Gerding, *Acta Chem. Scand.* **22**, 1283 (1968).

[b] Water, 3 M medium; Cl$^-$, Br$^-$, I$^-$, SCN$^-$: I. Leden, "Potentiometrisk undersökning av några kadmiumsalters komplexitet;" Diss., University of Lund, 1943; P. Gerding, *Acta Chem. Scand.* **20**, 79, 2771 (1966); CN$^-$: H. Persson, *ibid.* **25**, 543 (1971).

[c] Methanol, 2 M medium; Cl$^-$: Ya. I. Tury'an and B. P. Zhantalai, *Russ. J. Inorg. Chem. (Engl. Transl.)* **5**, 848 (1960); Br$^-$: O. I. Khotsyanovski and V. Sh. Telyakova, *Ukr. Khim. Zh.* **34**, 1126 (1968) (as cited in L. G. Sillén and A. E. Martell, "Stability Constants of Metal–Ion Complexes," Suppl. No. 1, Chem. Soc., London, 1971).

[d] From fast kinetics by means of a pressure jump technique, 20°C; H. Hoffmann, personal communication.

[e] S. Ahrland and N.-O. Björk, *Acta Chem. Scand., Ser. A* **30**, 257 (1976).

[f] S. Ahrland, N.-O. Björk, I. Persson, and R. Portanova, unpublished data.

[g] For CN$^-$ in DMSO, medium 0.1 M Et$_4$NClO$_4$, log $\beta_4 = 24.8$, no intermediate complexes can be proved; potentiometric measurements by means of cadmium electrode; M. Le Démézet, personal communication.

For both zinc and cadmium, the measurements in DMSO have been performed in ammonium perchlorate media, though the corresponding measurements in aqueous solution refer to sodium perchlorate media. These are not suitable in DMSO, however, due to the low solubility of NaCl. As very little is known about the influence of such medium changes in DMSO, however, it seems advisable to investigate this point further. This has been done for the cadmium halides where a 0.1 M as well as an 1 M ionic medium has been applied (see Table VIII). The chloride and, to a lower degree, the bromide complexes are all more stable in the 0.1 M medium. For the cationic and neutral complexes the stability increases are very considerable, much higher than for a corresponding medium change in aqueous solution. Moreover, even for the final anionic CdL_4^{2-} complexes increases are observed. These are certainly difficult to explain as due to ordinary activity changes. For the iodide, on the other hand, modest increases of stability are found for the first two complexes between 1 M and 0.1 M media, and a slight decrease for the final CdI_4^{2-} complex, i.e., a pattern quite similar to that found in aqueous solutions.[50]

The most reasonable interpretation of these facts is that the hydrogen bonding ligands Cl^- and, less marked, Br^-, in the absence of water interact with the hydrogens of NH_4^+, i.e., this ion competes with Cd^{2+} for the ligands, more severely the higher its concentration. The non-hydrogen-bonding I^-, on the other hand, is not particularly attracted by NH_4^+. Consequently, in an aprotic medium such as DMSO, the stabilities of chloride and bromide, but not of iodide complexes, are markedly depressed by an NH_4^+ medium. The stabilization of the chloride and bromide relative to the iodide complexes, which is in fact observed, is therefore less than would be found without this counteracting influence. At the lower concentration of 0.1 M, the residual influence is presumably modest, however, so the values found in this medium fairly faithfully reflect the conditions prevailing in a truly non-hydrogen-bonding medium.

For the cadmium halides also, one of the consecutive steps has a narrow range of existence in DMSO, though for these systems it is the second one (see Table VIII). The pattern is much the same in both media employed. The relative stabilities may thus stay fairly unaffected by even large changes of the medium. Especially in the iodide system, the second complex displays a narrow range of existence also in water (see Table VIII).

The cadmium thiocyanate complexes increase their stability from water to DMSO to about the same extent as the iodide complexes. They thus behave very much as the corresponding zinc complexes. Contrary to the halides, no step seems to be disfavored.

A closer examination of complex formation for all systems concerned is possible once the enthalpy and entropy changes involved have been determined (see Section V).

The mercury halide complexes are extremely stable in aqueous solutions, with a very marked (b) sequence (Table IX). In DMSO, a huge increase in the stability of the chloride complexes is observed, as is a somewhat smaller increase for the bromide complexes, while the stability of the iodide complexes stays almost the same as in water. As a consequence, the (b) sequence is practically leveled. For the higher complexes, measurements have also been performed in AN (Table IX). Very similar patterns are, as expected, found for the two aprotic solvents. The change from water even seems to be somewhat more marked for AN so that at least K_4 displays a mild (a) sequence in this solvent.

Very striking is of course the extremely wide range of existence of the second complex, reflected in very high ratios K_2/K_3 in water. In DMSO, separate values of K_1 and K_2 have so far not been definitely determined. Preliminary data indicate beyond doubt, however, that the values of K_2/K_3 are still quite high, though not as high as in water. As has long been recognized, this persistence of the second complex is connected with a change from linear to tetrahedral coordination.

Contrary to what is found in water, however, the third halide complex also has a rather wide range of existence in the two aprotic solvents, reflected in high ratios K_3/K_4 (Table IX). This step does not involve any major change of coordination, however, to judge from the structures determined for the HgI_3^- and HgI_4^{2-} complexes in DMSO.[4] Both are tetrahedral, though the HgI_3^- pyramid is somewhat flattened so that the mercury atom is closer to the base plane than expected for a regular tetrahedron. Also in this case, the enthalpy and entropy changes give a better insight in the processes involved (see Section V).

For all the mercury halide complexes, the variations in stability found between different media in a given solvent are small compared with the variations found between solvents of different types.

The stabilities of the mercury thiocyanate complexes increase little, or not at all, from water to DMSO (Table IX). In this respect, thiocyanate again behaves much like iodide. Contrary to what is found for the iodide system, however, the ratio K_3/K_4 has about the same low value in water and DMSO for the thiocyanate system.

3. COPPER(I), SILVER(I), AND GOLD(I)

The easy disproportionation of Cu^+ in water, and the low solubility of the copper(I) halides and pseudohalides in this solvent, makes the determination of K_1 for these systems a very difficult task. In fact no such values have been reliably established. A fairly stable second complex is formed while the higher complexes are formed rather reluctantly.[51] The second complexes thus have wide ranges of existence, cf., mercury(II). On the other

TABLE IX

STABILITIES OF MERCURY HALIDE AND PSEUDOHALIDE COMPLEXES IN PROTIC AND APROTIC SOLVENTS AT 25°C

Water, NaClO₄:

	0.5 M^a			1 M^b		3 M^c		0^d
	Cl^-	Br^-	I^-	Cl^-	SCN^-	Cl^-	Br^-	CN^-
$\log \beta_2$	13.22	17.33	23.82	13.23	16.86	13.98	17.98	32.75
$\log K_1$	6.74	9.05	12.87	6.72	9.08	7.07	9.40	17.00
$\log K_2$	6.48	8.28	10.95	6.51	7.78	6.91	8.58	15.75
$\log K_3$	0.85	2.41	3.78	1.00	2.84	0.75	2.76	3.56
$\log K_4$	1.00	1.26	2.23	0.97	1.97	1.38	1.49	2.66
K_1/K_2	1.8	5.9	83	1.6	20	1.4	6.6	18
K_2/K_3	4×10^5	7×10^5	1.5×10^7	3×10^5	9×10^4	1.4×10^6	7×10^5	1.5×10^{12}
K_3/K_4	0.7	14	35	1.1	7	0.23	19	8

DMSO, Et₄NClO₄:

	0.1 M^e				NaClO₄, 1 M^f		NH₄ClO₄, 1 M^g	
	Cl^-	Br^-	I^-	SCN^-	Br^-	I^-	Cl^-	Br^-
$\log \beta_2$	21.2	22.2	24.2	16.1	—	—	17.97	20.19
$\log K_3$	5.7	5.8	6.2	3.0	5.6	6.1	3.99	5.14
$\log K_4$	—	2.4	2.2	2.1	2.65	2.58	2.08	2.54
K_3/K_4	—	2.5×10^3	10^4	8	0.9×10^3	3.3×10^3	82	4.00

AN

	0^h		
	Cl^-	Br^-	I^-
$\log K_3$	6.00	6.00	5.95
$\log K_4$	2.23	2.04	1.60
K_3/K_4	5.9×10^3	9.1×10^3	2.2×10^4

^{a} L. G. Sillén, Acta Chem. Scand. 3, 539 (1949).
^{b} L. Ciavatta and M. Grimaldi, J. Inorg. Nucl. Chem. 30, 197 (1968) (for Cl^-); Inorg. Chim. Acta 4, 312 (1970) (for SCN^-).
^{c} R. Arnek, Ark. Kemi 24, 531 (1965).
^{d} J. J. Christensen, R. M. Izatt, and D. Eatough, Inorg. Chem. 4, 1278 (1965).
^{e} A. Foll, M. Le Demézét, and J. Courtot-Coupez, Bull. Soc. Chim. Fr., p. 1207 (1972).
^{f} R. Arnek and D. Poceva, Acta Chem. Scand., Ser. A 30, 59 (1976).
^{g} S. Ahrland, I. Persson, and R. Portanova, unpublished data.
^{h} G. Ellendt and K. Cruse, Z. Phys. Chem. (Frankfurt am Main) 201, 130 (1952).

hand, the third complexes also persist for quite wide ranges, which does not apply to mercury(II) in aqueous solution.

As already pointed out (Section IV,B,1), the disproportionation is much less severe in DMSO. The favorable solvation of Cu^+ by this solvent further makes all the halides and pseudohalides readily soluble.[5,6] The conditions for a complete determination of the complex formation in the halide systems are thus more favorable than in water. Consequently, values of K_1 has been reported. The same applies to AN where, however, only the chloride has been investigated[46] (Table X). In the aprotic solvents, the complexes beyond the second one are even more reluctantly formed than in water. No such complexes have in fact been found within the ligand concentrations applied (up to 0.25 M).[6]

In water, copper(I) shows a (b) sequence, though not a very marked one. The large increase in the stabilities of the chloride and also of the bromide complexes relative to the iodide ones brings about a change to an (a) sequence in DMSO (Table X). The iodide complexes are even somewhat less stable in DMSO than in water.

The silver halides and pseudohalides are considerably less soluble than the corresponding copper(I) compounds. In water, the solubility of the first complex, $C_s(AgL)$, which can be calculated according to $C_s(AgL) = K_1 K_{so}$, varies from $10^{-6.7}$ for the chloride to $10^{-9.5}$ M for the iodide (Table X). As very low solubilities of silver can be measured radiometrically, which is practically impossible for copper due to the lack of suitable active isotopes, it has nevertheless been possible to determine values of K_1 for the silver halides, especially as no disproportionation complicates the measurements. Mainly because of the much more favorable solvation of Ag^+, and certainly also of the complexes AgL_j^{j-1} (see Section III,B), the solubilities increase considerably in DMSO and also in AN. Due to the lack of values of K_1, no definite values of $C_s(AgL)$ can be given, but from the values of K_{so} and β_2 measured, it might be guessed that the solubilities are around 1000 times larger than in water. In spite of this, however, the silver halides and pseudohalides remain slightly soluble even in these strongly solvating solvents. The complex formation can therefore be investigated by solubility measurements in a great variety of solvents, and since various kinds of potentiometric measurements are also feasible, these systems have in the past certainly been the ones most thoroughly investigated in nonaqueous solvents. A selection of the results is presented in Table X.

In water, the silver halides display a (b) sequence more marked than that of the copper(I) halides. Following a now well-known pattern, the chloride and bromide complexes become much stronger relative to the iodide complexes in DMSO. Because the (b) sequence in water is so marked, however, these changes do not result in a switch to an (a) sequence, as for copper(I),

TABLE X

Stabilities of Copper(I), Silver(I), and Gold(I) Complexes in Protic and Aprotic Solvents at 25°C

Copper(I)

Water, NaClO₄:

	5 M[a]				0[b]				note[c]
	Cl⁻	Br⁻	I⁻	SCN⁻	Cl⁻	Br⁻	I⁻	SCN⁻	CN⁻
pK_{so}	7.38	8.89	12.72	14.78	6.50	8.23	11.96	13.4	—
$\log \beta_2$	6.06	6.23	8.69	—	5.31	5.89	8.76	—	16.26
$\log \beta_3$	5.94	7.44	10.41	11.60	—	—	—	—	21.65
$\log \beta_4$	—	—	9.53	12.03	—	—	—	—	24.27
$\log K_3$	−0.12	1.21	1.72	0.43	—	—	—	—	5.39
$\log K_4$	—	—	−0.88	—	—	—	—	—	2.62
K_3/K_4	—	—	400	—	—	—	—	—	600

AN, Et₄NClO₄:

	0.1 M[d]	DMSO, Et₄NClO₄, 0.1 M[e]			
	Cl⁻	Cl⁻	Br⁻	I⁻	SCN⁻
$\log \beta_2$	10.8	11.95	9.6	8.2	9.3
$\log K_1$	4.9	6	5.0	5.5	4.3
$\log K_2$	5.9	6	4.6	2.7	5.0
K_1/K_2	0.1	1	2.5	600	0.2

Silver(I)

Water, NaClO₄:

	5 M (4 M)[f]				0				CN⁻
	Cl⁻	Br⁻	I⁻	SCN⁻	Cl⁻	Br⁻	I⁻	SCN⁻	CN⁻
pK_{so}	10.10	12.62	16.35	12.11	9.75[g]	12.30[h]	16.08[g]	12.00[h]	15.92[i]
$\log \beta_2$	5.40	7.23	10.95	8.29	5.04[j]	7.34[k]	11.74[k]	8.23[j]	20.85[l]
$\log K_1$	3.08	4.2	—	4.59	3.04	4.38	6.58	4.75	—

log K_2	2.32	3.0	3.70	2.00	2.96	5.16	3.48	—
log K_3	0.75	1.85	1.77	0.00	0.66	1.94	1.22	0.95[l]
log K_4	−0.85	0.12	1.20	0.26	0.73	−0.58	0.22	−1.1[m]
K_1/K_2	5.7	16	7.8	11	26	26	19	—
K_2/K_3	37	14	85	100	200	1700	180	—
K_3/K_4	40	54	3.7	0.55	0.85	330	10	100

Methanol, LiClO$_4$:

1 M[n]	Cl⁻	Br⁻	I⁻
pK_{so}	13.0	15.2	18.2
log β_2	7.9	10.6	14.8

PC, Et$_4$NClO$_4$:

0.1 M[o]	Cl⁻	Br⁻	I⁻	SCN⁻
pK_{so}	20.0	20.5	21.8	16.4
log β_2	20.9	21.2	22.8	16.0

AN, Et$_4$NClO$_4$:

0.1 M[n]	Cl⁻	Br⁻	I⁻
pK_{so}	12.4	13.2	14.2
log β_2	13.1	13.8	15.2

DMSO, Et$_4$NClO$_4$:

0.1 M[p]	Cl⁻	Br⁻	I⁻	SCN⁻	CN⁻
pK_{so}	10.4	10.9	12.1	7.6	14.9
log β_2	11.7	12.0	13.0	8.4	23.4

Gold(I)

AN, Et$_4$NClO$_4$: 0.1 M[q]

	Cl⁻
log β_2	21.52
log K_1	12.63
log K_2	8.89
K_1/K_2	5.5 × 10³

DMSO, Et$_4$NClO$_4$:

0.1 M[r]	Cl⁻	Br⁻
log β_2	18.0	16.6
log K_1	12.6	10.6
log K_2	5.4	6.0
K_1/K_2	1.6 × 10⁷	4 × 10⁴

[a] S. Ahrland and B. Tagesson, *Acta Chem. Scand. Ser. A* **31** 615 (1977).
[b] W. M. Latimer, "The Oxidation States of the Elements and their Potentials in Aqueous Solution," 2nd ed. Prentice-Hall, Englewood Cliffs, New Jersey, 1952.

(continued overleaf)

(continued from previous page)

$I < 0.01$ M.

[c] C. Kappenstein and R. Hugel, J. Inorg. Nucl. Chem. 36, 1821 (1974). Refers to a perchloric acid medium of low and varying concentration, $I < 0.01$ M.

[d] S. E. Manahan and R. T. Iwamoto, Inorg. Chem. 4, 1409 (1965).

[e] A. Foll, M. Le Démézet, and J. Courtot-Coupez, J. Electroanal. Chem. 35, 41 (1972).

[f] Cl⁻: E. Berne and I. Leden, Sven. Kem. Tidskr. 65, 88 (1953); Br⁻: E. Berne and I. Leden. Z. Naturforsch., Teil A 8, 719 (1953); I⁻: I. Leden, Acta Chem. Scand. 10, 540, 812 (1956); SCN⁻: I. Leden and R. Nilsson, Z. Naturforsch. Teil A 10, 67 (1955). Medium 5 M for Cl⁻ and Br⁻, 4 M for I⁻ and SCN⁻.

[g] B. B. Owen and S. R. Brinkley, Jr., J. Am. Chem. Soc. 60, 2233 (1938).

[h] C. E. Vanderzee and W. E. Smith, J. Am. Chem. Soc. 78, 721 (1956).

[i] J. E. Ricci, J. Phys. Chem. 51, 1375 (1947).

[j] Values of β_2 and K_j for these systems: Leden et al., note f.

[k] Values of β_2 and K_j for the iodide system: K. H. Lieser, Z. Anorg. Chem. 292, 97 (1957), at 18°C.

[l] J. Zsakó and E. Petri, Rev. Roum. Chim. 10, 571 (1965), at 20°C.

[m] L. H. Jones and R. A. Penneman, J. Chem. Phys. 22, 965 (1954).

[n] D. C. Luehrs, R. T. Iwamoto, and J. Kleinberg, Inorg. Chem. 5, 201 (1966), at 23°C.

[o] J. Courtot-Coupez and M. L'Her, Bull. Soc. Chim. Fr. p. 675 (1969), at 22°C. Within the range of [L] employed (≤ 10 mM), a third complex was found only for SCN⁻, log $\beta_3 = 18.7$. For Cl⁻, the values of pK_{so} and log β_2 agree well with those found in the same medium at 25°C by J. N. Butler, Anal. Chem. 39, 1799 (1967), i.e., $pK_{so} = 19.87$, log $\beta_2 = 20.87$. In this study, the following values for the constants of the consecutive steps were also determined: log $K_1 = 15.15$, log $K_2 = 5.72$, log $K_3 = 2.52$. Hence $K_1/K_2 = 3 \times 10^9$ and $K_2/K_3 = 1.6 \times 10^3$. The fairly high value of log K_3 is certainly at variance with the result of Courtot-Coupez and L'Her. For Br⁻ and I⁻ in 0.1 M LiClO$_4$ at 25°C, J. N. Butler et al. have found values of pK_{so} and log β_2 which are not very different from those found in 0.1 M Et$_4$NClO$_4$, i.e., for Br⁻: $pK_{so} = 20.25$, log $\beta_2 = 20.95$; for I⁻: $pK_{so} = 20.5$, log $\beta_2 = 21.8$ [cited by M. Salomon, in "Physical Chemistry of Organic Solvent Systems" (A. K. Covington and T. Dickinson, eds.). p. 176, Plenum Press, New York, 1973].

[p] M. Le Démézet, C. Madec, and M. L'Her, Bull. Soc. Chim. Fr. p. 265 (1970), presumably at 22°C (cf. note o). For Cl⁻, Br⁻, and I⁻ reasonably concordant results have been reported by Luehrs et al. (note n). For Cl⁻, J. C. Synnott and J. N. Butler, J. Phys. Chem. 73, 1470 (1969), have found $pK_{so} = 10.28$ and log $\beta_2 = 11.73$. They also determined the following constants for the consecutive steps: log $K_1 = 6.8$, log $K_2 = 4.9$, log $K_3 = 1.4$. Hence $K_1/K_2 = 80$ and $K_2/K_3 = 3 \times 10^3$.

[q] O. Bravo and R. T. Iwamoto, Inorg. Chim. Acta. 3, 663 (1969).

[r] A. Foll, M. Le Démézet, and J. Courtot-Coupez, Bull. Soc. Chim. Fr. p. 408 (1972).

but stops at an almost complete leveling, as for mercury(II). The same applies to the other aprotic solvents investigated, e.g., AN and PC. For AN, which evidently has solvating properties similar to DMSO, the values of K_{so} and β_2 are of much the same magnitude as for DMSO. For the more poorly solvating PC, on the other hand (cf. Section III,A and Table IIIA), the solubilities are much lower, and the complex formation much stronger, but the leveling persists. Contrary to this, the protic methanol shows a marked (b) sequence similar to that of water. As is generally observed, however, the complexes are considerably more stable in methanol than in water.

Like copper(I), silver(I) is very reluctant to form halide and pseudohalide complexes beyond the second ones in aprotic solvents, though such complexes are fairly important in water. The existence of such complexes has been claimed for the thiocyanate and chloride systems in PC, and also for the chloride system in DMSO (see Table X, notes o and p). Especially for the chloride systems, the values of K_3 are small, and the results partly controversial.[52] This is certainly in marked contrast to mercury(II) where higher complexes are readily formed in DMSO and AN (see Table IX). Admittedly, also the first two complexes are much more stable in the case of mercury(II).

While in aqueous solution gold(I) is stable only in the presence of very strongly stabilizing ligands such as CN^-, it is the only stable oxidation state of gold in DMSO.[7] The same seems to apply to AN.[53] A direct comparison between protic and aprotic solvents is therefore not possible but the existing data, referring to the chloride and bromide complexes, nevertheless allow several interesting conclusions. Only two complexes are formed in these systems. The first complex is very stable and also, like the corresponding copper(I) complex, readily soluble. The second complex is much less stable, so that the ratios K_1/K_2 are quite large (see Table X). The overall constants β_2 are, as might be expected, much higher than for the copper(I) or silver(I) complexes, approaching the values found for the mercury(II) complexes (see Table IX). Only in DMSO are data available that bear on the affinity sequence. In this solvent, the chloride and bromide complexes are of about the same stability. In this respect, gold(I) thus seems to behave much like its lighter congeners, or as mercury(II).

4. COPPER(II) AND NICKEL(II)

In water, the halide complexes of copper(II) and nickel(II) are extremely weak and the thiocyanate complexes are of very modest strength (Table XI). Iodide complexes of copper(II) are moreover not stable with respect to the redox reaction resulting in free iodine and copper(I) iodide complexes (with the precipitation of the slightly soluble CuI, if the iodide concentration is not

TABLE XI

STABILITIES OF COPPER(II) AND NICKEL(II) COMPLEXES IN PROTIC AND APROTIC SOLVENTS

$Copper(II)$[a]	Water			DMSO[e]			DMSO[f]	DMSO[g]	AN[h]
	Cl⁻	Br⁻	SCN⁻	Cl⁻	Br⁻	SCN⁻	Cl⁻	Br⁻	Cl⁻
$\log K_1$	≈0[c]	≈0[c]	1.90[d]	4.5	3.4	3.2	4.4	1.5	9.7
$\log K_2$	—	—	1.10[d]	3.0	0.9	2.1	3.0	1.1	7.9
$\log K_3$	—	—	—	1.6	—	—	≈3	—	7.1
$\log K_4$	—	—	—	—	—	—	2	—	3.7
K_1/K_2	—	—	6	30	300	13	25	3	60
K_2/K_3	—	—	—	25	—	—	≈1	—	6
K_3/K_4	—	—	—	—	—	—	≈10	—	2.5×10^3

$Nickel(II)$[b]	Water		Methanol[j]		DMSO[k]	
	Cl⁻	SCN⁻	Cl⁻	SCN⁻	Cl⁻	SCN⁻
$\log K_1$	≈0[c]	1.76[i]	5.0	—	2.7	3.0
$\log K_2$	—	—	3.2	—	—	≈1.5

[a] At 25°C and $I = 0.1 M$, brought about by KNO_3 in water and by Et_4NClO_4 in DMSO and AN, except for one of the bromide measurements in DMSO, cf. note g.

[b] At 20°C and $I = 0$.

[c] L. G. Sillén and A. E. Martell, "Stability Constants of Metal Ion Complexes," Spec. Publ. Nos. 17 and 25. Chem. Soc. London, 1964 and 1971 resp.

[d] N. Tanaka and T. Takamara, J. Inorg. Nucl. Chem. 9, 15 (1959).

[e] A. Foll, M. Le Démézet, and J. Courtot-Coupez, J. Electroanal. Chem. 35, 41 (1972).

[f] T. E. Suarez, R. T. Iwamoto, and J. Kleinberg, Inorg. Chim. Acta 7, 292 (1973).

[g] At $I = 1 M$, brought about by NH_4ClO_4, calorimetric determination: S. Ahrland, B. Tagesson, and D. Tuhtar, unpublished data.

[h] S. E. Manahan and R. T. Iwamoto, Inorg. Chem. 4, 1409 (1965).

[i] T. Williams, J. Inorg. Nucl. Chem. 24, 1215 (1962).

[j] H. Hoffmann, personal communication.

[k] F. Dickert and H. Hoffmann, Ber. Bunsenges. Phys. Chem. 75, 1320 (1971).

high enough). Also the copper(II) thiocyanate system tends to an analogous reaction.[54] These redox reactions occur also in the aprotic solvents. The iodide system is thus excluded from investigation.

Chloride and bromide complexes are much more stable in DMSO than in water, and for copper(II), an (a) sequence can be discerned. As usual, the increase in stability relative to water is much smaller for the thiocyanate systems. The copper(II) complexes are stronger throughout than the corresponding nickel(II) complexes, as predicted by the Irving–Williams rule.[55] This order is in fact expected whether the ligands are coordinated by bonds of mainly electrostatic or mainly covalent character, though the differences should be larger in the latter case.[56] For the copper(II) bromide system in DMSO, a 1 M ammonium perchlorate medium suppresses complex formation quite considerably relative to media containing no hydrogen bonding cation, just as it does for cadmium(II).

In AN, copper(II) chloride complexes become very stable. The main reason is presumably the same as has been advanced to account for the stabilization of proton complexes in this solvent (see Section IV,A), i.e., a weaker solvation of the ions involved.

The random errors of K_j are seemingly so large that a discussion of the various ratios K_j/K_{j+1} is hardly worthwhile.

V. THERMODYNAMICS OF COMPLEX FORMATION REACTIONS IN THE SOLVENTS SELECTED

A. Enthalpy and Entropy Changes and Their Determination

As has already been stressed, the stability of a complex in solution is the net result of several competing interactions between the various components involved. These interactions all involve enthalpy changes which mainly reflect the energies spent or gained when existing bonds are broken and new bonds formed, as well as entropy changes connected with the changes of structural order during these processes. The stepwise stability constants K_j, discussed in Section IV,B, are connected with the net changes of free energy, ΔG_j^0, enthalpy, ΔH_j^0, and entropy, ΔS_j^0, of the respective steps according to the well-known formula

$$-RT \ln K_j = \Delta G_j^0 = \Delta H_j^0 - T\Delta S_j^0 \tag{7}$$

Measurements of ΔH_j^0, combined with the values of ΔG_j^0 found from the stability measurements, will thus provide complete sets of the thermodynamic functions mentioned for the formation of each consecutive complex. From these functions important conclusions may then be drawn about the various factors governing complex formation, such as the solvation effects, the character of the coordinate bond, and the changes of structure often taking place as complex formation proceeds.

In aqueous solution, reactions between typically hard acceptors and donors are often endothermic, especially as far as the first steps are concerned.[9,32a,33,57] In such cases complex formation is exclusively due to a large gain in entropy. Reactions between typically soft acceptors and donors are, on the other hand, always strongly exothermic while the entropy change is often negative or, if positive, contributes only little to the stability of the complexes.[9,32a,33] The formation of a stable soft–soft complex is therefore due exclusively, or at least mainly, to a large decrease in enthalpy.

Thus, in aqueous solution, typical hard–hard interactions are in general entropy controlled, while typical soft–soft interactions are enthalpy controlled. Moreover, that term which does not determine the course of the reaction quite often even acts to some extent in the opposite direction. For complexes formed between acceptors and donors on the borderline between hard and soft, on the other hand, the entropy and enthalpy terms often contribute about equally to the decrease in free energy.[32a]

As the changes of entropy and enthalpy evidently depend very much upon both the solvation of the reacting species and the structure of the solvent, large differences in these functions are to be expected between solvents as markedly different in these respects as the protic and the aprotic ones. The rules found for aqueous solutions are therefore likely to be considerably modified for an aprotic solvent such as DMSO. This of course applies especially to systems involving species which are very preferentially solvated by one solvent or the other.

To elucidate these points, the thermodynamics for the formation of halide and thiocyanate complexes of zinc,[49] cadmium,[58,59] and mercury(II)[59,60] in water and in DMSO will be compared. The stabilities of these complexes in solvents of various properties have been thoroughly discussed in Section IV,B. In Section V,B we will now consider how the stability differences expected and found between different solvents are related to changes in the entropy and enthalpy terms involved.

The enthalpy changes have been determined calorimetrically. This direct method is generally much faster and more precise than the calculation of ΔH_j^0 from the temperature dependence of K_j. and is therefore greatly preferred.[61–66] Only if the reactions are slow[67] (i.e., equilibrium not attained within $\approx \frac{1}{2}$ hour) or the concentrations available very low ($< 1mM$, which might be due either to scarce supply[62,63,68] or to low solubility[69]) the latter method is preferable, and may even be the only one possible, as is certainly true for the cases cited.

In principle, calorimetric measurements allow a simultaneous determination of ΔH_j^0 and K_j provided that K_j is not so large as to make the reaction virtually complete. Once this is the case, any sufficiently high value of K_j will of course satisfy the measurements. In practice, however, the values of ΔH_j^0 have to vary considerably between the consecutive steps if a reasonably precise determination of both ΔH_j^0 and K_j is to be made.[49,58,70] Generally, a separate determination of K_j by means of some well-established method is greatly recommended.

B. Zinc(II), Cadmium(II), and Mercury(II) Halides and Thiocyanates

As stated in Section IV,B, the halide complexes of the hard acceptor zinc(II) display an even more marked (a) sequence in DMSO than in water. Moreover, the overall stability of all the halide systems has increased considerably (see Table VII).

These stability changes are brought about by two very striking changes in the enthalpy and entropy terms of the complex formation reactions. As the terms connected with the individual steps are greatly influenced by those changes from octahedral to tetrahedral coordination which take place in all these systems (as will be further discussed below), the general trends are most clearly discerned in the sums $\Delta H_{\beta 3}^0$ and $\Delta S_{\beta 3}^0$ referring to the first three steps. This is as far as it has been possible to follow the complex formation. Fortunately, however, it seems as if the switch of coordination would be completed with this step for all the halides, so the sums mentioned should refer to the formation of complexes of the same coordination.

The first very striking change from water to DMSO is that the trend of $\Delta H_{\beta 3}$ between the halides is reversed from $Cl^- > Br^-$ to $Cl^- \ll Br^- (\ll I^-)$ (Table XII). This reversal evidently reflects that no hydrogen bonds have to be broken in DMSO. The resulting changes of $\Delta H_{\beta 3}^0$ very much favor the formation of chloride complexes in DMSO relative to water and also, though to a lesser extent, the formation of bromide complexes. The (a) sequence thus tends to become more marked in DMSO.

Also in this solvent, however, $\Delta H_{\beta 3} > 0$ for all the halides. The complexes would therefore still be quite weak, as in water, were it not for the very positive values of the entropy terms $\Delta S_{\beta 3}^0$ (Table XII). This certainly depends on the fact that the total entropy gain due to the desolvation of cations and anions on complex formation is much larger in a relatively unstructured solvent as DMSO than it is in a highly structured one such as water. In DMSO, the solvent molecules leave well-ordered solvates and enter a fairly unordered bulk solvent. In water they leave solvates which might be somewhat less well ordered than in DMSO (e.g., Zn^{2+}, I^-; see Table IIIA) or somewhat more well ordered (e.g., Cl^-, Br^-) but in any case they enter

TABLE XII

THERMODYNAMICS OF ZINC HALIDE AND THIOCYANATE COMPLEXES IN WATER
AND DMSO AT 25°C IN PERCHLORATE MEDIA OF THE COMPOSITION INDICATED[a]

Water, NaClO$_4$:	$3M$[b]		$1\ M$[c]	DMSO, NH$_4$ClO$_4$, $1\ M$[d]			
	Cl$^-$	Br$^-$	SCN$^-$	Cl$^-$	Br$^-$	I$^-$	SCN$^-$
ΔG_1^0	1.1	3.3	−4.0	−11.1	−4.8	4.0	−7.9
ΔG_2^0	2.4	4.7	−1.9	−22.2	−16.5	−8.0	−8.1
ΔG_3^0	−4.3	−3	−0.8	−12.9	−7.7	−0.9	−13.7
ΔG_4^0	—	—	−1.9	—	—	—	−9.4
ΔH_1^0	5.5	1.5	−5.8	22.3	27.8	19.0	5.5
ΔH_2^0	38	42	−2	0.8	9.1	29.4	23.5
ΔH_3^0	0	−8	−1	−10.2	−4.2	12.7	−17.8
ΔH_4^0	—	—	−8	—	—	—	−10.7
ΔS_1^0	15	−5.9	−6	112	110	50	45
ΔS_2^0	120	125	0	77	85	127	105
ΔS_3^0	8	−17	0	9.1	12	45	−13
ΔS_4^0	—	—	−20	—	—	—	−4
$\Delta G_{\beta 3}^0$	−0.8	5	−6.7	−46.2	−29.0	−4.9	−29.7
$\Delta H_{\beta 3}^0$	44	36	−9	12.9	32.7	61.1	11.2
$\Delta S_{\beta 3}^0$	143	102	−6	198	207	222	137

[a] ΔG_j and ΔH_j in kJ mole^{-1}, ΔS_j in JK^{-1} mole^{-1}.
[b] L. G. Sillén and B. Liljeqvist, *Sven. Kem. Tidskr.* **56**, 85 (1944); P. Gerding, *Acta Chem. Scand.* **23**, 1695 (1969).
[c] S. Ahrland and L. Kullberg, *Acta Chem. Scand.* **25**, 3692 (1971).
[d] S. Ahrland, N.-O. Björk, and R. Portanova, *Acta Chem. Scand., Ser. A* **30**, 270 (1976).

a bulk solvent which is definitely much better ordered than DMSO. Consequently, much more entropy will be gained in DMSO than in water. By structuring the solvent, hydrogen bonding between the solvent molecules in protic solvents thus tends to make the complexes less stable than they are in aprotic solvents.

For zinc thiocyanate in DMSO, the value of $\Delta H_{\beta 3}^0$ is about the same as for chloride, but the value of $\Delta S_{\beta 3}^0$ is much lower than for any of the halides. As will be seen, this is also a general feature, certainly due to the extra loss of conformational entropy suffered by the triatomic SCN$^-$ on complex formation. This ligand has more degrees of freedom to lose than the monoatomic halide ions.

If complex formation only involves the stepwise displacement of one solvent molecule by one ligand anion, the entropy gain due to desolvation is expected to decrease smoothly from step to step, as a consequences of that continual decrease of solvate order which accompanies the lowering of the effective charge on the metal atom. In systems where such simple substitutions certainly occur, this smooth decrease of ΔS_j^0 is also encountered and, moreover, both for hard–hard and soft–soft interactions. Several good examples are found among the numerous systems investigated in aqueous solution.[32a] Thus the fluoride systems of the typically hard Al^{3+} and Be^{2+} (where the hydrate and the fluoride complexes are all octahedral for Al^{3+} and all tetrahedral for Be^{2+}) behave in this way, as does also the chloride system of the soft Pd^{2+} (where the hydrate and the complexes are all square planar). On the other hand, if the complex formation is no simple substitution reaction but involves a change of the coordination figure, such a smooth decrease of ΔS_j^0 cannot be expected. Especially if the switch wholly or mainly occurs at a certain step, the consecutive values of ΔS_j^0 are bound to vary in a seemingly irregular fashion, reflecting the sudden rupture of the existing solvate pattern.[32a]

When no complexing agents are present, Zn^{2+} certainly exists as an octahedral hexasolvate both in water and in DMSO.[49] In water, the halide and also the thiocyanate complexes are too unstable to allow any safe conclusions about any coordination changes that might possibly occur. In DMSO, the complexes are much more stable (see Section IV,B,2 and Table VII). Higher complexes are consequently easily reached, in the thiocyanate system even $Zn(SCN)_4^{2-}$. This complex clearly constitutes the upper limit of complex formation, however. In the chloride and bromide systems, complex formation is even reluctant to proceed beyond the third step, while the iodide complexes are too unstable even in DMSO to allow any conclusions on this point.[71] In no case, however, are complexes formed beyond the fourth one. In view of the fact that not only $ZnCl_4^{2-}$ but also $ZnBr_3(H_2O)^-$ have been found as discrete tetrahedral groups in solid compounds,[72] it seems most probable that the higher halide and pseudohalide complexes formed in DMSO are in fact tetrahedral. This is also in line with the generally found preference for tetrahedral coordination in solid zinc chloride, bromide, and iodide compounds.[72]

The coordination is thus bound to change somewhere between the octahedral solvate and the tetrahedral higher complexes. If the switch takes place at a certain step, the rupture of the solvate shell results in the liberation of more solvate molecules than would occur in the case of a simple substitution. Consequently this step will be characterized by an abnormally large entropy gain.[32a] On the other hand, extensive desolvation will absorb extra energy which results in an abnormally unfavorable enthalpy change, counteracting the entropy gain. Whether the complex formed at the switch will in the end be more or less stable than its neighbors depends upon the balance between these two influences.[58]

For the zinc iodide and thiocyanate systems, the value of ΔS_2^0, and also of ΔH_2^0, are abnormally high in DMSO (see Table XII), indicating that the switch from octahedral to tetrahedral coordination at least mainly takes place at the formation of the second complex. In the chloride and bromide systems, on the other hand, ΔS_1^0 and ΔH_1^0 are largest, which would imply that in these cases the first step is already involved in the switch. Since the values of ΔS_2^0 are still quite large, however, the switch might well extend over the first two steps, especially in the case of bromide.

The balance between the entropy and the enthalpy terms is different in the different systems so that the first complex has an especially narrow range of existence for chloride, bromide, and iodide while the second is disfavored for thiocyanate, though not at all to the same extent (see Table VII). For the halides, the second complex is in fact rather favored in the sequence $Cl^- > Br^- > I^-$. Evidently, abnormal ratios of K_j/K_{j+1} do appear at co-ordination changes, but on account of the subtle balance between the entropy and enthalpy terms involved, one cannot tell from stability measurements alone whether the switch takes place at the jth or at the $(j+1)$th step.[58] In order to decide this, one has to know the entropy and enthalpy changes involved.

The mild (b) sequence displayed by the cadmium halide systems in water is switched into an (a) sequence in DMSO (see Table VIII). As for zinc, the overall stabilities of the complexes are moreover considerably enhanced in DMSO. This of course applies especially when the concentrations of cations competing with Cd^{2+} for the ligands are kept low (in the present investigation such ions are exemplified by NH_4^+, cf. Section IV,B,2).

These stability changes are brought about by the same factors that are at work, with analogous results, at the zinc halide systems. First, the trend of $\Delta H_{\beta 3}^0$ switches from $Cl^- > Br^- > I^-$ in water to $Cl^- < Br^- < I^-$ in DMSO (see Table XIII) reflecting the absence of hydrogen bonding in DMSO which so greatly favors the chloride and also the bromide complexes relative to the iodide ones. As for zinc, however, all values of $\Delta H_{\beta 3}^0$ are >0 in DMSO, and the complexes would thus be very unstable, were it not for the very positive values of $\Delta S_{\beta 3}^0$. These are somewhat lower than for zinc, evidently on account of the weaker solvation of Cd^{2+} relative to Zn^{2+} in DMSO ($\Delta H_{sv}^0 = -1898$ and -2123 kJ, respectively, see Table IIIA), but nevertheless large enough to bring about fairly stable complexes (see Tables VIII and XIII). For the same reason, the values of $\Delta H_{\beta 3}^0$ are all more favorable for cadmium than for zinc complexes. The overall result is that in DMSO $ZnCl_3^-$ is slightly more stable than $CdCl_3^-$, while $ZnBr_3^-$ is less stable than $CdBr_3^-$ and ZnI_3^- much less stable than CdI_3^- (compare the values of $\Delta G_{\beta 3}^0$ in Tables XII and XIII). Thus, the softer the ligand, the stronger are the Cd^{2+} complexes relative to the Zn^{2+} complexes which evidently reflects the stronger capacity for covalent bonding of Cd^{2+}.

For the thiocyanate system, the value of $\Delta S_{\beta 3}^0$ is again much lower than for any of the halides. The difference is so large that the stability also becomes much lower, in spite of the relatively favorable value of $\Delta H_{\beta 3}^0$ (Table XIII).

Also in the cadmium halide systems, a switch of coordination certainly takes place, both in aqueous solution and in DMSO. In both solvents, Cd^{2+} exists as an octahedral hexasolvate in noncomplexing solutions, as has been proved by x-ray diffraction measurements.[73,74] Also, the solid solvates $(NH_4)_2[Cd(H_2O)_6](SO_4)_2$ and $[Cd(DMSO)_6](ClO_4)_2$ contain this structural unit.[75,76] The upper limit of complex formation in both water and DMSO is the fourth complex,[59] CdL_4^{2-}, in all the halide systems. Again by x-ray diffraction, CdI_4^{2-} has been proved[73,77] to be a regular tetrahedron in both solvents and there is certainly every reason to believe that the same is true also for $CdCl_4^{2-}$ and $CdBr_4^{2-}$.

The abnormally high values of ΔS_2^0 found for the cadmium halides in DMSO indicate that for all these systems the switch from octahedral to tetrahedral coordination mainly takes place as the second complex is formed (Table XIII). As would be expected, the corresponding values of ΔH_2^0 are also abnormally high. In water, where the values of ΔS_j^0 are smaller throughout, the switch is not as clearly marked but no doubt it mainly takes place at the third step, as indicated by the abnormal values of ΔS_3^0. The entropy and enthalpy changes thus reveal that the change of coordination takes place at different steps in water and in DMSO, in spite of the fact that in both solvents the same complex, i.e., the second one, has an especially narrow range of existence (see Table VIII). This is another example of the fact that determination of the complete thermodynamics of complex formation is necessary in order to ensure a reliable interpretation of the reactions involved. Interestingly enough, the measurements performed on the chloride system in DMSO containing only 0.1 M NH_4ClO_4 show that ΔS_2^0 and ΔH_2^0 are not as much larger than their neighbors in this medium as they are in 1 M NH_4ClO_4 (see Table XIII). This change of medium actually seems to postpone the switch of coordination so that, to a substantial extent, it takes place at the third step also. If so, at least the second step must involve an equilibrium between octahedral and tetrahedral complexes.

In the chloride and bromide systems, the switch of coordination in DMSO thus takes place at a lower step for Zn^{2+} than for Cd^{2+}. This might be connected with the more severe crowding of DMSO around Zn^{2+} which should encourage an earlier rupture of the solvate. From this point of

TABLE XIII

Thermodynamics of Cadmium Halide and Thiocyanate Complexes in Water and DMSO[a]

	Water, NaClO$_4$: 3 M[b]				1 M[c]			DMSO, NH$_4$ClO$_4$: 1 M[d]				0.1 M[e]
	Cl⁻	Br⁻	I⁻	SCN⁻	Cl⁻	I⁻	SCN⁻	Cl⁻	Br⁻	I⁻	SCN⁻	Cl⁻
ΔG_1^0	-9.0	-10.0	-11.9	-8.0	-7.7	-10.7	-7.5	-18.4	-16.7	-12.5	-10.3	-24.6
ΔG_2^0	-3.7	-3.3	-4.0	-4.8	-2.5	-4.6	-3.8	-11.2	-10.9	-7.8	-5.2	-16.5
ΔG_3^0	-1.0	-5.6	-12.2	-1.3	2.1	-9.6	-0.3	-14.7	-15.7	-16.8	-1.2	-19.5
ΔG_4^0	—	-2.1	-9.2	0.0	—	-7.6	0.9	-10.0	-9.5	-6.7	—	-13.0
ΔH_1^0	-0.4	-4.1	-9.5	-8.1	0.5	-10.2	-9.6	-6.3	-3.9	2.4	-3.0	-5.0
ΔH_2^0	0.1	-2.4	-0.8	-7.2	2.1	-2.1	-8.1	15	17	27	-2.8	12.9
ΔH_3^0	7.7	7.2	-3.1	-6.6	8	-5.9	-9	1	2	-5	4.2	6.0
ΔH_4^0	—	1.3	-15.9	-4	—	-16.8	—	-12.2	-13.0	-9.5	—	-15.9
ΔS_1^0	29	20	8	0	28	2	-7	41	43	50	25	66
ΔS_2^0	13	3	10	-8	15	8	-14	88	94	117	8	98
ΔS_3^0	29	43	31	-18	18	12	-30	53	59	40	18	86
ΔS_4^0	—	11	-23	-15	—	-33	—	-7	-12	-10	—	-9
$\Delta G_{\beta 3}^0$	-13.7	-18.9	-28.1	-14.1	-8.1	-24.9	-11.6	-44.3	-43.3	-37.1	-16.7	-60.6
$\Delta H_{\beta 3}^0$	7.4	0.7	-13.4	-21.9	11	-18.2	-27	10	15	24	-2	13.9
$\Delta S_{\beta 3}^0$	77	66	49	-26	61	22	-51	182	196	207	51	250

[a] At 25°C, in perchlorate media of the composition indicated. ΔG_j and ΔH_j in kJ mole⁻¹, ΔS_j in JK⁻¹ mole⁻¹.

[b] I. Leden, "Potentiometrisk undersökning av några kadmiumsalters komplexitet," Diss., University of Lund, 1943.

[c] C. E. Vanderzee and H. J. Dawson, J. Am. Chem. Soc. 75, 5659 (1953); P. Gerding and I. Jönsson, Acta Chem. Scand. 22, 2247 (1968); P. Gerding and B. Johansson, ibid. p. 2255.

[d] S. Ahrland and N.-O. Björk, Acta Chem. Scand., Ser. A 30, 257 (1976).

[e] S. Ahrland, N.-O. Björk, I. Persson, and R. Portanova, unpublished data.

view it seems of course strange that I^-, in spite of being the largest ligand, causes the solvate to break at a later step than does Cl^- and Br^-. The reason might be that the lower charge density of I^- does not bring about the necessary decrease of the effective charge on the zinc atoms.

The cadmium thiocyanate system does not display any markedly abnormal values, either in DMSO or in water, of ΔS_j^0 or ΔH_j^0, though a slight increase relative to the previous step is observed both for ΔS_3^0 and ΔH_3^0 (see Table XIII). It might be that the switch occurs mainly at the highest, i.e., the fourth step, which is not reached in the present measurements.

The complete thermodynamics for the formation of mercury halide and thiocyanate complexes in DMSO is not yet known, but sufficient data have nevertheless been accumulated to allow several interesting conclusions. Thus it may be inferred that the almost complete leveling in DMSO of the very marked (b) sequence found in water is again due to that drastic change in the relative magnitudes of ΔH_j^0 which is caused by the absence of hydrogen bonding in the aprotic solvent. For the very soft acceptor Hg^{2+} the values of ΔH_j^0 are as a rule strongly negative, reflecting the very covalent character of the metal to ligand bonds. In water, these values become more negative for all steps in the sequence $Cl^- < Br^- < I^-$ (Table XIV) just as the values of $\Delta H_{\beta 3}^0$ for Zn^{2+} and Cd^{2+} (see Tables XII and XIII). In DMSO, the differences between the various Hg^{2+} systems are much smaller; in some instances they have even disappeared (Table XIV). Contrary to what is found for Cd^{2+}, however, a switch to an opposite sequence does not take place in the case of Hg^{2+}. This is again a consequence of the capacity of Hg^{2+} for covalent bonding which favors the coordination of softer ligands relative to harder ones. It should be remembered that the same tendency, though to a lesser extent, is observed for Cd^{2+} relative to Zn^{2+}.

Like Zn^{2+} and Cd^{2+}, Hg^{2+} is hexasolvated in both water and DMSO, with the solvent molecules in a regular octahedral arrangement, as has been found by x-ray diffraction measurements.[74,78] The evidence is even more conclusive in the case of DMSO where not only the Hg–O but also the Hg–S distances can be found. The results also prove beyond doubt that DMSO is indeed coordinated via the oxygen atom (see Section III,B).

In DMSO, separate values for the stabilities of the first two complexes are so far not known, so no values of ΔS_1^0 or ΔS_2^0 can be calculated. The values of β_2 determined in 1 M NH_4ClO_4 (Table IX) allow the calculation of $\Delta G_{\beta 2}^0 = -102.5$ and -115.2 kJ mole^{-1} for the chloride and bromide systems, however. Combined with the values of $\Delta H_{\beta 2}^0 = \Delta H_1^0 + \Delta H_2^0$ (Table XIV) these yield $\Delta S_{\beta 2}^0 = 179$ and 196 JK^{-1}, respectively. These values are much higher than the values of $\Delta S_{\beta 2}^0$ found for the corresponding cadmium complexes (Table XIII). They are in fact about as high as for

TABLE XIV

THERMODYNAMICS OF MERCURY(II) HALIDE COMPLEXES IN WATER AND DMSO[a]

	Water					DMSO			
	NaClO$_4$, 0.5 M^b			3 M^c		NaClO$_4$, 1 M^d		NH$_4$ClO$_4$, 1 M^e	
	Cl⁻	Br⁻	I⁻	Cl⁻	Br⁻	Br⁻	I⁻	Cl⁻	Br⁻
ΔG_1^0	−38.5	−51.7	−73.5	−40.3	−53.7	—	—	—	—
ΔG_2^0	−37.0	−47.3	−62.5	−39.4	−49.0	—	—	—	—
ΔG_3^0	−5.4	−13.8	—	−4.3	−15.8	−32.0	−34.8	−22.8	−29.3
ΔG_4^0	−6.0	−7.2	—	−7.9	−8.5	−15.1	−14.7	−11.9	−14.5
ΔH_1^0	−24.7	−42.2	−75.3	−24.2	−40.0	—	—	−20.4	−24.8
ΔH_2^0	−28.9	−44.8	−67.8	−27.2	−40.2	—	—	−28.6	−32.2
ΔH_3^0	−9.2	−12.0	—	−4.3	−10.8	−26.9	−26.8	−19.5	−27.3
ΔH_4^0	0.4	−17.2	—	−6.2	−18.6	−22.7	−22.8	−13.5	−18.7
ΔS_1^0	46	32	−6	54	46	—	—	—	—
ΔS_2^0	27	8	−18	41	30	—	—	—	—
ΔS_3^0	−13	6	—	0	17	17	27	11	7
ΔS_4^0	21	−33	—	5	−34	−25	−27	−5	−14
$\Delta G_3^0 + \Delta G_4^0$	−11.4	−21.0	−34.5	−12.2	−24.3	−47.1	−49.5	−34.7	−43.8
$\Delta H_3^0 + \Delta H_4^0$	−8.8	−29.2	−42.0	−10.5	−29.4	−49.6	−49.6	−33.0	−46.0
$\Delta S_3^0 + \Delta S_4^0$	8	−27	−25	5	−17	−8	0	6	−7

[a] At 25 C, in perchlorate media of the composition indicated. ΔG_j and ΔH_j in kJ mole⁻¹, ΔS_j in JK⁻¹ mole⁻¹.

[b] L. G. Sillén, *Acta Chem. Scand.*, **3**, 539 (1949); Y. Marcus, *ibid.* **11**, 599 (1957); P.K. Gallagher and E. L. King, *J. Am. Chem. Soc.* **82**, 3510 (1960); M. Björkman and L. G. Sillén, *Trans. Royal Inst. Technol. Stockholm*, No. 199, 1963; J. J. Christensen, R. M. Izatt, L. D. Hansen, and J. D. Hale, *Inorg. Chem.* **3**, 130 (1964).

[c] R. Arnek, *Ark. Kemi* **24**, 531 (1965).

[d] R. Arnek and D. Poceva, *Acta. Chem. Scand., Ser. A* **30**, 59 (1976).

[e] S. Ahrland, I. Persson, and R. Portanova, unpublished data.

the zinc halides (Table XII), in spite of the fact that Zn^{2+} is more strongly solvated than Hg^{2+} in DMSO by over 200 kJ (see Table III).

The obvious explanation is that, with the formation of the second linear Hg^{2+} complex, all the solvent molecules coordinated in the octahedral solvate have been relegated to an outer coordination sphere, while the formation of the second tetrahedral Zn^{2+} complex leaves two of them in the inner coordination sphere (see Section III,B). The former process evidently involves an extra entropy gain which more than compensates the stronger bonding of the DMSO molecules in the initial Zn^{2+} solvate.

That this difference in structure between the ZnL_2 and HgL_2 complexes indeed involves energy differences comparable to those found between $\Delta H^0_{sv}(Zn^{2+})$ and $\Delta H^0_{sv}(Hg^{2+})$ is evident from the values of $\Delta H^0_{sv}(ZnL_2)$ and $\Delta H^0_{sv}(HgL_2)$ listed in Table IV. The former are more strongly solvated by 115 and 120 kJ mole^{-1} for the chloride and the bromide, respectively, which compensates much of the difference between $\Delta H^0_{sv}(Zn^{2+})$ and $\Delta H^0_{sv}(Hg^{2+})$. For Cd^{2+}, the values of $\Delta S^0_{\beta2}$ for the chloride and bromide systems in DMSO are only $\simeq 130$ JK^{-1} mole^{-1} (see Table XIII), i.e., much lower than for both Hg^{2+} and Zn^{2+}. This is to be expected as Cd^{2+} has the lowest solvation enthalpy of the three acceptors (see Table III) and this weaker solvation is not compensated by any extensive desolvation on the formation of the second complex. On the contrary, the switch to tetrahedral coordination occurs later in the Cd^{2+} than in the Zn^{2+} systems and it might well be that the CdL_2 complexes in fact represent equilibria between octahedral and tetrahedral structures. This would of course tend to suppress $\Delta S^0_{\beta2}$ even more.

In water, the thermodynamics for the formation of the first two complexes are well known for all the mercury(II) halide systems. On the other hand, those determined for the zinc halides are very uncertain, on account of the extremely low stability of the complexes (see Table XII). A valid comparison is therefore possible only with the cadmium halides. Preferably, that medium where data exist for both acceptors, i.e., 3 M sodium perchlorate, should then be considered (Tables XIII and XIV). As in DMSO, the values of $\Delta S^0_{\beta2}$ for the mercury(II) chloride and bromide, 95 and 76 JK^{-1} mole^{-1}, respectively, are much higher than the corresponding values for the cadmium systems, 42 and 23 JK^{-1} mole^{-1}, and evidently for the same reason as in DMSO. Since the change from octahedral to tetrahedral coordination in the cadmium systems mainly takes place at the third step in water but at the second step in DMSO, it is also understandable that the difference between the two acceptors is proportionally even larger in water.

Especially in DMSO, the values of ΔS^0_3 and ΔS^0_4 for the mercury(II) halides systems are low relative to $\Delta S^0_{\beta2}$ which reflects the weak solvation of the second and third complexes as compared with first and second ones. The rather large difference found between ΔS^0_3 and ΔS^0_4 in DMSO is presumably connected with the change from linear to tetrahedral coordination which takes place on the formation of the third complex. As the formation of the fourth complex does not involve any profound change of coordination (cf. Section IV,B,2), the considerable difference between the entropy changes connected with the two steps is in fact not surprising.

In DMSO, the wide ranges of existence of the second as well as of the third complex in the mercury(II) halide systems are thus largely entropy effects. The effects due to differences in ΔH_j^0 are surprisingly small at the formation of both the third and the fourth complex (Table XIV). This is in marked contrast to the conditions in water where the preponderance of the second complex is mainly an enthalpy effect (Table XIV). The third complex has no particularly wide range of existence in water, due to counteracting influences from the enthalpy and the entropy terms.

These conclusions again demonstrate the deeper insight to be gained from the determination of all the thermodynamic functions involved in formation reactions.

C. Silver Complexes of Monodentate Donors Coordinating via N, P, As, Sb, and Bi

Ligands with low affinities for solvent molecules are generally not soluble in a highly structured solvent like water, where the strong bonds between the solvent molecules virtually prevent any solvate formation. In less well-structured solvents like DMSO, on the other hand, the specific interactions between the solvent molecules are much weaker, and the formation of ligand solvates becomes a more likely reaction. Consequently, many ligands which are not at all soluble in water dissolve readily in DMSO. Among these are the simple triphenyl compounds Ph_3X formed by the donor atoms of the nitrogen group; X = N, P, As, Sb, and Bi. These donors coordinate metal ions in a most discriminating way and the investigation of their complex equilibria is therefore of great interest. As both the ligands and most of the complexes formed are soluble in DMSO, extensive quantitative studies of the thermodynamics of complex formation have now become relatively easy to perform.

Early preparative work indicated that hard acceptors, with the stability sequence $F^- \gg Cl^- > Br^- > I^-$ toward the halide ions (see Section IV,B,1), displayed an analogous sequence $N \gg P > As > Sb$ toward ligands coordinating via donor atoms of the nitrogen group. The affinities of typically hard acceptors for the heavy donor atoms of this group are indeed very weak. In sharp contrast to this, typically soft acceptors, with the halide sequence $F^- \ll Cl^- < Br^- < I^-$, show strong affinities for the heavy donors of the nitrogen group, though the affinity sequence is a different one: $N \ll P > As > Sb > Bi$.[79] In aqueous solution, this sequence has been confirmed quantitatively for the soft acceptor Ag^+, as far as the first three donor atoms are concerned.[80] This was achieved by means of ligands which had been sulfonated in order to become sufficiently soluble. The complex formation of monosulfonated triphenyl phosphine in aqueous solution with several other acceptors has since been determined. These measurements confirm that really strong complexes are formed only with very soft acceptors, i.e., also copper(I),[81] gold(I),[82] mercury(II),[83] palladium(II),[84] and platinum(II),[84] besides silver(I), while complexes formed by borderline acceptors, such as cadmium(II)[85] and lead(II),[81] are already quite weak. Similar results have also been found for an aliphatic phosphine, triethyl-phosphine, made soluble in water by the introduction of an alcoholic group.[86] The aliphatic phosphine is, however, a stronger donor than the aromatic one. For one thing it is a fairly strong base (log K_1 = 8.1, at 22°C in 1 M potassium nitrate)[86] while the basic properties of the aromatic phosphine are extremely weak (log K_1 = 0.63, at 25°C in 1 M perchloric acid).[87]

To modify the ligands in order to reach a sufficiently high solubility is a time-consuming and rather difficult procedure and, in the case of antimony and bismuth compounds, it has not even been achieved. The solubilities obtained are moreover too low for precise calorimetric measurements. The use of the aprotic DMSO instead of water for the investigations of reactions of this type therefore brings many advantages.

Stabilities in DMSO have now been determined[88] for the silver(I) systems of all the Ph_3X ligands. The affinity sequence postulated before for typically soft acceptors has been fully confirmed (Table XV). The phosphine complexes are far more stable than the arsine or stibine ones which in turn do not differ very much from each other. The triphenylbismuth and, especially, the triphenylamine form only weak complexes. As a consequence no complexes beyond the first one are formed in these systems even at the highest ligand concentrations that can be reached.

A most interesting feature common to the three systems where complex formation proceeds beyond the first step is that the first and third complexes are particularly stable relative to their neighbors. The second complex, so prominent in all the silver(I) halide systems (cf. Section IV,B), plays only a subordinate role here. The fourth complex is not formed in appreciable amounts within the present concentration range. At least for the phosphine system, this means that the fourth complex is formed very reluctantly as is immediately evident from the complex formation functions drawn in Fig. 1. These functions give the ligand number, $\bar{n}$, defined as the average number of ligands per acceptor ion, as a function of the log of the free ligand concentrations, [L]. For the phosphine system, $\bar{n}$ clearly approaches 3 without any tendency to exceed this value within the range of [L] available. The strong predominance of the first complex is evident from the marked inflexions at $\bar{n}$ = 1 for all the three systems and also, of course, from the high values of K_1/K_2 relative to K_2/K_3 (Table XV). Also in this respect, the phosphine system stands out before the arsine and stibine ones which again do not differ very much among themselves.

If an excess of ligand is added at sufficiently high silver(I) concentrations

TABLE XV

EQUILIBRIUM CONSTANTS (K_j IN M^{-1}) AND THERMODYNAMICS OF SILVER(I) COMPLEXES OF LIGANDS COORDINATING VIA NITROGEN-GROUP DONORS IN DMSO AND WATER

| | DMSO; ligands Ph$_3$X; 0.1 M NH$_4$ClO$_4^b$ | | | | | H$_2$O | | |
	X = N	P	As	Sb	Bi	Dpm^c	Dop^d	Asm^c
log K_1	0.19	6.58	3.56	3.16	0.80	8.15	11.83	5.36
log K_2	—	4.15	1.81	1.45	—	5.95	9.02	—
log K_3	—	2.44	1.31	1.45	—	5.40	4.86	—
K_1/K_2	—	269	56	51	—	160	650	—
K_2/K_3	—	52	3.2	1.0	—	3.5	14500	—
$-\Delta G_1^0$	1.1	37.6	20.3	18.1	4.6	46.5	66.8	30.6
$-\Delta G_2^0$	—	23.7	10.4	8.3	—	34.0	51.0	—
$-\Delta G_3^0$	—	13.9	7.5	8.3	—	30.8	27.4	—
$-\Delta H_1^0$	1	51.8	34.5	32.1	0.5	—	80.8	—
$-\Delta H_2^0$	—	38.1	19.4	8.6	—	—	69.0	—
$-\Delta H_3^0$	—	36.3	44.5	57.1	—	—	38.1	—
$-\Delta S_1^0$	0	48	48	47	-14	—	47	—
$-\Delta S_2^0$	—	48	30	1	—	—	60	—
$-\Delta S_3^0$	—	75	124	164	—	—	36	—
$-\Delta G_{\beta 3}^0$	—	75.2	38.1	34.6	—	111.3	145.2	—
$-\Delta H_{\beta 3}^0$	—	126.2	98.4	97.8	—	—	187.9	—
$-\Delta S_{\beta 3}^0$	—	171	202	212	—	—	143	—

^{a} At 25°C (Dop 22°C). ΔG_j^0 and ΔH_j^0 in kJ mole^{-1}, ΔS_j^0 in JK^{-1} mole^{-1}.

^{b} S. Ahrland, T. Berg, and P. Trinderup, *Acta. Chem. Scand. Ser. A* **31** (1977). In press.

^{c} Dpm = Ph$_2$PC$_6$H$_4$SO$_3^-$ (m–); 0.1 M NaClO$_4$; Asm = As(C$_6$H$_4$SO$_3^-$)$_3^{3-}$ (m–); 0.2 M NaClO$_4$. For the phosphine system, the ratio K_3/K_4 is certainly >4000. For the arsine system, no complex beyond the first one can be proved in the range of [L] available; the ratio K_1/K_2 is certainly >20,000. S. Ahrland, J. Chatt, N. R. Davies, and A. A. Williams, *J. Chem. Soc.* p. 276 (1958).

^{d} Dop = Et$_2$PCH$_2$CH$_2$OH; 1 M KNO$_3$. For this system, a ratio K_3/K_4 > 4000 has also been found. M. Meier, "Phosphinokomplexe von Metallen," Diss. No. 3988. Eidgenössische Technische Hochschule, Zürich, 1967.

solid perchlorates Ag(Ph$_3$X)$_4$ClO$_4$, X = P, As, Sb, precipitate.[88–90] A preliminary structure determination[91] indicates that these compounds are isostructural, with the ligands in a regular tetrahedral coordination around Ag$^+$. The fourth complex is thus finally formed but precipitates readily in the medium used as the fairly insoluble perchlorate.

The complexes formed by the Ph$_3$P and Ph$_3$As ligands in DMSO are considerably less stable than those formed by the sulfonated phosphine, Dpm,

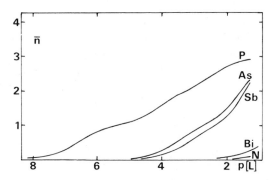

FIG. 1. Average number of Ph_3X ligands, where X = N, P, As, Sb, and Bi per silver(I) ion $\bar{n}$ as a function of the free ligand ion concentration, [L], in DMSO solution. Medium, $0.1 \, M \, NH_4ClO_4$; 25°C.

and arsine, Asm, in water (Table XV) in spite of the fact that sulfonation ought to weaken the donor properties. This lower stability in DMSO is certainly mainly due to the stronger solvation of Ag^+ in this solvent, reflected in the very negative $\Delta H_{tr}^0(H_2O \rightarrow DMSO) = -54 \, kJ \, mole^{-1}$ (see Table III and Sections III,A and IV,B). Due to the stronger donating properties of the aliphatic phosphine Dop, the silver(I) complexes formed by this ligand in aqueous solution are even stronger than those formed by the aromatic Dpm (Table XV).

The first and third complexes are also predominating in the aromatic phosphine system in water. For the arsine Asm, the first complex has an even wider range of existence. No subsequent complex is formed within the range of [L] available which means that $K_1/K_2 \gtrsim 20\,000$. The aliphatic phosphine behaves similarly insofar as both the first and the third complex have unusually wide ranges of existence (Table XV). In this system, however, this applies even more to the second complex which of course resembles the conditions in the silver(I) halide systems (see Section IV,B). Also, in most other silver(I) systems, the linear second complex is of course the favored one. Good examples are the pseudohalides and all amines, including ammonia.[92]

The interactions involving the soft acceptor Ag^+ and the soft donor atoms P, As, or Sb are all strongly exothermic both in DMSO and water (Table XV). This is certainly expected from the strongly covalent Ag—X bond and the weak solvation, especially of the ligands. The weak solvation also causes unfavorable entropy terms, however, so that the stabilities of the complexes are rather much lower than would be expected from the very considerable enthalpy decreases (Table XV). The value of $-\Delta H_{\beta 3}$, referring to the highest complex actually found in the solutions, decreases in the order P > As > Sb, reflecting the declining strength of the Ag—X bond, while the value of $-\Delta S^0_{\beta 3}$ increases in the same order, reflecting the fact that the ligands have to conform more strictly as they grow more bulky. For both $\Delta H^0_{\beta 3}$ and $\Delta S^0_{\beta 3}$, however, the big jump is between P and As, while the difference between As and Sb is modest, especially for $\Delta H^0_{\beta 3}$.

For the individual steps, quite different but very consistent patterns are found for ΔH^0_j in the different systems (Table XV). Certainly, a gradual change of the character of the Ag—X bond with the donor atom must be one of the main causes for the variations observed. At present, however, it seems wise to postpone any elaborate interpretation until experimental data have also been gathered for other acceptors coordinating these donors.

The weak complex formation between silver(I) and the ligands Ph_3N and Ph_3Bi is characterized by quite low values of both ΔH^0_1 and ΔS^0_1 (Table XV). Evidently, these reactions involve only very modest changes of bonding strength and structural order.

REFERENCES

1. R. Fernandez-Prini, in "Physical Chemistry of Organic Solvent Systems" (A. K. Covington and T. Dickinson, eds.), p. 571 et seq. Plenum, New York, 1973.
2. S. Ahrland and N.-O. Björk, Coord. Chem. Rev. 16, 115 (1975).
3. S. Ahrland and N.-O. Björk, Acta Chem. Scand., Ser. A 28, 823 (1974).
4. F. Gaizer and G. Johansson, Acta Chem. Scand. 22, 3013 (1968).
4a. M. Sandström and G. Johansson, Acta Chem. Scand., Ser. A 31, 132 (1977).
5. A. Foll, M. Le Démézet, and J. Courtot-Coupez, J. Electroanal. Chem. 35, 41 (1972).
6. T. E. Suarez, R. T. Iwamoto, and J. Kleinberg, Inorg. Chim. Acta 7, 292 (1973).
7. A. Foll, M. Le Démézet, and J. Courtot-Coupez, Bull. Soc. Chim. Fr. p. 408 (1972).
8. J. W. Vaughn, in "The Chemistry of Nonaqueous Solvents" (J. J. Lagowski, ed.), Vol. 2, p. 192. Academic Press, New York, 1967.
9. G. Schwarzenbach, Pure Appl. Chem. 24, 307 (1970).
10. J. J. Lagowski and G. A. Moczygemba, in "The Chemistry of Nonaqueous Solvents" (J. J. Lagowski, ed.), Vol. 2, p. 319. Academic Press, New York, 1967.
11. R. Alexander, D. A. Owensby, A. J. Parker, and W. E. Waghorne, Aust. J. Chem. 27, 933 (1974).
12. J. A. V. Butler, Annu. Rep. Chem. Soc. 34, 75 (1937) (as cited by W. J. Moore, "Physical Chemistry," 5th ed., p. 915. Longmans, Green, New York, 1972).
13. R. M. Noyes, J. Am. Chem. Soc. 84, 513 (1962).
14. H. F. Halliwell and S. C. Nyburg, Trans. Faraday Soc. 59, 1126 (1963).
15. D. F. C. Morris, Struct. Bonding (Berlin) 4, 63 (1968); 6, 157 (1969).
16. M. Salomon, J. Phys. Chem. 74, 2519 (1970).
17. K. Fajans, Verh. Dsch. Phys. Ges. 21, 709 (1919) (as cited by Halliwell and Nyberg[14] and by Morris[15]).
18. E. Grunwald, G. Baughman, and G. Kohnstam, J. Am. Chem. Soc. 82, 5801 (1960).
19. R. Alexander and A. J. Parker, J. Am. Chem. Soc. 89, 5549 (1967).
20. C. V. Krishnan and H. L. Friedman, J. Phys. Chem. 73, 3934 (1969).
21. I. M. Kolthoff and M. K. Chantooni, Jr., J. Phys. Chem. 76, 2024 (1972).

22. B. G. Cox and A. J. Parker, *J. Am. Chem. Soc.* **95**, 402 (1973).
23. J. Barthel, "Ionen in nichtwässerigen Lösungen," p. 74 *et seq.* Dietrich Steinkopff Verlag, Darmstadt, 1976.
24. Y.-C. Wu and H. L. Friedman, *J. Phys. Chem.* **70**, 501 (1966).
25. E. M. Arnett and D. R. McKelvey, *J. Am. Chem. Soc.* **88**, 2598 (1966).
26. H. F. Halliwell and S. C. Nyburg, *J. Chem. Soc.* p. 4603 (1960).
27. C. M. Criss and M. Salomon, *in* "Physical Chemistry of Organic Solvent Systems" (A. K. Covington and T. Dickinson, eds.), p. 254. Plenum, New York, 1973.
28. M. Spiro *in* "Physical Chemistry of Organic Solvent Systems" (A. K. Covington and T. Dickinson, eds.), p. 636. Plenum, New York, 1973.
29. J. F. Coetzee and G. P. Cunningham, *J. Am. Chem. Soc.* **87**, 2529 (1965).
30. J. Kenttämaa, *Suom. Kemistil. B* **33**, 179 (1960).
31. S. Ahrland, L. Kullberg, and R. Portanova, *Acta Chem. Scand., Ser. A* **32**, (1978). In press.
32. K. Hartley, H. Pritchard, and H. Skinner, *Trans. Faraday Soc.* **47**, 254 (1951).
32a. S. Ahrland, *Struct. Bonding (Berlin)* **15**, 167 (1973).
33. S. Ahrland, *Struct. Bonding (Berlin)* **5**, 118 (1968).
34. J. Courtot-Coupez and M. Le Démézet, *Bull. Soc. Chim. Fr.* p. 1033 (1969).
35. R. Domain, M. Rinfret, and R. L. Benoit, *Can. J. Chem.* **54**, 2101 (1976).
36. E. M. Arnett, T. C. Moriarty, L. E. Small, J. P. Rudolph, and R. P. Quirk, *J. Am. Chem. Soc.* **95**, 1492 (1973).
37. E. J. King, *in* "Physical Chemistry of Organic Solvent Systems" (A. K. Covington and T. Dickinson, eds.), p. 331. Plenum, New York, 1973.
38. W. S. Matthews, J. E. Bares, J. E. Bartmess, F. G. Bordwell, F. J. Cornforth, G. E. Drucker, Z. Margolin, R. J. McCallum, G. J. McCollum, and N. R. Vanier, *J. Am. Chem. Soc.* **97**, 7006 (1975).
39. I. M. Kolthoff and M. K. Chantooni, Jr., *J. Phys. Chem.* **72** 2270 (1968).
40. R. L. Benoit and C. Buisson, *Electrochim. Acta* **18**, 105 (1973).
41. C. McCallum and A. D. Pethybridge, *Electrochim. Acta* **20**, 815 (1975).
42. I. M. Kolthoff, S. Bruckenstein, and M. K. Chantooni, Jr., *J. Am. Chem. Soc.* **83**, 3927 (1961).
43. G. Klopman, *J. Am. Chem. Soc.* **90**, 223 (1968).
44. R. G. Pearson, *J. Chem. Educ.* **45**, 581 and 643 (1968).
45. R. G. Pearson, ed., "Hard and Soft Acids and Bases." Dowden, Hutchinson & Ross, Stroudsburg, Pennsylvania, 1973.
46. S. E. Manahan and R. T. Iwamoto, *Inorg. Chem.* **4**, 1409 (1965).
47. S. Ahrland and J. Rawsthorne, *Acta Chem. Scand.* **24**, 157 (1970).
48. S. Fronaeus, *Tech. Inorg. Chem.* **1**, 1 (1963).
49. S. Ahrland, N.-O. Björk, and R. Portanova, *Acta Chem. Scand., Ser. A* **30**, 270 (1976).
50. P. Gerding, *Acta Chem. Scand.* **22**, 1283 (1968).
51. S. Ahrland and B. Tagesson, *Acta Chem. Scand., Ser. A* **31**, 615 (1977).
52. J. Courtot-Coupez and M. L'Her, *Bull. Soc. Chim. Fr.* p. 675 (1969).
53. O. Bravo and R. T. Iwamoto, *Inorg. Chim. Acta* **3**, 663 (1969).
54. D. Krüger, W. Büssem, and E. Tschirch, *Ber. Dsch. Chem. Ges. B* **69**, 1601 (1936).
55. H. Irving and R. J. P. Williams, *J. Chem. Soc.* p. 3192 (1953).
56. G. Schwarzenbach, *Chimia* **27**, 1 (1973).
57. I. Grenthe and H. Ots, *Acta Chem. Scand.* **26**, 1229 (1972).
58. S. Ahrland and N. O. Björk, *Acta Chem. Scand., Ser. A* **30**, 257 (1976).
59. S. Ahrland, I. Persson, and R. Portanova, unpublished data.
60. R. Arnek and D. Poceva, *Acta Chem. Scand., Ser. A* **30**, 59 (1976).
61. A. McAuley and G. H. Nancollas, *J. Chem. Soc.* p. 989 (1963).
62. A. D. Jones and G. R. Choppin, *Actinides Rev.* **1**, 311 (1969).

63. G. R. Choppin and J. K. Schneider, *J. Inorg. Nucl. Chem.* **32**, 3283 (1970).
64. R. A. Day, Jr. and R. M. Powers, *J. Am. Chem. Soc.* **76**, 3895 (1954).
65. S. Ahrland and L. Kullberg, *Acta Chem. Scand.* **25**, 3471 and 3677 (1971).
66. J. J. Christensen and R. M. Izatt, "Handbook of Metal Ligand Heats." Dekker, New York, 1970.
67. L.-I. Elding, *Acta Chem. Scand.* **24**, 1331 (1970).
68. R. G. de Carvalho and G. R. Choppin, *J. Inorg. Nucl. Chem.* **29**, 737 (1967).
69. J. H. Jonte and D. S. Martin, Jr., *J. Am. Chem. Soc.*, **74**, 2052 (1952).
70. L. Kullberg, *Acta Chem. Scand.*, *Ser. A* **28**, 829 and 897 (1974).
71. S. Ahrland and N.-O. Björk, *Acta Chem. Scand.*, *Ser. A* **30**, 265 (1976).
72. A. F. Wells, "Structural Inorganic Chemistry," 4th ed., pp. 390 and 394. Oxford Univ. Press, (Clarendon), London and New York, 1975.
73. H. Ohtaki, M. Maeda, and S. Ito, *Bull. Chem. Soc. Jpn.*, **47**, 2217 (1974).
74. S. Ahrland, I. Persson, and M. Sandström, unpublished data.
75. H. Montgomery and E. C. Lingafelter, *Acta Crystallogr.* **20**, 728 (1966).
76. M. Sandström, personal communication.
77. G. Johansson and S. Pocev, personal communication.
78. G. Johansson, *Acta Chem. Scand.* **25**, 2787 (1971).
79. S. Ahrland, J. Chatt, and N. R. Davies, *Qt Rev.*, *Chem. Soc.* **11**, 265 (1958).
80. S. Ahrland, J. Chatt, N. R. Davies, and A. A. Williams, *J. Chem. Soc.* p. 276 (1958).
81. R. George and J. Bjerrum, *Acta Chem. Scand.* **22**, 497 (1968).
82. C. J. Hawkins, O. Mønsted, and J. Bjerrum, *Acta Chem. Scand.* **24**, 1059 (1970).
83. B. Salvesen and J. Bjerrum, *Acta Chem. Scand.* **16**, 735 (1962).
84. J. C. Chang and J. Bjerrum, *Acta Chem. Scand.* **26**, 815 (1972).
85. S. Ahrland, J. Chatt, N. R. Davies, and A. A. Williams, *J. Chem. Soc.* p. 1403 (1958).
86. M. Meier, "Phosphinokomplexe von Metallen," Diss. No. 3988. Eidgenössische Technische Hochschule, Zürich, 1967.
87. G. Wright and J. Bjerrum, *Acta Chem. Scand.* **16**, 1262 (1962).
88. S. Ahrland, T. Berg, and P. Trinderup, *Acta Chem. Scand. Ser. A* **31** (1977). In press.
89. F. A. Cotton and D. M. L. Goodgame, *J. Chem. Soc.* p. 5267 (1960).
90. R. H. Nuttall, E. R. Roberts, and D. W. A. Sharp, *J. Chem. Soc.* p. 2854 (1962).
91. A. Cassel, personal communication.
92. L. G. Sillén and A. E. Martell, "Stability Constants of Metal-Ion Complexes," Special Publ. Nos. 17 and 25. Chem. Soc., London, 1964 and 1971 resp.

~ 2 ~

Solvent Basicity

∾

ROBERT L. BENOIT AND CHRISTIAN LOUIS

Département de Chimie, Université de Montréal
Montréal, Quebec, Canada

I. INTRODUCTION

When one considers the large number of solvents and the many definitions of basicity, this survey of the subject of solvent basicity seems to be a rather ambitious undertaking. Yet this very ambition emphasizes the importance of the work. An excellent paper on basicity itself has recently been published by Arnett *et al.*[1] Basicity is an essential solvent property often used to account for the influence of the solvent on chemical phenomena, as will be apparent from the following. For example, it has recently been proposed that the solvent effect on various physicochemical properties related to ion–solvent and ion–ion interactions depends directly on only two parameters, the Lewis basicity and the Lewis acidity of the solvent.[2] The dissociation of an acid HA in a solvent is another physicochemical process affected by the basicity of the solvent toward both species, HA and H^+.[3] The problem of relating acidity scales in different solvents is also highly relevant.[4] Furthermore, the formation of hydrogen-bonded species such as $(HA)_n$, $A^- \cdots HA$ or $BH^+ \cdots B$ is highly dependent on the basic properties of the solvent.[5] Examples pertaining to the influence of the basicity of the solvent on kinetics and thermodynamics of chemical reactions can be found in reviews dealing with solvent effects such as that of Amis and Hinton.[6] In addition, correlations have often been sought between spectroscopic properties of acid solutes, whether molecular[7] or cationic,[8] and solvent basicity.

It can be stated that most molecular solvents possess some basic properties in that the solvent molecules, due to the availability of electrons on some of their atoms, can usually be considered as Lewis bases. The need has obviously arisen to classify solvents according to their basicity so that quantitative aspects of basicity effects can be rationalized. That the basic strength of solvents varies widely is readily apparent. Nitromethane and hexamethylphosphotriamide, on the one hand, and fluorosulfuric acid and liquid ammonia on the other differ by some 30 and 50 kcal/mole, respectively, in their free energy of solvation of the gaseous proton. However, when one attempts to set up a quantitative or even a qualitative basicity scale, difficulties arise. For example, while in the gas phase, pyridine is more basic than ammonia, in that its reaction with the proton is some 18 kcal/mole more exothermic, in aqueous solution the reverse holds true since the free energy of protonation of ammonia is some 5 kcal/mole higher than that of pyridine. Although such an apparent contradiction has been nicely explained in terms of solvation effects,[9] the question remains: which solvent, ammonia or pyridine, is more basic and by how much? We will attempt to answer such questions.

Despite the fact that it has been proposed to utilize core-level binding-energy shifts as one operational measure of the Lewis basicity,[10] so that basicity may be viewed as an intrinsic molecular property, basicity scales are usually established by comparing parameters of reactions involving the various bases with a common reference acid. This approach implies some built-in problems because of the multiplicity of reaction parameters. First, the common acids are many and may belong to the Brønsted or Lewis class. Then, the nature of the acid–base reaction will differ according to whether it involves the transfer of a proton or the formation of an ion pair or a molecular adduct. Finally, the medium used to carry out the acid–base reaction is of importance. While the basic solvent itself should be the preferred reaction medium for the purpose of setting up solvent basicity scales, other media have also been used, e.g., bulk acids, such as fluorosulfuric acid; inert media, such as carbon tetrachloride or 1,2-dichloroethane; and more recently the gas phase. It is then readily apparent that some of the discrepancies observed between different basicity scales could simply reflect the fact that, only when the previous terms of reference and the measured acid–base reaction parameters, whether thermodynamic or spectroscopic, are well defined, has basicity then an operational meaning.

Our survey of solvent basicity, which does not pretend to be exhaustive, will be conducted with these limitations in mind. In Section II we will consider data relating to the basicity of the bulk solvent S using either the proton H^+ or Brønsted acids such as p-fluorophenol as the common acid. Then, in Sections III, IV, and V, we will examine results derived from the reaction between isolated basic solvent molecules B and a common acid in various media.

First, in Section III, we will look at data pertaining to the reaction of B in an acidic medium of reference, usually a strong acid in water. This leads to the familiar pK_a scale. In order to accommodate weaker bases, more acidic media are used to extend the previous scale with the H_o Hammett acidity function. Bulk fluorosulfuric acid is also used to establish a scale based on enthalpies of protonation instead of free energies.

We will then turn our attention in Section IV to the many thermodynamic and spectroscopic determinations using an inert medium such as CCl_4 or $1,2\text{-}C_2H_4Cl_2$ for the reaction, at high dilution, of B with a common acid such as antimony pentachloride, or Brønsted acids such as phenol.

Finally, in Section V we will review recent results relating to gas phase basicity of B through measurements of gas phase proton affinities. In the concluding section, Section VI, we will attempt to find out to what extent the somewhat limited results on bulk basicity can be confirmed or supplemented by more extensive data on basicity of isolated solvent molecules and we will also compare results on the basicity of bulk solvents.

In order to limit our survey, we will consider only solvents of practical interest. This gives rise to the problem of classifying the solvents for the purpose of presenting data in table form. Rather than using a classification

such as that proposed by Brønsted and used by others,[5,11] we will group
solvents somewhat arbitrarily under the following convenient types[4,12]:
nitrogen bases (amines, nitriles); oxygen bases (alcohols, ethers, aldehydes,
ketones, carboxylic acids, esters); other bases (amides, nitro compounds,
sulfoxides, sulfones, inorganic compounds); hydrocarbons, and derivatives.
Liquid metals and molten salts will not be dealt with here, although
attempts have been made by others to relate basicity in the latter media and
molecular solvents.[13] Solvent mixtures will also be left out because of their
number and diversity. As a final remark on the limitations of our survey, we
will consider experimental methods in some detail only when they have
some direct bearing on the quality of the results or when they are relatively
unfamiliar to solution chemists. A case in point would be the mass-
spectrometric methods used to obtain proton affinities.

II. BASICITY OF BULK SOLVENTS

The establishment of the simplest solvent basicity scale would involve the
choice of one common reference acid and the determination of either the free
energy or the enthalpy change for the reaction between this acid and bulk
solvents. The first acid which comes to mind, because of its ubiquitous role,
is the proton. However, the determination of the enthalpy or the free energy
of solvation of a single ion, H^+, can only be done with the help of extrapolations or extrathermodynamic assumptions, procedures which are open to
criticism. The use of an uncharged Brønsted acid (HA) such as phenol does
not offer such problems, and the determinations of solubility or heat of
solution of the acid in various solvents lead to free energy and enthalpy
values. Lewis acids such as $SbCl_5$ or I_2 may not be suitable because they give
additional complex ionization reactions with the more basic solvents.

A. Free Energy and Enthalpy of Solvation of the Gaseous
Proton in Bulk Solvents

The process of proton solvation can be represented by the equation

$$H^+(g) \rightarrow H^+(s) \tag{1}$$

The enthalpy of solvation of H^+, $\Delta H^0_{g \rightarrow s}(H^+)$, is defined as the enthalpy
change occurring when 1 mole of proton is transferred from the gas phase at
low pressure to infinite dilution in the liquid phase. The free energy of
solvation of H^+, $\Delta G^0_{g \rightarrow s}(H^+)$, is defined as the free energy change occurring
when 1 mole of H^+ is transferred from a standard gaseous state to a stan-

dard state in solution. The choice of the standard state when defining ΔG^0 values has been discussed by Friedman and Krishnan.[14] ΔG^0 values quoted in this chapter will refer, unless otherwise specified, to a 1 atm gas phase standard state and to a 1 M solution standard state.

$\Delta G^0_{g \to s}(H^+)$ and $\Delta H^0_{g \to s}(H^+)$ account for the interactions between solvents S and the common acid H^+ and give an indication of the relative basic strengths of the bulk solvents S. However, these thermodynamic parameters have large numerical values which cannot be obtained with any degree of precision and their variations from solvent to solvent are more meaningful. Emphasis has therefore been placed on relative or transfer solvation enthalpies and free energies [$\Delta H^0_{w \to s}(H^+)$ and $\Delta G^0_{w \to s}(H^+)$], water (w) being often selected as the reference solvent. The corresponding process

$$H^+(w) \to H^+(s) \tag{2}$$

is better suited to experimental investigation than is the reaction depicted by Eq. 1. In place of $\Delta G^0_{w \to s}(H^+)$, the free energy of transfer of H^+ from water to solvent, the transfer activity coefficient $^w\gamma^s(H^+)$ is often used. The relation between these two quantities is

$$\Delta G^0_{w \to s}(H^+) = RT \ln {}^w\gamma^s(H^+) \tag{3}$$

Equation 4 gives the relationship between the thermodynamic parameters P for reactions given by Eqs. 1 and 2.

$$\Delta P^0_{g \to s}(H^+) = \Delta P^0_{w \to s}(H^+) + \Delta P^0_{g \to w}(H^+) \tag{4}$$

where P stands for free energy or enthalpy. $\Delta G^0_{g \to w}(H^+)$ and $\Delta H^0_{g \to w}(H^+)$, the strongly exothermic free energy and enthalpy of hydration of the proton have been evaluated by a number of authors.[15] Although some of the results differ by as much as 50 kcal/mole, the most recent estimates[14,16-19] agree on values of -261 ± 2 and -270 ± 2 kcal/mole, respectively.

1. EXPERIMENTAL DETERMINATIONS LEADING TO SOLVATION ENTHALPIES AND FREE ENERGIES OF H^+

It is physically impossible to transfer protons from one phase to another without simultaneously transferring an equivalent number of anions in the same direction or an equivalent number of cations in the opposite direction. The experimental techniques described here therefore yield free energies and enthalpies of solvation of pairs of ions. To separate these experimental values into contributions for single ions some extrathermodynamic assumption is needed. Some of these assumptions are discussed in Section II,A,2.

a. *Free Energy of Transfer of H^+ from Water to Solvent.* If the dissociation constant K_d of a weak acid HA is known in water and in the solvent S, then the thermodynamic quantity sought $\Delta G^0_{w \to s}(H^+)$ is calculated from

$$\Delta G^0_{w \to s}(H^+) + \Delta G^0_{w \to s}(A^-) = RT \ln (_wK_d/_sK_d) + \Delta G^0_{w \to s}(HA) \qquad (5)$$

Dissociation constants may be obtained from

1. Conductometric measurements. The electrical conductivity of HA solutions of different concentrations is measured. Treatment of the data by a method such as that of Shedlovsky yields K_d. It should be noted that the presence of basic impurities in the solvent may lead to errors in the interpretation of the results.

2. Potentiometric measurements. The potential of an electrode sensitive to proton activity (glass electrode, hydrogen electrode) is measured in solutions of HA. Calibration of the electrode in solutions, preferably buffered, containing known concentrations of solvated protons is required for K_d to be obtained. This calibration often presents problems as the slope is rarely RT/F.

3. Spectrophotometric measurements. If the species HA and A^- have distinct absorption bands, their concentrations can easily be determined spectrophotometrically and K_d calculated.

The free energy of transfer of the uncharged compound HA, $\Delta G^0_{w \to s}(HA)$, can be determined from its solubilities (sol) in water (w) and solvent (s)—if it is slightly soluble and not appreciably ionized—from its partition coefficient between w and s, $K_{w \to s}$, if w and s are immiscible or if a third solvent is available which is immiscible with w and s, or from its partial vapor pressure over its solutions in w and s.

$$\Delta G^0_{w \to s}(HA) = RT \ln(_w\text{sol}/_s\text{sol}) \qquad (6)$$

$$= -RT \ln(K_{w \to s}) \qquad (7)$$

$$= RT \ln(_wK_H/_sK_H) \qquad (8)$$

where K_H is the proportionality constant between vapor pressure and concentration at infinite dilution.

Free energies of solvation of pairs of ions including H^+ can also be obtained from standard potentials (potentials with respect to the standard hydrogen electrode, SHE, in the medium studied) of reversible electrodes in w and in s. For example, if the redox process studied in both solvents is

$$M^+(l) + e^- \rightleftharpoons M(\text{solid}) \qquad (9)$$

then

$$\Delta G^0_{w \to s}(M^+) - \Delta G^0_{w \to s}(H^+) = F(_sE^0 - {}_wE^0) \qquad (10)$$

Although the species M and H_2 participate in the electrode process defining E^0, they are both in standard states, solid and gaseous (1 atm) respectively, and their solvation free energies do not appear in Eq. 10. If the reduction of M^+ at a dropping mercury electrode is reversible, E^0 can be approximated by $E_{1/2}$, the corresponding polarographic half-wave reduction potential with respect to SHE. Other voltammetric techniques may be preferable in certain circumstances.

b. *Enthalpy of Transfer of H^+ from Water to Solvent.* $\Delta H^0_{w \to s}$ values for pairs of ions can be obtained from the variation of the corresponding free energy values with temperature. More accurate values are obtained, however, from calorimetric measurements. For example, if there exists a Brønsted acid, HA, which dissociates completely in the different solvents investigated

$$HA(pure) \to H^+(s) + A^-(s) \tag{11}$$

then its heats of solution at infinite dilution, ΔH^0_s and ΔH^0_w, can be measured and enthalpies of transfer calculated from

$$\Delta H^0_{w \to s}(H^+) + \Delta H^0_{w \to s}(A^-) = \Delta H^0_s - \Delta H^0_w \tag{12}$$

$HSbCl_6$ has been used as a strong acid in related determinations.[3]

2. EXTRATHERMODYNAMIC ASSUMPTIONS FOR THE ESTIMATION OF SOLVATION ENTHALPIES AND FREE ENERGIES OF SINGLE IONS

Because several reviews and papers have adequately covered this subject[13,20-23] our discussion here will be brief.

a. *Use of Modified Born Equations.* If an ion is assumed to be a hard sphere of radius r with a centrally symmetrical charge distribution, and a solvent to be a fluid with a uniform dielectric constant D_s, then the contribution of electrostatic interactions to the free energy of solvation or transfer of the ion can be expressed by the Born equation

$$\Delta G^{el}_{g \to s} = -(6.023 \times 10^{23} z^2 e^2 / 2r)[1 - (1/D_s)] \tag{13}$$

or

$$\Delta G^{el}_{w \to s} = -(6.023 \times 10^{23} z^2 e^2 / 2r)[(1/D_s) - (1/D_w)] \tag{14}$$

where ze is the charge of the ion considered. ΔH^{el}, the enthalpy change analogous to ΔG^{el}, can be obtained by differentiation of Eqs. 13 and 14 and from values of $d(D_s)/d(T)$. The nonelectrostatic part of the free energy or of the enthalpy of ion solvation (ΔG^{neut} or ΔH^{neut}) can be equated to the free energy or to the enthalpy of solvation of an uncharged analog of the ion considered (the analog should be isoelectronic and isostructural with the ion). The free energy or enthalpy of solvation of the ion is then obtained by adding the electrostatic and nonelectrostatic contributions.

The validity of this method for obtaining individual ionic solvation energies can be checked by comparing the sum of the calculated energies for a pair of ions with the corresponding experimental value. In practice, experimental and calculated values compare well only for a few pairs of ions. This is not surprising since Eqs. 13 and 14 are based on a very simplified model for ion solvation. Several modifications of the Born equation have been proposed[14,22] to take into account such factors as the variation of the dielectric constant of the solvent in the neighborhood of the ion (dielectric saturation effects) and the molecular nature of the solvent. But the use of these new equations does not improve significantly the correlation between experimental and calculated values.[22] However, the fact that the Born equation is expected to hold better for large and spherically symmetrical ions provides a way of getting around this problem. If the calculated value is assumed to be correct for such an ion, then solvation energies for all other ions are obtained from experimental data for pairs of ions.

   b. *Extrapolation Methods.* Most of these methods are based on the assumption that the solvation energy of an ion can be expressed as a function of $1/r^n$, where the integer n can have values of 1, 2, 3, 4, or 6. Thus, the solvation energy of an infinitely large ion $(1/r \to 0)$ is assumed to be zero. On this basis Izmailov[24] has obtained values of $\Delta G^0_{g \to s}(H^+)$ for several solvents by plotting experimental values for $[-\Delta G^0_{g \to s}(M^+) + \Delta G^0_{g \to s}(H^+)]$ and $[\Delta G^0_{g \to s}(H^+) + \Delta G^0_{g \to s}(X^-)]$ against $1/r(M^+)$ and $1/r(X^-)$ respectively $(M^+$ being alkali ions and X^- halide ions). Both of these functions extrapolate to give $\Delta G^0_{g \to s}(H^+)$ at $1/r \to 0$. By plotting energies of transfer from water, instead of absolute solvation energies, Feakins and Watson[25] have obtained values of $\Delta G^0_{w \to s}(H^+)$ by the same method. Alfenaar and De Ligny[26] have proposed an improved method which takes into account the fact that only the electrostatic part of $\Delta G^0_{g \to s}$ is a function of $1/r$. On the basis of the relationship

$$\Delta G^0_{g \to s}(H^+) + \Delta G^0_{g \to s}(A^-) - \Delta G^{neut}_{g \to s}(A^-)$$
$$= \Delta G^0_{g \to s}(H^+) + a/r(A^-) + b/r(A^-)^2 + \cdots \quad (15)$$

they have plotted the left hand side expression (values obtained experimentally) against $1/r(A^-)$ to obtain $\Delta G^0_{g \to s}(H^+)$ by extrapolation to $1/r(A^-) \to 0$.

   Another extrapolation method, that of Izmailov,[27] is based on the assumption that solvation energies of cations and anions can be expressed as a function of $1/n^2$, where n is the principal quantum number of the lowest vacant orbital of the ion. If M^+ and X^- are isoelectronic alkali and halide ions respectively, then the function $[-\Delta G^0_{g\to s}(H^+) + \{\Delta G^0_{g\to s}(M^+) - \Delta G^0_{g\to s}(X^-)\}/2 = f(1/n^2)$ was assumed to yield $-\Delta G^0_{g\to s}(H^+)$ when extrapolated to $1/n^2 \to 0$.

The results obtained by such extrapolation methods are not too reliable because of the limited number of experimental points and the hazardous nature of the extrapolations.

c. *Negligible Liquid Junction Assumption.* The potential difference, ΔE, measured across a cell such as

Pt,H$_2$(1 atm fugacity)/H$^+_{(w)}$(1 M activity)//H$^+_{(s)}$(1 M activity)/H$_2$(1 atm fugacity), Pt

is related to the free energy of transfer of the proton from w to s by the equation

$$\Delta E = \Delta G^0_{w\to s}(H^+)/F + E_j \tag{16}$$

where E_j is the liquid junction potential across the aqueous–nonaqueous boundary. It has been proposed[28,29] that the interposition of a concentrated salt bridge between the two phases makes E_j negligible so that $\Delta G^0_{w\to s}(H^+)$ can be calculated directly from ΔE. Saturated aqueous KCl solutions[28] and saturated tetraethylammonium picrate in several solvents[29] have been used as salt bridges.

Another approach[30] used to evaluate $\Delta G^0_{w\to s}(H^+)$ from Eq. 16, has been to calculate approximate values for E_j from expressions such as

$$E_j = -K \int_w^s \sum (t_i/z_i) \, d(\log M_i) \tag{17}$$

where t_i, z_i, and M_i are respectively the Hittorf transference numbers, the charges, and the molarities of the ionic species i in the transition layers between the two phases w and s.

The advantage of the negligible junction potential assumption is the ease with which it can be applied to yield numerical results. Its fault is that junction potentials may not always be negligible even with the interposition of concentrated salt bridges,[13,31] while Eq. 17 yields only very approximate values of E_j.[22]

d. *The Rubidium Ion Assumption.* This assumption, advanced by Pleskov,[32] was one of the first to be formulated to allow the calculation of $\Delta G^0_{w\to s}$ values for single ions. It stipulates that, owing to its large radius, its low charge and polarizability, and its minimal tendency toward specific solvation, the rubidium ion should have small and nearly constant solvation energies in all solvents. Thus $\Delta G^0_{w \to s}(Rb^+) = 0$ and all other $\Delta G^0_{w \to s}$ values could be calculated from experimental data. However, Rb^+ is probably not nearly large enough ($r = 1.48$ Å) to justify this assumption and its interest is mainly historical.

e. *The Ferrocene–Ferricinium Assumption.* Ferrocene [iron(II) dicyclopentadienyl], Fc^+ can be oxidized to give the ferricinium cation, Fc^+, according to

$$Fc(s) \rightleftharpoons Fc^+(s) + e^-$$ (18)

The assumption proposed by Strehlow[33] states that the absolute standard electrode potential of this redox system can be considered invariable irrespective of the solvent. The difference between the standard potentials of this redox system in w and s is related to solvation energies by

$$F(_sE^0 - {}_wE^0) = \Delta G^0_{w \to s}(Fc^+) - \Delta G^0_{w \to s}(Fc) - \Delta G^0_{w \to s}(H^+)$$ (19)

Assuming that the ferrocene molecule is an excellent neutral analog of the ferricinium ion, it follows that

$$\Delta G^0_{w \to s}(Fc^+) - \Delta G^0_{w \to s}(Fc) = \Delta G^0_{w \to s}(Fc^+) - \Delta G^{neut}_{w \to s}(Fc^+)$$

$$= \Delta G^{el}_{w \to s}(Fc^+)$$ (20)

Strehlow's assumption would then be equivalent to saying that, because of the large size of the ferricinium ion ($r = 3.8$ Å), its structure and its low surface charge density, $\Delta G^{el}_{w \to s}(Fc^+)$ should be negligible so that

$$F(_sE^0 - {}_wE^0) = \Delta G^0_{w \to s}(H^+)$$ (21)

The reasoning leading to this assumption is sound, although one questionable part is that a 3.8 Å ionic radius may not be large enough to allow dielectric saturation problems to be disregarded. However, one drawback of the method is that the determination of the redox potential of the system proves difficult in some solvents and voltammetric techniques are often preferred over potentiometry. It must also be kept in mind that ferrocene and ferricinium ions may react with some solvents either through protonation or redox processes.[34]

f. *Assumption Related to Hammett Acidity Function.* Hammett bases, B, have been used to measure acidity levels in aqueous and nonaqueous solutions. Equation 22 expresses the difference between the pK_d of a cationic acid, BH^+, in water and in a solvent as an algebraic sum of free energies of solvation.

$$RT[_s(pK_d) - {}_w(pK_d)] = \Delta G^0_{w \to s}(H^+) + \Delta G^0_{w \to s}(B) - \Delta G^0_{w \to s}(BH^+)$$ (22)

If the quantity $[\Delta G^0_{w \to s}(B) - \Delta G^0_{w \to s}(BH^+)]$ is assumed negligible, then $\Delta G^0_{w \to s}(H^+)$ can be calculated directly from $_s(pK_d)$ and $_w(pK_d)$. This assumption can be justified in the same way as Strehlow's ferrocene assumption. The quality of the results it yields will depend on how good a neutral analog of BH^+ B is and on how large and spherically symmetrical BH^+ is. The validity of the assumption is readily tested by comparing the results obtained using different large cationic acids. Significant differences[23] have been observed between such results so that this method may only be useful in estimating orders of magnitude for $\Delta G^0_{w \to s}(H^+)$.

g. *The Tetraphenylborate Assumption.* This assumption, proposed by Grunwald et al.,[35] states that for an electrolyte, Y^+X^-, composed of two large symmetrical ions which are similar in size and structure, the experimental values for $\Delta G^0_{w \to s}(Y^+X^-)$ or $\Delta H^0_{w \to s}(Y^+X^-)$ can be divided equally between the anion and the cation:

$$\Delta P^0_{w \to s}(X^-) = \Delta P^0_{w \to s}(Y^+) = \tfrac{1}{2}\Delta P^0_{w \to s}(Y^+X^-) \qquad (23)$$

It has been argued that because of the similarity in size and structure for Y^+ and X^- both the electrostatic and the nonelectrostatic contributions to their solvation energies should be equal. The reference electrolyte which has been mostly used in conjunction with this method is tetraphenylarsonium tetraphenylborate (Ph_4AsBPh_4).[36] Other reference electrolytes include the tetraphenylphosphonium[35] and triisoamyl-n-butylammonium[37] tetraphenylborates. Once the solvation energies of the ions of the Y^+X^- electrolyte have been established, other ionic solvation energies, including that of H^+, are obtained from the solvation energies of electrolytes involving either X^- or Y^+.

h. *Other Assumptions.* Researchers have discussed other more rarely used assumptions.[20-22] More interestingly, it has recently been proposed[34] that polynuclear alternate aromatic hydrocarbons such as perylene and their cation or anion radicals form redox systems which are analogous but more useful than the ferrocene–ferricinium system since they constitute a homologous series with E^0's spanning some 5 volts. Since some of these systems are stable in oxidizing or reducing solvents and in molten salt, they allow the basicity of these media to be estimated.

3. DATA FOR FREE ENERGY AND ENTHALPY OF TRANSFER OF H^+

Tables I and II show selected values for the transfer parameters $\Delta G^0_{w \to s}(H^+)$ and $\Delta H^0_{w \to s}(H^+)$, respectively, obtained or calculated from data in the literature. Corresponding $\Delta G^0_{g \to s}(H^+)$ and $\Delta H^0_{g \to s}(H^+)$ can be simply deduced using Eq. 3 with $\Delta G^0_{g \to w}(H^+) = -261$ kcal/mol and

TABLE I

Free Energies[a] of Transfer of H^+ from Water to Solvent at 25°C (kcal/mole)

Solvent	$\Delta G^0_{w \to s}(H^+)$ Modified Born equation	Extrapolation method	Negligible liquid junction	Pleskov hypothesis	Hammett-type hypothesis	Ferrocene assumption	Tetraphenylarsonium tetraphenylborate assumption
Ammonia		-21.6[d,p]				-24[q,r]	
Hydrazine		-18.0[d,p]					
Pyridine						-13.4[k], -5.3[s]	
Acetonitrile	3.3[c]	7.7[d]		5.9[j]	7.0[g]	8.5[k]	11.6[i], 11.1[l]
Water	O	O	O	O	O	O	O
Methanol	0.3[c]	4.4[d]		0.6[j]	0.6[h]	-2.4[k]	2.4[i], 2.6[i]
Ethanol		5.5[d]	3.6[e]				2.7[n,o]
n-Butanol		5.9[d]					
Isoamyl alcohol		5.9[d]					
Acetone		4.7[d]					
Formic acid	10.9[c]	13.5[d]		12.1[f]	6.4[h]	9.7[k], 5.9[w]	
Acetic acid						22.2[k]	
Trifluoroacetic acid						-3.2[k]	
Formamide	1.5[c]				0.8[h]		-2.0[l]
N,N-Dimethylformamide					-1.9[j]	-6.8[k]	-4.1[l], -3.4[l]
N,N-Dimethylacetamide						-8.8[k]	-5.8[l,m]
N-Methylpyrrolidone						-8.1[k]	-7.1[l]
Nitromethane						20.8[k]	23.8[l,m]
Dimethyl sulfoxide					-2.2[j]	-7.6[k]	-5.0[l], -4.5[l]
Tetramethylene sulfone[b]						14.0[k]	13.9[l]
Fluorosulfuric acid					20.5[u]	27.8[k,r]	
Sulfuric acid					16.2[u]		
Hydrogen fluoride					13.2[v]		

[a] Except where otherwise mentioned, ΔG^0 values are on the molar scale.

[b] Values at 30°C.

[c] O. Popovych, *Crit. Rev. Anal. Chem.* **7**, 73 (1970).

[d] N. A. Izmailov, *Dokl. Akad. Nauk SSSR* **149**, 884, 1103, and 1364 (1963).

[e] N. Bjerrum and E. Larsson, *Z. Phys. Chem. (Leipzig)* **127**, 358 (1927).

[f] V. A. Pleskov, *Usp. Khim.* **16**, 254 (1947).

[g] H. Strehlow, *in* "The Chemistry of Nonaqueous Solvents" (J. J. Lagowski, ed.), Vol. I, Chapter 4. Academic Press, New York, 1966.

[h] H. Strehlow, *Z. Elektrochem.* **56**, 827 (1952).

[i] G. Demange-Guérin, *Talanta* **17**, 1075 (1970).

[j] J. Courtot-Coupez, M. Le Démézet, A. Laouenan, and C. Madec, *J. Electroanal. Chem.* **29**, 21 (1971).

[k] Calculated from $\Delta G^0_{w\rightarrow s}(H^+) = 23.05 \, ({}_wE^0_{Fc} - {}_sE^0_{Fc})$. The values of the standard potentials of the ferrocene-ferricinium electrode were taken from A. Foucault, Doctoral Thesis, University of Paris VI, Paris (1975).

[l] Calculated from $\Delta G^0_{w\rightarrow s}(H^+) = 23.05 \, ({}_wE^0_{Ag} - {}_sE^0_{Ag}) + \Delta G^0_{w\rightarrow s}(Ag^+)$. The standard potentials of the silver electrode were calculated from the results of R. Alexander, A. J. Parker, J. H. Sharp, and W. E. Waghorne, *J. Am. Chem. Soc.* **94**, 1148 (1972); values of $\Delta G^0_{w\rightarrow s}(Ag^+)$ were taken from B. G. Cox, G. R. Hedwig, A. J. Parker, and D. W. Watts, *Aust. J. Chem.* **27**, 477 (1974) except in the case of nitromethane and *N,N*-dimethylacetamide.

[m] Values of $\Delta G^0_{w\rightarrow s}(Ag^+)$ taken from I. M. Kolthoff and M. K. Chantooni, Jr., *J. Phys. Chem.* **76**, 2024 (1972).

[n] O. Popovych and A. J. Dill, *Anal. Chem.* **41**, 456 (1969).

[o] Value obtained using the TAB BPh$_4$ assumption.

[p] ΔG^0 value on the molal scale.

[q] D. Bauer and J. P. Beck, *Bull. Soc. Chim. Fr.* p. 1252 (1973).

[r] Values obtained using polynuclear alternate aromatic hydrocarbon systems as reference electrode.

[s] L. M. Mukherjee, *J. Phys. Chem.* **76**, 243 (1972).

[t] I. M. Kolthoff and M. K. Chantooni, Jr., *J. Phys. Chem.* **76**, 2024 (1972).

[u] R. J. Gillespie and T. E. Peel, *J. Am. Chem. Soc.* **95**, 5173 (1973).

[v] G. Dallinga, J. Gaaf, and E. L. Mackor, *Recl. Trav. Chim. Pays-Bas* **89**, 1068 (1970).

[w] G. Petit and J. Bessière, *J. Electroanal. Chem.* **34**, 489 (1972).

TABLE II

ENTHALPIES OF TRANSFER OF H^+ FROM WATER TO SOLVENT

Solvent	$\Delta H^0_{w \to s}(H^+)^a$
Acetonitrile	13.4^b
Water	0
Propylene carbonate	10.5^b
N,N-Dimethylformamide	-6.7^b
Dimethyl sulfoxide	-6.1^b
Tetramethylene sulfone^c	17.4^b

[a] Values obtained using the tetraphenylborate assumption.
[b] R. Domain, M. Rinfret, and R. L. Benoit, Can. J. Chem. **54**, 2101 (1976).
[c] At 30°C.

$\Delta H^0_{g \to w}(H^+) = -270$ kcal/mole. The merits of the various extrathermodynamic assumptions leading to single-ion transfer free energies have been compared by several authors.[20-22] At present, the tetraphenylborate assumption appears to be the preferred one although the ferrocene assumption is still often used by electrochemists for reasons of convenience. Most of the values given in Tables I and II were obtained with these assumptions. However, it has been remarked that, for a number of solvents, the value of $\Delta G^0_{w \to s}(H^+)$ based on the Strehlow assumption is 2–3 kcal/mole more negative than the value based on the tetraphenylborate assumption.[38] This discrepancy may be related to shortcomings of the ferrocene assumption with water[39] and protic solvents, since single-ion free energies of transfer between dipolar aprotic solvents obtained with both assumptions are reasonably close. The values reported by different authors are at times in poor agreement, as in the case of pyridine (see Table I). There are, of course, many experimental difficulties in the determination of $\Delta G^0_{w \to s}(H^+)$, such as the reaction of H^+ with basic impurities in weakly basic solvents. Also, the fact that the steps required to reach transfer values for H^+ from values for BPh_4^- or $AsPh_4^-$ may often be numerous gives rise to large errors. Solvents of lower dielectric constant, ammonia and pyridine, for example, present special problems which have not always been taken into account. Electrolytes are largely present in such solvents as ion pairs, including SH^+A^-, and the corresponding dissociation constants should be taken in account to calculate $\Delta G^0_{w \to s}(H^+)$. In order to get around these complicating factors, the use of a background electrolyte has been suggested together with some simplifying assumptions concerning the dissociation constants.[40]

At present the available values of $\Delta H^0_{w \to s}(H^+)$ are limited to those of Domain[3] for a few dipolar aprotic solvents. $\Delta H^0_{w \to s}(H^+)$ appears to follow the corresponding $\Delta G^0_{w \to s}(H^+)$ values. Too little is so far known about single ion $\Delta S^0_{w \to s}$ values[41–43] to warrant any further discussion.

B. Free Energy and Enthalpy of Transfer of Reference Brønsted and Lewis Acids to Bulk Solvents

The process of transfer of a reference acid such as a Brønsted acid, HA, from a reference phase RP to a bulk solvent is represented by Eq. 24

$$HA(RP) \to HA(s) \tag{24}$$

Values of enthalpies of transfer have been determined for HA = alcohols, phenols, and HCl, and these values have been used to compare basicities of bulk solvents. Although free energies of transfer can also be used for the same purpose, available data are less numerous, possibly because they are more difficult to obtain experimentally.

1. ALCOHOLS

Krishnan and Friedman[44] have studied the transfer of normal alcohols, ROH, from the gas phase to infinite dilution in a series of solvents. Their calorimetric data on the enthalpies of solution of ROH were plotted as a function of m, the number of carbon atoms in the alkyl group R. The plots were linear except when the solvent was water. The extrapolated enthalpy values for $m = 0$, $\Delta H^0_{K,F}$, are a measure of the interaction of the hypothetical reference acid H $\cdots$ OH with the solvent. Such values are given in Table III.

2. PHENOLS

Arnett et al., and Duer and Bertrand have investigated the transfer of phenol,[7,45] p-fluorophenol,[7,46] and n-butanol[7] from infinite dilution in CCl_4 to infinite dilution in solvents

$$HA(CCl_4) \to HA(s) \tag{25}$$

Calorimetric data on the heats of solution of HA in CCl_4 and in S at different concentrations were extrapolated to infinite dilution to obtain $\Delta H^0_s(HA)$ and $\Delta H^0_{CCl_4}(HA)$. The enthalpy change for Eq. 25 was calculated from

$$\Delta H^0_t(HA) = \Delta H^0_s(HA) - \Delta H^0_{CCl_4}(HA) \tag{26}$$

$\Delta H^0_t(HA)$ is a measure of the interactions between HA and S, namely hydrogen bonding and nonspecific interactions. Arnett[46] has proposed that

TABLE III

ENTHALPIES OF TRANSFER OF HA FROM A REFERENCE PHASE R TO A SOLVENT (kcal/mole)

Solvents	HA: Phenol[a] R: CCl$_4$	p-Fluorophenol[b] CCl$_4$	n-Butanol[c] CCl$_4$	Acetic acid[d] Gas phase	HCl Gas phase	H···OH Gas phase	H$_2$O Liquid phase	CF$_3$CO$_2$H[p] Gas phase
Triethylamine	−8.85[e]	−8.92[f]	−5.53[e]	—	—	—	—	—
Pyridine	−7.34[e]	−7.40[f]	−4.47[e]	—	—	—	—	—
Acetonitrile	—	−4.2	—	−3.1[i]	−5.7[h]	−6.13[i]	+1.98[n]	−6.6[g]
Water	—	—	—	—	—	−7.68[i]	—	—
Methanol	—	—	—	—	—	−7.87[i]	—	—
Ethanol	—	—	—	—	—	−7.82[i]	—	—
n-Butanol	—	—	—	—	—	−7.75[i]	—	—
Diethyl ether	−5.45[e]	−5.57[f]	−2.96[e]	—	—	—	—	—
Anisole	−3.08[e]	−3.13[e]	−1.69[e]	—	—	—	—	—
Tetrahydrofuran	−5.75[e]	−5.75[f]	−3.06[e]	—	—	—	—	—
Dioxane	−5.11[e]	−5.10[f]	−3.11[e]	—	—	—	—	—
Diethyl sulfide	—	−3.63[f]	—	—	—	—	—	—
Tetrahydrothiophene	−5.3[j]	−3.71[f]	—	—	—	—	+1.05[o]	—
Acetone	−5.23[e]	−5.59[f]	—	—	—	—	—	—
2-Butanone	—	−5.20[e]	−3.05[e]	—	—	—	—	—
Cyclohexanone	—	−5.66[f]	—	—	—	—	—	—
Ethyl acetate	−4.75[e]	−4.74[f]	−2.43[e]	—	−6.3[h]	−6.20[i]	+1.93[n]	—
Propylene carbonate	—	−4.53[f]	—	—	—	−7.58[i]	—	—
Formamide	—	—	—	—	—	—	—	—
N-Methylformamide	−6.38[e]	−6.44[f]	−3.99[e]	—	—	—	—	—
N,N-Dimethylformamide	−6.86[e]	−6.97[f]	−4.14[e]	−6.4[g]	−17.8[h]	−7.85[i]	−0.9[h]	−11.7[g]
N,N-Dimethylacetamide	−7.36[e]	−7.44[f]	−4.43[e]	—	—	−8.18[i]	—	—
Hexamethylphosphoramide	—	−8.72[f]	—	—	—	—	—	—
N-Methyl-2-pyrrolidone	—	−7.38[f]	—	—	—	−5.52[i]	—	—
Nitromethane	—	—	—	—	—	—	2.4[o]	—

Solvent								
Nitrobenzene	—	—	—	—	-4.1[k]	—	—	—
Thionyl chloride	-7.21[e]	-1.17[f]	—	—	—	—	—	—
Dimethyl sulfoxide	—	-7.21[f]	-4.66[e]	-6.7[g]	-18.3[h]	-8.13[i]	-1.29[n]	-11.5[g]
Tetramethylene sulfone	—	-4.25[f]	—	-3.4[l]	-5.8[h]	-6.15[i,m]	1.63[n]	-6.4[l]
Phosphorus oxychloride	—	-3.71[f]	—	—	—	—	—	—
Trimethyl phosphate	—	-6.44[f]	—	—	—	—	—	—
Benzene	-1.18[e]	-1.23[f]	-0.50[e]	—	—	—	—	—
Toluene	-1.33[e]	-1.27[f]	-0.50[e]	—	—	—	—	—
n-Butyl chloride	-1.71[e]	-1.93[e]	—	—	—	—	—	—
1,2-Dichloroethane	—	—	—	—	-4.6[k]	—	—	—
o-Dichlorobenzene	—	-0.50[f]	—	—	—	—	—	—

[a] Values corrected for the enthalpies of transfer of anisole.
[b] Values corrected for the enthalpies of transfer of p-fluoroanisole.
[c] Values corrected for the enthalpies of transfer of n-butyl chloride.
[d] Values corrected for the enthalpies of transfer of methyl acetate.
[e] E. M. Arnett, L. Joris, E. Mitchell, T. S. S. R. Murty, J. M. Gorrie, and P. v. R. Schleyer, J. Am. Chem. Soc. 92, 2365 (1970).
[f] E. M. Arnett, T. S. S. R. Murty, P. v. R. Schleyer, and L. Joris, J. Am. Chem. Soc. 89, 5955 (1967).
[g] C. Louis, Doctoral Thesis, University of Montreal, Montreal (1974).
[h] R. Domain, Doctoral Thesis, University of Montreal, Montreal (1974).
[i] C. V. Krishnan and H. L. Friedman, J. Phys. Chem. 75, 3598 (1971).
[j] W. C. Duer and G. L. Bertrand, J. Am. Chem. Soc. 92, 2587 (1970).
[k] R. V. Choudhari and L. K. Doraiswamy, J. Chem. Eng. Data 17, 428 (1972).
[l] S. Y. Lam, personal communication.
[m] Value for 3-methyl sulfolane.
[n] R. L. Benoit and S. Y. Lam, J. Am. Chem. Soc. 96, 7385 (1974).
[o] S. Y. Lam and R. L. Benoit, Can. J. Chem. 52, 718 (1974).
[p] Values corrected for the enthalpies of transfer of methyl trifluoroacetate.

the nonspecific interactions can be approximated by the enthalpy of transfer of M, a nonhydrogen-bonding analog of HA, from infinite dilution in CCl_4 to infinite dilution in S.

$$M(CCl_4) \rightarrow M(s) \tag{27}$$

Anisole, p-fluoroanisole, and n-butyl chloride were used as analogs for phenol, p-fluorophenol, and n-butanol, respectively. $\Delta H_t^0(M)$ was obtained by an experimental method similar to that used for $\Delta H_t^0(HA)$. ΔH_h^0 (ΔH_f^0 in the paper by Arnett et al.), the enthalpy change corresponding to the acid–base interaction between HA and bulk solvent was obtained from

$$\Delta H_h = \Delta H_t^0(HA) - \Delta H_t^0(M) \tag{28}$$

The values of ΔH_h obtained by Arnett et al.[7,46] and Duer and Bertrand[45] are reported in Table III for selected solvents. We also give some values determined in our laboratory using acetic and trifluoroacetic acids as the common acid, HA, the reference phase being the gas phase; the methyl esters were selected as the nonhydrogen-bonding analogs.

3. HYDROCHLORIC ACID

Gerrard et al.[47] have studied the transfer of HCl as the common acid from the gas phase to a series of bulk solvents, mainly alcohols, ethers, and esters. Unfortunately, the reported HCl solubilities are usually high and the units used (mole of HCl per mole of solvent) are such that free energies of solution are not easily attainable. We include in Table III some enthalpy values we determined for the transfer of HCl from the gas phase to high dilution in some dipolar aprotic solvents. Some vapor pressure results have also been obtained[3] in the same solvents. Although we did not carry out systematic studies, it would appear that the enthalpies of solution of gaseous CH_3Cl (which unfortunately may not be a very good analog of HCl) do not vary appreciably for the type of solvents we examined. Infrared shifts of the HCl band in a few solvents have been reported by Gerrard et al.[47]

4. OTHER BRØNSTED ACIDS

Available data from the literature on heats of mixing for binary mixtures where one component would be a common Brønsted acid such as chloroform[48,49] could, in principle, be used to set up a basicity scale. However, the treatment of such data involves some assumptions concerning the ideality of the mixtures.[48] Finding a model molecule to account for nonspecific interactions between the acid solute and the solvent also presents some problems. For example, it has been suggested that CH_3CCl_3 and not CCl_4 should be used for the determination of the magnitude of the nonspecific contribution to the properties of mixtures of $CHCl_3$ and of the proton acceptor.[50] Nevertheless, uncorrected heats of mixing of a common proton donor, at low concentration with different solvents, reflect in some way a basicity order, as indicated by the values given in Table III for the enthalpy of mixing of water at infinite dilution with some dipolar aprotic solvents.

5. LEWIS ACIDS

Table IV gives some as yet unpublished data we obtained on heats of solution of $SbCl_5$, I_2, and SO_2 at high dilution (0.01–0.3 M) in several dipolar aprotic solvents. However, such values should be used with caution to compare basicities since complex ionization reactions between the first

TABLE IV

ENTHALPIES OF SOLUTION OF ANTIMONY PENTACHLORIDE, IODINE, AND SULFUR DIOXIDE IN BULK SOLVENTS

Solvent	ΔH^0 (kcal/mole)		
	$SbCl_5(g)^a$	$I_2(g)^b$	$SO_2(g)$
Dimethylaniline	—	—	−15.0
Pyridine	—	—	−11.6
Acetonitrile	−26.8	−10.3	−6.7
Tetrahydrofuran	—	—	−9.4
Water	—	—	−6.3
Ethyl acetate	—	—	−7.1
Propylene carbonate	−29.9	−12.4	−7.5
Trimethyl phosphate	—	—	−8.9
Dimethylformamide	−44.6^c	−14.85^c	−10.9
Nitrobenzene	−21.2	−10.2	−6.3
Nitromethane	—	−8.9	−6.2
Dimethyl sulfoxide	−46.5^c	—	−11.5
Tetramethylene sulfone	−26.9	—	−7.3
1,2-Dichloroethane	—	—	−5.5
Benzene	—	—	−5.8
Carbon tetrachloride	—	—	−5.7
n-Heptane	—	—	−3.8
Cyclohexane	—	—	−3.5
Isooctane	—	—	−3.5

^{a} Heat of vaporization of $SbCl_5(l)$, $SbCl_5(l) = SbCl_5(g)$, $\Delta H^0_{vap.} = +11.5$ kcal/mole.
^{b} Heat of sublimation of I_2, $I_2(c) = I_2(g)$, $\Delta H^0_{subl.} = +14.9$ kcal/mole.
^{c} Possible complex ionization reactions between acid and solvent.

two acids and the more basic solvents are likely. Nevertheless, these values will be of interest when we attempt later, in Section VI, to look for correlations.

6. Influence of the Reference Brønsted Acid

Solvent basicity scales based on the enthalpies of transfer of reference Brønsted acids HA are obviously easier to establish experimentally than is the scale based on $\Delta H^0_{g \to s}(H^+)$. But for such scales to be really effective they should be relatively independent of the acid HA. With this purpose in mind, Arnett et al.[7] have compared sets of ΔH^0 values obtained for a series of 27 bases with three different acids, phenol, p-fluorophenol, and n-butanol. Good linear correlations were observed. However, it must be noted that phenol and p-fluorophenol are closely related acids giving nearly equal ΔH^0 values, and that the series of bases included only one hydrogen-bonded base, N-methylformamide. The ΔH^0 values for HA = HCl and HA = p-fluorophenol give the same order of basic strength for the few dipolar aprotic solvents studied with both acids. Data in Table III indicate that the spread of ΔH^0 values increases in the order alcohol < phenol $\sim CH_3CO_2H < CF_3CO_2H < HCl$ so that the latter acid has a larger discriminating factor to compare basicities. Unfortunately, HCl, also protonates the stronger bases DMF and DMSO so that experimental values had to be corrected to obtain $\Delta H^0_{g \to s}(HCl)$ quoted in Table III. It is therefore not practical to select too strong a reference acid HA. There is, however, an obvious need to attempt to extend correlations between ΔH^0 values for different types of acids HA such as $HCCl_3$ and HCO_2R. It is interesting to note that, for the two solvents of similar basicities, DMF and DMSO, the ΔH^0 values of Domain[3] for HCl, uncorrected for nonspecific interactions, are respectively -17.8 and -18.3 kcal/mole, indicating DMSO to be the stronger base. This is in agreement with the values obtained by Krishnan and Friedman[44] for $H \cdots OH$, -7.85 and -8.13 kcal/mole, and the corrected values of Arnett et al.[7] for p-fluorophenol -6.97 and -7.21 kcal/mole. However, the uncorrected values of Arnett et al.[7] were in the opposite order, -8.13 and -7.66 kcal/mole. The latter results suggest that the benzene ring interacts more strongly with DMF than with DMSO, as also indicated by values of the heat of solution of anisole and of benzene in the two solvents. It may be that by selecting a proper acid, other than phenols, a basic strength order could be established more simply, without having to correct for nonhydrogen-bonding interactions with analogs of HA.

III. BASICITY OF SOLVENTS AS SOLUTES
IN REFERENCE ACID MEDIA

Solvents as molecular species, B, can be classified according to their base strength by comparing free energies or enthalpies of their protonation reaction in a common acid medium M either

$$B(M) + H^+(M) \rightleftharpoons BH^+(M) \tag{29}$$

or

$$B(M) + HA(M) \rightleftharpoons BH^+(M) + A^-(M) \tag{30}$$

Thermodynamic parameters for the reactions represented by Eqs. 29 and 30 have been determined in three instances for a sufficiently large number of basic solvents to make the basicity scales meaningful.

A. Protonation of Solvents in Mixtures of Water and Strong Acids

Dilute solutions of strong acids (H_2SO_4, $HClO_4$, HCl) in water have long been used to characterize the basic strength of molecular species. The thermodynamic parameter used for this purpose is usually the equilibrium constant K_{BH^+} corresponding to Eq. 29 written as

$$K_{BH^+} = [B][H^+]/BH^+ \times f_B f_{H^+}/f_{BH^+} \tag{31}$$

where f_B, f_{H^+}, and f_{BH^+} are activity coefficients. For dilute aqueous solutions, the activity coefficients can be calculated and K_{BH^+} obtained from pH, conductometric, or spectrophotometric measurements. However, many of the bases commonly used as solvents are not protonated in dilute acid solutions so that solutions with variable concentrations of a strong acid have to be used. In such media the activity coefficients are largely unknown and K_{BH^+} values cannot be calculated without formulating some extrathermodynamic assumption. Hammett's postulate is that for some organic uncharged indicator bases, In, the ratio of the activity coefficients of In and InH^+ has the same values for all uncharged indicator bases in the same solution.[23] This hypothesis was found to hold satisfactorily for substituted primary anilines in mixtures of various acids with water. An acidity function, H_0, was thus defined based on the degree of protonation of such indicator bases

$$H_0 = -\log a_{H^+} \times f_{In}/f_{InH^+} = pK_{InH^+} - \log[InH^+]/[In] \tag{32}$$

The technique used to establish the H_0 scale is well known. Starting with a basic indicator In_1 such as p-nitroaniline whose $pK_{In_1H^+}$ is known in aqueous dilute acid medium, H_0 values for more acidic media are obtained by determining spectrophotometrically the ratio $I = [In_1 H^+]/[In_1]$ in these solutions. Then a slightly less basic indicator, In_2, is used in these previous media and from H_0 and I, $pK_{In_2 H^+}$ is calculated. In_2 is then used to obtain the H_0 values for more acidic media. The stepwise procedure is repeated with increasingly less basic overlapping indicators and increasingly more acidic media. The value of pK_{BH^+} for a basic solvent molecule B is thus in principle obtained by simply measuring the corresponding I value in an acidic solution whose H_0 is known. However, it has been found that for some bases, log I is not a linear function with unit slope of H_0, as indicated by Eq. 32. This means that the assumption concerning f_{InH^+}/f_{In} does not hold for all bases. Nevertheless, bases can be divided into groups for which the assumption is valid so that a number of acidity functions have been proposed, each corresponding to a different family of bases (e.g., the H_A function based on protonation of substituted amides). In order to obtain pK_{BH^+} for a new base it is therefore necessary to first establish which acidity function it follows. More details about acidity functions and the various experimental methods for obtaining I are found in reviews such as those by Arnett[51] and Liler.[52] Table V gives selected pK_{BH^+} values, most of them from Arnett's review, for a number of bases used as solvents. For some bases, divergent values are available so that a choice had to be made.

B. Protonation of Solvents in Acetonitrile-Dilute Methanesulfonic Acid

The use of a solvent less basic than water would seem to be preferable to carry out the protonation of weak bases, because large concentrations of an acid such as H_2SO_4 would not be required as in the previous case. Changing solvation effects on BH^+ and B when passing gradually from pure water to concentrated sulfuric acid would thus be eliminated, the medium remaining essentially the pure solvent. With this purpose in mind, Kolthoff et al.[53] have selected acetonitrile as solvent, and dilute methanesulfonic acid as acid. The method used to study protonation equilibria (Eq. 30) was conductometry. Some selected values of K_{BH^+} obtained under these conditions are given in Table V.

C. Protonation of Solvents in Bulk Fluorosulfuric Acid

Arnett et al.[1,54] have used fluorosulfuric acid as a protonating medium to establish a basicity scale. In such a strong acid medium, most of the usual bases are completely protonated, as shown by IR and NMR spectroscopy, so that enthalpy is the most convenient parameter to characterize the protonation reaction given by Eq. 30. Arnett et al. further chose as a reference

TABLE V

THERMODYNAMIC PARAMETERS FOR PROTONATION OF SOLVENT
MOLECULES IN REFERENCE ACID MEDIA

Solvent	pK_{BH^+} in water[a]	pK_{BH^+} in acetonitrile[b]	ΔH_i^0 in fluorosulfuric acid[c]
Ammonia	9.2[d]	16.5[d]	43.3[u]
Butyl amine	10.6[d]	18.3[d]	—
Pyrrolidine	11.3[e]	—	—
Piperidine	11.1[e]	—	—
Pyridine	5.2[d]	12.3[d]	−38.6[n]
Quinuclidine	11.0[e]	—	−45.8[n]
Guanidine	13.6[e]	—	—
Acetonitrile	−10.0[f]	—	−13.6[n]
Benzonitrile	−10.3[f]	—	—
Water	−2.3[g]	2.2[r]	−16.5[u]
Methanol	−4.9[h] or −2.1[h]	2.4[r]	−17.1[u]
Ethanol	−4.8[i]	—	−18.7[u]
n-Propanol	−4.7[h]	—	—
Isopropanol	−4.7[h]	—	—
n-Butanol	−2.3[f]	—	—
Isobutanol	−2.2[f]	—	—
tert-Butanol	−2.6[f]	2.4[r]	—
Ethylene glycol	—	1.5[r]	—
Diethyl ether	−3.6[j]	0.4[r]	−19.5[n]
Anisole	−6.5[k]	−0.54[r]	—
Tetrahydrofuran	−2.1[l]	1.1[r]	−19.6[c]
Tetrahydropyran	−2.8[l]	—	—
Dioxane	−3.2[m]	—	−21.5[n]
Diethyl sulfide	−6.8[n]	—	−19.0[n]
Tetrahydrothiophene	−7[n]	—	−19.7[n]
Acetone	−7.2[o]	−0.1[r]	−18.3[n]
Acetophenone	−6.7[f]	−0.14[r]	−18.9[w]
Benzophenone	−6.0[c]	—	−16.9[w]
Cyclohexanone	−6.8[e]	—	−18.1[n]
Acetaldehyde	−10.2[x]	—	—
Benzaldehyde	−7.6[f]	—	−16.1[w]
Acetic acid	−6.8[f]	1.1[r]	—
Ethyl acetate	−6.9[p]	−0.7[r]	−17.4[n]
Propylene carbonate	—	—	−17.8[n]
Formamide	−2.7[m]	—	—
N-Methylformamide	—	—	−29.6[n]
N,N-Dimethylformamide	−0.7[q]	6.1[r]	−29.5[n]
N,N-Dimethylacetamide	0.1[q]	—	−32.0[n]
N-Methyl-2-pyrrolidone	0.2[q]	—	—
Tetramethylurea	0.4[q]	—	−37.6[n]
Nitromethane	−11.9[m]	—	—

(*Continued*)

TABLE V—*continued*

Solvent	pK_{BH^+} in water[a]	pK_{BH^+} in acetonitrile[b]	ΔH_i^0 in fluorosulfuric acid[c]
Nitrobenzene	-12.1^f	—	-6.6^r
Sulfur dioxide	-13.3^f	—	—
Dimethyl sulfoxide	-1.8^r	5.8^r	-28.6^n
Tetramethylene sulfone	—	—	-11.8^n
Phosphorus oxychloride	—	—	-4.8^n
Trimethyl phosphate	—	—	-20.4^n
Benzene	-9.2^s	—	—
o-Dichlorobenzene	—	—	-14.5^n
Acetyl chloride	-10.9^t	—	—
Benzoyl chloride	-11.1^t	—	-6.0^r

[a] K_{BH^+} is the equilibrium constant of the reaction: $BH_{(w)}^+ \rightleftharpoons B_{(w)} + H_{(w)}^+$ expressed in mole liter^{-1}

[b] K_{BH^+} is the equilibrium constant of the reaction: $BH_{(AN)}^+ \rightleftharpoons B_{(AN)} + H_{(AN)}^+$ expressed in mole liter^{-1}.

[c] ΔH_i^0 is the enthalpy of transfer of the solvent from infinite dilution in CCl_4 to infinite dilution in HSO_3F expressed in kcal/mole.

[d] I. M. Kolthoff, M. K. Chantooni, Jr., and S. Bhowmik, *J. Am. Chem. Soc.* **90**, 23 (1968).

[e] D. D. Perrin, "Dissociation Constants of Organic Bases in Aqueous Solution." Butterworth, London, 1965.

[f] M. Liler, "Reaction Mechanisms in Sulfuric Acid." Academic Press, New York, 1971.

[g] E. M. Arnett, J. J. Burke, J. V. Carter, and C. F. Douty, *J. Am. Chem. Soc.* **94**, 7837 (1972).

[h] R. E. Weston, S. Ehrenson, and K. Heinzinger, *J. Am. Chem. Soc.* **89**, 481 (1967) or P. Bonivicini, A. Levi, V. Lucchini, G. Modera, and G. Scorrano, *ibid.* **84**, 1674 (1962).

[i] J. T. Edward, J. Leane, and I. Wang, *Can. J. Chem.* **40**, 1521 (1972).

[j] E. M. Arnett and C. Wu, *J. Am. Chem. Soc.* **84**, 1680 (1962).

[k] E. M. Arnett, C. Wu, J. Anderson, and R. Bushick, *J. Am. Chem. Soc.* **84**, 1674 (1962).

[l] E. M. Arnett and C. Wu, *J. Am. Chem. Soc.* **84**, 1684 (1962).

[m] E. M. Arnett, *Prog. Phys. Org. Chem.* **1**, 223 (1963).

[n] E. M. Arnett, E. J. Mitchell, and T. S. S. R. Murty, *J. Am. Chem. Soc.* **96**, 3875 (1974).

[o] N. C. Deno and M. Wisotsky, *J. Am. Chem. Soc.* **85**, 1735 (1963).

[p] C. A. Lane, *J. Am. Chem. Soc.* **86**, 2521 (1964).

[q] R. L. Alderman, *J. Org. Chem.* **29**, 1837 (1964).

[r] I. M. Kolthoff and M. K. Chantooni, Jr., *J. Am. Chem. Soc.* **95**, 8539 (1973).

[s] D. M. Bronwer, E. L. Mackor, and C. Maclean, *in* "Carbonium Ions" (G. A. Olah and P. v. R. Schleyer, eds.), Vol. 2, p. 837. Wiley (Interscience), New York, 1970.

[t] M. Liler, *J. Chem. Soc. B* p. 205 (1966).

[u] E. M. Arnett and J. F. Wolf, *J. Am. Chem. Soc.* **95**, 978 (1973).

[v] E. M. Arnett, R. P. Quirk, and J. J. Burke, *J. Am. Chem. Soc.* **92**, 1260 (1970).

[w] E. M. Arnett, R. P. Quirk, and J. W. Larsen, *J. Am. Chem. Soc.* **92**, 3977 (1970).

[x] G. C. Levy, J. D. Cargioli, and W. Racela, *J. Am. Chem. Soc.* **92**, 6238 (1970).

state for B a dilute solution in CCl_4, thus eliminating variable solvation effects on B which would be operative in both previous methods. From the enthalpies of protonation determined by Arnett et al. for a number of bases,[1,54] we have taken those corresponding to commonly used solvents. Values are listed in Table V.

D. Correlation of Protonation Data in Water, Acetonitrile, and Fluorosulfuric Acid

We have plotted in Fig. 1 the pK_{BH^+} values obtained for nine bases in acetonitrile by Kolthoff et al.[53] against the corresponding values in $H_2O-H_2SO_4$ media. The points fall close to a straight line of slope 1, the correlation extending over some 18 pK units. This linear relationship would imply that the difference between the free energies of transfer of BH^+ and of B from $H_2O-H_2SO_4$ to acetonitrile is independent of B (Eq. 22). This is somewhat unexpected considering the widely different nature of the bases considered. Kolthoff et al.[53] have discussed further the relationship between pK_{BH^+} values in both media.

Arnett et al.[1] have compared two other thermodynamic parameters for the protonation of an extensive series of bases. These authors have obtained

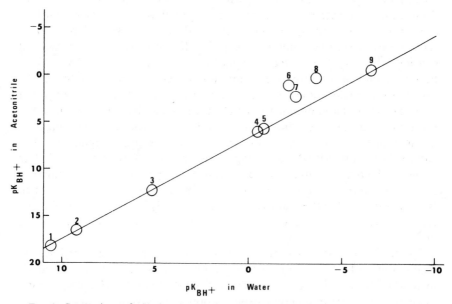

FIG. 1. Comparison of pK_{BH^-} values measured in $H_2O-H_2SO_4$ media and in acetonitrile for nine bases. 1. butylamine; 2, ammonia; 3, pyridine; 4, dimethylformamide; 5, dimethyl sulfoxide; 6, tetrahydrofuran; 7, methanol; 8, diethyl ether; 9, anisole.

a good linear correlation between the enthalpies of protonation in bulk HSO_3F and the corresponding aqueous pK_{BH^+} values. As Arnett *et al.* point out, it is surprising that the correlation, which holds over 22 pK_{BH^+} units, is this good, considering the different entropy factors which might be expected in these two solvents. The different reference states for B, in the first case CCl_4 and in the second case $H_2O-H_2SO_4$, add to the complexity of the relationship. It would be of interest to know whether the above correlation extends to common hydrogen-bonding bases which were not included in the series, except for *N*-methylformamide and some anilines.

IV. BASICITY OF SOLVENTS AS REACTANTS WITH BRØNSTED AND LEWIS REFERENCE ACIDS IN INERT MEDIA

The base strength of isolated solvent molecules B can be compared on the basis of the free energy or enthalpy changes for the adduct formation reaction, given by Eq. 33, between B and a reference Lewis or Brønsted acid AR in an inert medium M:

$$B(M) + AR(M) \rightleftharpoons BAR(M) \qquad (33)$$

An inert medium which gives minimum interactions with both B and AR is chosen and experimental conditions are selected so that B and AR molecules do not autoassociate and no complexes of stoichiometry higher than 1 : 1 are formed. Thermodynamic data, ΔH_f^0 and ΔG_f^0, for Eq. 33 have been reported for a very large number of acids AR and inert media M. We will consider the very frequently used molecular acids, $SbCl_5$, I_2, phenol, *p*-fluorophenol, chloroform, methanol, water, and as inert media, CCl_4, cyclohexane, dichloromethane, and dichloroethane. A charged Brønsted acid, the tributylammonium ion Bu_3NH^+, has also been used as acid, the medium being chlorobenzene or *o*-dichlorobenzene. The corresponding data are included here since they provide some interesting comparisons. The nature of the adduct BAR may change within a series of bases or acids.[11] For example, the reaction given by Eq. 33 with HA-type acids may only involve hydrogen bonding or it may also involve proton transfer as is the case when B = pyridine and AR = carboxylic acids in CCl_4.[55] This will make the establishment of correlations between a series of bases or acids more difficult.

Spectroscopic data have been widely used to characterize, through the reaction given by Eq. 33, the basicity of molecular species B. When AR is a hydroxylic Brønsted acid, $\Delta\nu_{OH}$, which is the shift of the infrared stretching frequency of $-OH$ upon complex formation, is a convenient parameter. $\Delta\delta$, the increase of the NMR chemical shift of the acidic proton upon complex formation, is another parameter used.

A. Experimental Methods

Two main experimental methods have been used to establish thermody-namic data for the reaction represented by Eq. 33.

1. CALORIMETRY

Calorimetric methods used to investigate hydrogen bond and charge transfer complexes have been recently and comprehensively discussed by Fenby and Hepler.[56] Earlier reviews by Lamberts[57] and Christensen et al.[58] provide supplementary material. Two related types of determinations have been made. Some workers have measured the heat change produced by dissolving the pure base in a dilute solution of the reference acid. The heat change due to the reaction given by Eq. 33 is then obtained by correcting for the heat of solution of the base in the inert solvent. Other investigators have measured the heat change produced by mixing dilute solutions of the base and of the acid in the inert solvents. If necessary, corrections are then made for the heats of dilution of the base and of the acid. In both cases, the enthalpy change for the reaction (Eq. 33) ΔH_f^0, can be simply calculated from the observed heat change if a value of K_f, the corresponding equilib-rium constant has been determined independently, for example from IR or NMR measurements. ΔH_f^0 values are sensitive to errors in K_f, particularly when K_f is small.[7] Both K_f and ΔH_f^0 can also be obtained simultaneously by calorimetric measurements at a single temperature.[56] Trial-and-error methods of calculation such as those of Lamberts and Zeegers-Huyskens[59] and Bolles and Drago[60] are used. Cabani and Gianni[61] have discussed the limitations coming from the propagation of errors in such methods. A new and potentially better procedure has been proposed by Fenby and Hepler[56] to avoid trial-and-error calculations. Obviously, weak complexing makes the determination of both K_f and ΔH_f^0 a particularly difficult problem.

2. SPECTROSCOPIC METHODS

The equilibrium constant K_f (Eq. 33) can be determined by IR, UV, and NMR spectroscopies. The corresponding enthalpy, ΔH_f^0, is obtained from the variation of K_f with temperature.

IR spectroscopy is a popular method when phenol is the proton donor.[7,62] In the presence of a base B, the IR spectra of solutions of phenol show two bands due to stretching of the O–H bond: one is due to free phenol, the other to bonded phenol. The intensity of the band due to free phenol is used to monitor the acid–base reaction. Little overlap of the free and bonded bands is observed and the concentration of free phenol in solution can be obtained from the Beer equation. From the initial concentrations of AR and B, and the equilibrium concentration of AR, the concentrations of all species at equilibrium can be calculated and K_f obtained. The same type of procedure has been used[63] to determine K_f values with two other proton donors, CH_3OH[64] (free OH band used to monitor concentration) and $CHCl_3$ (CH-stretching band used to monitor concentration).

UV spectrophotometry is a convenient method when BAR is a charge-transfer complex. For example, when iodine is the reference acid,[65,66] the intensity of the charge-transfer band observed between 240 and 300 nm[67] has been used to monitor the complex concentration in solution. When neither the pure base nor the pure reference acid absorbs at the wavelength selected for measurements, the Benesi–Hildebrand equation can be used to compute K_f

$$C_{AR}^0/A = 1/K_f\varepsilon_c C_B^0 + 1/\varepsilon_c \qquad (34)$$

where C_{AR}^0 and C_B^0 are total concentrations, ε_c is the extinction coefficient of the complex and A the absorbance per unit path length. When the base and the pure acid also absorb at the wavelength selected, Eq. 34 is replaced by a more general equation.[68] At any rate, both equations contain two unknowns: K_f and ε_c. A weighted linear least-squares method is recommended to obtain reliable values of K_f.[69]

NMR spectroscopy has been used with $CHCl_3$, H_2O (H-NMR),[70-72] and p-fluorophenol (F-NMR)[73] as reference acids. The experimental procedure involves measuring chemical shifts ($C\underline{H}Cl_3$, $\underline{H}_2O$, p-$\underline{F}C_6H_4OH$) in solutions containing a constant total concentration of reference acid AR and varying amounts of base B. The chemical shift measured is a weighted mean of the chemical shift of the pure acid, δ_{AR}, and of that of the complex, δ_{BAR}. The resulting equation can be combined with the mass balance equations to allow calculation of K_f. Two unknowns have to be determined simultaneously (δ_{BAR} and K_f) and a procedure similar to that previously indicated can be employed. However, in some cases it is possible to determine δ_{BAR} independently by having a large excess of base and/or by lowering the temperature, thereby favoring the complete conversion of AR into BAR.[73]

B. Influence of Inert Media and Reference Acids on Acid-Base Reaction Parameters

Selected thermodynamic and spectroscopic data for the reaction given by Eq. 33 have been gathered in Tables VI to XII. Such data are dependent on two main factors, the inert medium and the reference acid, which will be considered in the following two sections. We will also discuss in a third section the different parameters used to characterize this reaction.

1. THE INERT MEDIUM

Because of solubility limitations, several inert media M, some of them more polar than others, have been used to study the reaction (Eq. 33). As the species AR, B, and BAR are unlikely to interact in the same way with the different media, the values of the thermodynamic parameters determined for a given adduct formation vary from one reaction medium to another. Several authors[74–79] have tried to rationalize the influence of the reaction medium on these parameters in order to allow the prediction of values in one medium from values in another.

Drago et al.[76–78] have proposed a simple method to correlate ΔH_f^0 values determined in two different reaction media. This method, called elimination of solvent procedure, is based on the assumption that for a series of adducts of a given reference acid in a particular solvent, the difference between the solvation energies of the adduct and of the base is a constant.

$$\Delta H^1_{\text{olv}}(\text{BAR}) - \Delta H^1_{\text{solv}}(\text{B}) = \text{const.} \qquad (35)$$

The index 1 refers to the reaction medium. This assumption leads to the following relationship between enthalpies of adduct formation with a given reference acid in two reaction media

$$\Delta H_f^1 = \Delta H_f^2 + \text{const.} \qquad (36)$$

The validity of the procedure has been tested for the reference acids: m-fluorophenol[76,77] and hexafluoro-2-propanol[78] and the following reaction media: CCl_4, n-hexane, benzene, o-dichlorobenzene, nitrobenzene, nitromethane, cyclohexane, and 1,2-dichloroethane. The method was found to allow the evaluation of reaction media effects to within 0.2 kcal/mole. However, Olofsson and Olofsson[80] have shown that the method does not account for differences in ΔH_f^0 values which they found with the stronger reference acid, $SbCl_5$, in CCl_4 and 1,2-dichloroethane. These authors have shown that ΔH_f^0 values obtained with this acid in both media are linearly related, but that the slope of the correlation is not unity as predicted by Eq. 36. Arnett et al.[1] have plotted ΔH_f data from Drago for phenol and m-fluorophenol in different solvents and also obtained similar linear correlations. Allerhand and Schleyer[79] have shown that Δv values obtained with a series of bases with the same reference acid in two different reaction media are linearly related.

Christian et al.[74,75] have offered another method to correlate adduct formation energies in different solvents. Their method is based on the assumption that the free energy or enthalpy of transfer of the adduct BAR

TABLE VI

REACTION PARAMETERS FOR THE INTERACTION OF PHENOL WITH SOLVENT MOLECULES IN INERT MEDIA (M)

Solvent	ΔH_f^{0}[a] (kcal/mole)		ΔG_f^{0}[b] (kcal/mole)	$\Delta \nu^c$ (cm^{-1})	$\Delta \delta^d$ (ppm)
	M: CCl$_4$	Dichloromethane	CCl$_4$	CCl$_4$	Dichloromethane
Ammonia	−8.2[e]				
Pyridine	−7.5[f]	−8.1[g]	−2.46[j]	465[f]	6.12[g]
Acetonitrile	−4.65[h]	−3.5[g]	−0.93[h]	179[j]	2.49[g]
Diethyl ether	−5.41[j]	−5.1[g]	−1.29[j]	276[j]	3.71[g]
Anisole	−3.27[c]			161[i]	
Tetrahydrofuran	−5.29[j]	−5.5[g]	−1.65[j]	285[c]	4.02[g]
Tetrahydropyran	−5.19[j]		−1.16[j]		
Dioxane	−4.41[c]	−4.4[g]		245[i]	3.51[g]
Diethyl sulfide	−4.6[k]				
Tetrahydrothiophene	−3.0[l]		−0.2[l]	257[c]	
Acetone	−4.83[m]	−1.37[g]	−1.47[h]	225[i]	2.91[g]
2-Butanone	−4.87[m]				
Acetophenone	−4.65[m]			193[j]	
Benzophenone	−4.48[m]			193[j]	
Cyclohexanone	−4.65[m]			242[i]	
Ethyl acetate	−4.62[m]	−3.3[g]	−1.17[h]	179[j]	2.47[g]
N,N-Dimethylformamide	−6.30[m]	−5.3[g]	−2.46[n]	294[n]	3.82[g]
N,N-Dimethylacetamide	−6.48[m]	−6.2[g]	−2.90[n]	342[n]	4.36[g]
Nitromethane	−1.88[c]				
Nitrobenzene	−2.04[c]				
Dimethyl sulfoxide	−6.5[k]	−6.4[g]	−3.08[k]	364[i]	4.54[g]
Tetramethylene sulfone	−4.9[k]	−3.5[g]	−1.68[k]		2.69[g]
Trimethylphosphate	−5.3[f]	−5.7[g]	−3.06[f]	315[f]	4.29
Benzene	−1.5[k]			55[i]	
Toluene	−1.51[c]			61[i]	

[a] Enthalpy change for the reaction: $HA(M) + S(M) \rightarrow HAS(M)$.

[b] Free energy change (molar scale) for the reaction given above.

[c] Shift in the OH stretching frequency (infrared) of phenol on complex formation.

[d] NMR shift of the acid proton of phenol on complex formation.

[e] R. S. Drago and K. F. Purcell, *Prog. Inorg. Chem.* **6**, 271 (1964).

[f] G. A. Aknes and T. Gramstad, *Acta Chem. Scand.* **14**, 1485 (1960).

[g] D. P. Eyman and R. S. Drago, *J. Am. Chem. Soc.* **88**, 1617 (1966).

[h] T. D. Epley and R. S. Drago, *J. Am. Chem. Soc.* **89**, 5770 (1967).

[i] L. J. Bellamy and R. J. Pace, *Spectrochim. Acta, Part A* **25**, 319 (1969).

[j] R. West, D. L. Powell, M. K. T. Lee, and L. S. Whatley, *J. Am. Chem. Soc.* **86**, 3227 (1964).

[k] R. S. Drago, B. Wayland, and R. L. Carlson, *J. Am. Chem. Soc.* **85**, 3125 (1963).

[l] J. N. Spencer, R. S. Harner, L. I. Freed, and C. D. Penturelli, *J. Phys. Chem.* **79**, 332 (1975).

[m] D. Neerinck, A. van Audenhaege, and L. Lamberts, *Ann. Chim. (Paris)* [14] **4**, 43 (1969).

[n] M. D. Joesten and R. S. Drago, *J. Am. Chem. Soc.* **84**, 2696 (1962).

93

TABLE VII

REACTION PARAMETERS FOR THE INTERACTION OF p-FLUOROPHENOL
WITH SOLVENT MOLECULES IN CCl_4

Solvent	ΔH_f^0 (kcal/mole)	ΔG_f^0 (kcal/mole)	$\Delta\nu^a$ (cm^{-1})	$\Delta\delta^b$ (ppm)
Pyridine	-7.1^c	$-2.56^{d,e}$	485^e	2.49^d
Quinuclidine	-9.5^c	-3.58^e	814^e	—
Acetonitrile	-4.2^c	$-1.43^{d,e}$	184^e	1.88^d
Benzonitrile	—	-1.08^d	—	1.71^d
Diethyl ether	-5.6^c	-1.38^e	285^e	1.88^d
Anisole	-3.1^c	-0.49^e	169^e	—
Tetrahydrofuran	-5.6^c	-1.70^e	292^e	2.00^d
Dioxane	-4.6^d	-0.97^d	252^e	1.45^d
Diethyl sulfide	—	-0.15^e	263^e	—
Tetrahydrothiophene	—	—	$282^{e.}$	—
Acetone	—	-1.60^e	232^e	—
2-Butanone	-4.8^d	-1.63^e	221^e	2.02^d
Acetophenone	—	-1.54^d	—	1.92^d
Cyclohexanone	-5.8^d	-1.79^e	229^e	2.12^d
Benzaldehyde	—	-1.13^d	—	1.70^d
Ethyl acetate	-4.7^d	-1.49^e	199^e	1.85^d
N-Methylformamide	-5.5^c	-2.67^e	271^e	—
N,N-Dimethylacetamide	—	-3.29^e	356^e	2.86^d
N,N-Dimethylformamide	-6.6^c	$-2.81^{d,e}$	305^e	2.72^d
Hexamethylphosphoramide	-8.0^c	-4.85^d	479^e	3.71^d
N-Methyl-2-pyrrolidone	-7.0^c	-3.22^e	339^e	—
Tetramethylurea	-7.8^c	-3.30^d	350^e	3.00^d
Dimethyl sulfoxide	-6.6^c	-3.46^e	367^e	2.71^d
Tetramethylene sulfone	—	—	186^e	—
Trimethylphosphate	-6.5^c	-3.37^e	323^e	2.74^d
Benzene	—	—	49^e	—
Toluene	—	—	57^e	—
Butyl chloride	—	—	62^e	—

^{a} Shift in the OH stretching frequency (IR) of p-fluorophenol on complex formation.
^{b} NMR shift of the fluorine atom of p-fluorophenol on complex formation.
^{c} E. M. Arnett, E. J. Mitchell, and T. S. S. R. Murty, *J. Am. Chem. Soc.* **96**, 3875 (1974).
^{d} D. Gurka and R. W. Taft, *J. Am. Chem. Soc.* **91**, 4794 (1969).
^{e} E. M. Arnett, L. Joris, E. M. Mitchell, T. S. S. R. Murty, T. M. Gorrie, and P. v. R. Schleyer, *J. Am. Chem. Soc.* **92**, 2365 (1970).

from solvent 1 to solvent 2 is directly proportional to the sum of the transfer
energies of AR and B. For example,

$$\Delta G^0_{1\to 2}(BAR) = \alpha[\Delta G^0_{1\to 2}(AR) + \Delta G^0_{1\to 2}(B)] \qquad (37)$$

This assumption leads to the following relationship between ΔG_f^0 (or ΔH_f^0)
values in media 1 and 2

$$\Delta G_f^2 - \Delta G_f^1 = (\alpha - 1)[\Delta G^0_{1\to 2}(AR) + \Delta G^0_{1\to 2}(B)] \qquad (38)$$

TABLE VIII

REACTION PARAMETERS FOR THE INTERACTION OF WATER AND METHANOL WITH SOLVENT MOLECULES IN INERT MEDIA (M)

Solvent	ΔH_f^0 (kcal/mole) HA: H_2O M: CCl_4	ΔH_f^0 (kcal/mole) H_2O C·H^a	ΔG_f^0 (kcal/mole) H_2O CCl_4	ΔG_f^0 (kcal/mole) H_2O C·H^a	$\Delta\delta^b$ (ppm) H_2O C·H^a	ΔH_f^0 (kcal/mole) CH_3OH CCl_4
Pyridine	—	—	—	—	—	-4.4[f]
Tetrahydrofuran	-3.5[d]	-2.4[c]	—	-0.14[c]	1.40[c]	—
Dioxane	—	—	-1.5[d]	—	—	—
Acetone	—	-2.4[c]	—	-0.49[c]	1.67[c]	—
Benzophenone	—	—	—	—	—	-1.83[e]
Ethyl acetate	—	—	—	—	—	-2.55[e]
Dimethylformamide	—	—	—	—	—	-3.80[e]
Dimethylacetamide	—	-3.6[c]	—	-0.05[c]	1.83[c]	—
Tributyl phosphate	-4.1[d]	—	-2.46[d]	—	—	—

[a] Cyclohexane.
[b] NMR shift of the water protons on complex formation.
[c] F. Takahashi and N. C. Li, J. Am. Chem. Soc. 88, 1117 (1966).
[d] N. Muller and P. Simon, J. Phys. Chem. 71, 568 (1967).
[e] D. Neerinck, A. van Audenhaege, and L. Lamberts, Ann. Chim. (Paris) [14] 4, 43 (1969).
[f] T. Kitao and C. H. Jarboe, J. Org. Chem. 32, 407 (1967).

TABLE IX

REACTION PARAMETERS FOR THE INTERACTION OF CHLOROFORM
WITH SOLVENT MOLECULES IN INERT MEDIA (M)

Solvent	M:	ΔH_f^0 (kcal/mole) C · H[b]	ΔG_f^0 (kcal/mole) C · H[b]	ΔG_f^0 (kcal/mole) CCl$_4$	$\Delta\delta^a$ (ppm) C · H[b]	$\Delta\delta^a$ (ppm) CCl$_4$
Pyridine		—	—	−0.38[f]	—	2.27[f]
Acetonitrile		—	—	−0.08[f]	—	0.97[f]
Diethyl ether		—	−0.78[f]	−0.22[f]	0.91[f]	1.27[f]
Dioxane		−2.56[c]	0.31[c]	—	0.69[c]	—
Diethyl sulfide		−1.70[c]	0.88[c]	—	0.93[c]	—
Tetrahydrothiophene		−2.28[c]	0.76[c]	—	0.87[c]	—
Acetone		−2.34[c]	0.15[c]	−0.43[f]	0.97[c]	1.42[f]
Cyclohexanone		−2.44[c]	−0.02[c]	—	0.98[c]	—
Ethyl acetate		−2.51[c]	0.18[c]	—	0.84[c]	—
Hexamethylphosphoramide		—	−1.54[g]	—	2.02[g]	—
N-Methyl-2-pyrrolidone		−3.99[c]	−0.74[c]	—	1.39[c]	—
Nitroethane		−1.53[c]	0.59[c]	—	0.50[c]	—
Tributylphosphate		−4.3[d]	−0.93[d]	—	1.37[d]	—
Benzene		−1.97[e]	−0.03[e]	—	−1.91[e]	—

[a] NMR shift of the proton of chloroform (CHCl$_3$) on complex formation.
[b] Cyclohexane.
[c] G. R. Wiley and S. I. Miller, *J. Am. Chem. Soc.* **94**, 3287 (1972).
[d] S. Nushimura, C. H. Ke, and N. C. Li, *J. Phys. Chem.* **72**, 1297 (1968).
[e] C. J. Creswell and A. L. Allred, *J. Phys. Chem.* **66**, 1469 (1962).
[f] B. B. Howard, C. F. Jumper, and M. T. Emerson, *J. Mol. Spectrosc.* **10**, 117 (1963).
[g] T. Olsen, *Acta Chem. Scand.* **24**, 3081 (1970).

The Christian method has only been tested for a few systems. Its disadvantage is that precise transfer energies for AR and B have to be known in order to evaluate the influence of the solvent on ΔH_f^0 or ΔG_f^0. It seems that further systematic studies are needed to account fully for the effect of the medium on the thermodynamics of the reaction given by Eq. 33.

2. THE REFERENCE ACID

Attempts have long been made to find correlations between ΔH_f^0 (or ΔG_f^0) values obtained for a series of bases with pairs of reference acids. For example, Arnett et al[1] have shown that for 18 bases there is a linear correlation between their ΔH_f^0 values with p-fluorophenol as reference acid and those of Gutmann[80a] with SbCl$_5$ as reference acid, the average deviation being ±0.4 kcal/mole. Although the same trend is observed, the correlations

TABLE X

THERMODYNAMIC PARAMETERS FOR THE REACTION OF Bu_3NH^+ [a] WITH SOLVENT MOLECULES IN INERT MEDIA (M)

Solvent	ΔH_f^o (kcal/mole)		ΔG_f^o (kcal/mole)	
	M: Chlorobenzene	o-Dichlorobenzene	Chlorobenzene	o-Dichlorobenzene
Pyridine	−7.43[b]	−6.30[b]	−4.27[b]	−4.16[b]
Acetonitrile	−4.51[b]	−4.01[b]	−3.33[b]	−3.25[b]
Benzonitrile	—	—	—	−2.93[d]
Methanol	—	−4.38[c]	—	−2.56[c]
n-Propanol	—	−4.54[c]	—	−2.53[c]
Isopropanol	—	−5.14[c]	—	−2.53[c]
tert-Butyl alcohol	—	−6.56[c]	—	−2.42[c]
Diethyl ether	—	−5.71[c]	—	−1.30[c]
Tetrahydrofuran	−5.50[b]	−4.82[b]	−2.95[b]	−2.91[b]
Acetone	−4.52[b]	−4.32[b]	−3.32[b]	−3.26[b]
N,N-Dimethylacetamide	—	—	—	−6.12[d]
Hexamethylphosphoramide	—	—	—	−8.66[d]
Dimethyl sulfoxide	—	—	—	−6.23[d]
Tetramethylene sulfone	—	—	—	−3.88[d]

[a] Tributylammonium ion.
[b] M. L. Junker and W. R. Gilkerson, J. Am. Chem. Soc. 97, 493 (1975).
[c] H. B. Flora II and W. R. Gilkerson, J. Am. Chem. Soc. 92, 3273 (1970).
[d] H. W. Aitken and W. R. Gilkerson, J. Am. Chem. Soc. 95, 8551 (1973).

TABLE XI

ENTHALPIES OF ADDUCT FORMATION
BETWEEN SbCl$_5$ AND SOLVENT
MOLECULES IN 1,2-DICHLOROETHANE

Solvent	ΔH_f^0 (kcal/mole)
Ammonia	-59.0^a
Hydrazine	-44.0^a
Ethylenediamine	-55.0^a
Pyridine	-33.1^b
Acetonitrile	-14.1^b
Benzonitrile	-11.9^b
Water	-18.0^b, -24.3^c
Methanol	-19.0^b
Diethyl ether	-19.2^b
Tetrahydrofuran	-20.0^b
Acetone	-17.0^b
2-Butanone	-17.4^d
Cyclohexanone	-17.8^d
Ethyl acetate	-17.1^b
Propylene carbonate	-15.1^b
N,N-Dimethylformamide	-26.6^b
N,N-Dimethylacetamide	-27.8^b
Hexamethylphosphoramide	-38.8^b
N-Methyl-2-pyrrolidone	-27.3^b
Nitromethane	-2.7^b
Nitrobenzene	-4.4^b, -8.1^e
Thionyl chloride	-0.4^b
Sulfuryl chloride	-0.1^b
Dimethyl sulfoxide	-29.8^b
Tetramethylene sulfone	-14.8^b
Phosphorus oxychloride	-11.7^b
Trimethyl phosphate	-23.0^b
Tributyl phosphate	-23.7^b
Benzene	-0.1^b
Acetyl chloride	-0.7^b
Benzoyl chloride	-2.3^b
Acetic anhydride	-10.5^b

[a] M. Herlem and A. I. Popov, *J. Am. Chem. Soc.* **94**, 1431 (1972).
[b] V. Gutman and R. Schmid, *Coord. Chem. Rev.* **12**, 263 (1974).
[c] G. Olofsson, *Acta Chem. Scand.* **22**, 1352 (1968).
[d] G. Olofsson, *Acta Chem. Scand.* **22**, 377 (1968).
[e] G. Olofsson and I. Olofsson, *J. Am. Chem. Soc.* **95**, 7231 (1973).

TABLE XII

ENTHALPIES AND FREE ENERGIES OF ADDUCT
FORMATION BETWEEN IODINE AND
SOLVENT MOLECULES IN CCl_4

Solvent	ΔH_f^0 (kcal/mole)	ΔG_f^0 (kcal/mole)
Ammonia	-4.8^a	—
Pyridine	-7.8^a	-3.3^a
Acetonitrile	-2.3^b	$+0.54^b$
Methanol	-1.9^a	$+0.45^a$
Ethanol	-2.1^a	$+0.47^a$
Diethyl ether	-4.2^b	-0.09^b
Tetrahydrofuran	-5.3^a	—
Dioxane	-3.5^a	—
Diethyl sulfide	-7.8^b	-3.17^b
Acetone	—	$+0.10^c$
Ethyl acetate	-3.06^a	—
N,N-Dimethylformamide	-3.7^b	-0.63^b
N,N-Dimethylacetamide	-4.0^b	-1.14^b
Dimethyl sulfoxide	-4.4^b	-1.45^b
Tetramethylenesulfone	-2.2^b	$+0.19^b$
Benzene	-1.5^b	$+1.12^b$
Toluene	-1.8^a	—

[a] R. S. Drago and K. F. Purcell, *Prog. Inorg. Chem.* **6**, 271 (1964).
[b] R. S. Drago, B. Wayland, and R. L. Carlson, *J. Am. Chem. Soc.* **85**, 3125 (1963).
[c] R. L. Middaugh, R. S. Drago, and R. Niedyielski, *J. Am. Chem. Soc.* **86**, 388 (1964).

are poorer when pairs of reference acids like p-fluorophenol and $CHCl_3$[1] or phenol and I_2[65] are compared. This may be due in part to difficulties in getting some reliable values with a weak acid such as $CHCl_3$.[72] Nevertheless, in some cases, clear-cut inversions in basicity order are observed; for example, ΔH_f^0 values with phenol as acid indicate that $(C_2H_5)_2S$ is a weaker base than $(C_2H_5)_2O$ while ΔH_f^0 values with I_2 as acid indicate that $(C_2H_5)_2S$ is a stronger base than $(C_2H_5)_2O$. Such differences can be accounted for in a qualitative way[65] by using the concept of hard and soft acids and bases.[81] While soft bases and acids have molecules which are easily distortable, the opposite holds true for hard base or acid molecules. Thus, I_2 and $CHCl_3$ are soft acids, while $SbCl_5$ and phenol are hard acids. Acid-base interactions of the type soft–soft or hard–hard are strong while others are weak. Drago[82] has further proposed a quantitative empirical treatment to account for donor–acceptor interactions in weakly solvating solvents or in the gas phase. Each acid A and base B is characterized by two parameters, E and C. E is the susceptibility of the acid or base to undergo electrostatic interaction, and C, the susceptibility of the acid or base to form a covalent bond. ΔH_f^0, the enthalpy of adduct formation between a base and an acid, is then given by a four-parameter equation:

$$-\Delta H_f^0 = E_A E_B + C_A C_B \qquad (39)$$

Iodine was selected as a reference acid and assigned $E_A = C_A = 1$. Values of E and C have been calculated from experimental ΔH_f^0 values and tabulated for a large number of acids and bases,[82] although some values are still being refined.[72] However, Arnett et al.[1] have reported deviations as large as 1 kcal/mole between calculated and experimental ΔH_f^0 values of p-fluorophenol and fifteen bases. They have therefore expressed doubts about the usefulness of Eq. 39 to give more than reasonable estimates of the enthalpies of hydrogen-bond formation.

Association constants for the reaction of aprotic bases with the charged Brønsted acid, Bu_3NH^+, and with p-fluorophenol have been compared.[82a] The correlation is unexpectedly strong (Fig. 2) considering the different nature of both acids, and of the inert media used, o-dichlorobenzene and CCl_4, respectively. The corresponding ΔH_f^0 correlation is not as good but this may be due to the fact that the enthalpy values for Bu_3NH^+ were derived from the van't Hoff equation. Gilkerson et al.[82a] have discussed the significance of these relationships.

HCl has been used as reference acid to investigate the basicity of a series of ethers.[83] Christian and Keenan[83] have also reported data which provide comparisons with two other acids, phenol and HNCS, and two inert media, CCl_4 and heptane.

3. PARAMETERS USED TO CHARACTERIZE BASICITY

The recent and extensive thermodynamic and spectroscopic data of Arnett et al.[1] provide the best basis for considering various correlations of parameters characterizing the reaction given by Eq. 33. The same authors have also reviewed previous work on this subject. Arnett et al.[1] have plotted ΔH_f^0 values for the reaction of a series of bases with the reference acid p-fluorophenol against the independently determined corresponding ΔG_f^0 values. They have found that these parameters correlate linearly for bases having the same functional group (pyridines, sulfoxides, amides, etc.). Each straight line obtained has a slope approximately equal to unity. This result indicates that the entropy change ΔS_f^0 accompanying the adduct formation reaction (Eq. 33) is essentially constant within each series of bases ΔS_f^0.

In as much as spectroscopic data could be used to predict ΔH_f^0 values, it is interesting to examine the correlations between IR shifts and ΔH_f^0. As

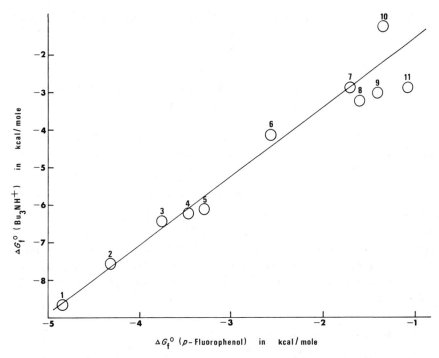

FIG. 2. Comparison of free energies of adduct formation for the reference acids p-fluorophenol and Bu_3NH^+ with bases. 1, hexamethylphosphoramide; 2, triphenylphosphine oxide; 3, pyridine N-oxide; 4, dimethyl sulfoxide; 5, dimethylacetamide; 6, pyridine; 7, tetrahydrofuran; 8, acetone; 9, acetonitrile; 10, diethyl ether; 11, benzonitrile.

early as 1937, Badger and Bauer[84] had proposed that the infrared shift Δv produced by hydrogen bonding might be linearly related to the corresponding enthalpy change ΔH_f^0. Since then, evidence has been produced for[66,85-87] and against[88] the existence of such a relationship. The success or failure of the attempted correlations seems to depend on the nature of the bases examined. Arnett $et\ al.$[1,7] have shown that although no general correlation can be established between Δv_{OH} and ΔH_f^0, for a series of bases interacting with p-fluorophenol, a plot of Δv_{OH} versus ΔH_f^0 is linear for bases possessing the same functional group. Olofsson[89] has established a similar relationship between v_{CO} and ΔH_f^0 for a series of esters interacting with $SbCl_5$.

Correlations between $\Delta \delta$ and ΔH_f^0 or ΔG_f^0 have also been attempted. Eyman and Drago[90] have reported a linear relationship between $\Delta \delta$ (H-NMR) and ΔH_f^0 values for adduct formation between a series of bases and phenol. Gurka and Taft,[73] on the other hand, have reported a linear relationship between $\Delta\delta$ (F-NMR) and ΔH_f^0 values for adduct formation between a series of bases and p-fluorophenol.

V. BASICITY OF SOLVENTS AS REACTANTS WITH REFERENCE ACIDS IN THE GAS PHASE

An extension of the previous methods is to eliminate the medium entirely when reacting the basic solvent molecule B with a reference acid AR. By carrying out the reaction given by Eq. 40 in the gas phase

$$B(g) + AR(g) \rightleftharpoons BAR(g) \tag{40}$$

solvation effects disappear and an " intrinsic basicity order " with respect to AR can be established on the basis of corresponding thermodynamic data. Unfortunately, available gas-phase results have been rather limited up until the present time because of experimental difficulties. We will now consider in turn the two main AR reference acids used: the proton and boron compounds.

A. The Proton as Reference Acid

Mass spectroscopic techniques have recently been developed to study the thermodynamics of ion–molecule reactions in the gas phase (Eq. 40, where $AR = H^+$). Judging from present activity in this area it is expected that such data will become available in ever-increasing volume and quality. Regrettably, molecules B so far investigated do not include many of a type which correspond to widely used solvents.

1. PROTON AFFINITY

The proton affinity of the species B, $PA(B)$, is defined as the enthalpy change for the gas phase endothermic deprotonation reaction

$$BH^+(g) \rightarrow B(g) + H^+(g) \tag{41}$$

Proton affinities expressed in kcal/mole vary widely from a low of 112 for HF to a high of 235 for ethylenediamine; this corresponds to a change of about 90 pK units. Because of difficulties in obtaining absolute PA values, most of the values have been derived from proton transfer reactions.

a. *Absolute Proton Affinity.* Two methods have been used to determine absolute PA values. The first method involves the use of crystal lattice energies and thermochemical cycles. Sherman[91] has thus obtained

$PA(NH_3) = 207$ kcal/mole from the lattice energies of ammonium halides. The same author also calculated $PA(H_2O) = 182$ kcal/mole by assuming that the lattice energies of the isomorphous crystals NH_4ClO_4 and H_3OClO_4 were equal. Sokolov et al.[92] have corrected this value for the energy contribution of hydrogen bonding in the H_3OClO_4 crystal and obtained $PA(H_2O) = 168 \pm 7$ kcal/mole.

The second method for obtaining PA values involves the measurement of the appearance potential (AP) of the BH^+ ion in the source of a mass spectrometer.[93][96] AP is taken as equal to the endothermicity of the reaction producing the ion. If that reaction can be identified and if the enthalpies of formation, ΔH_F^0, of the other species involved are known, $\Delta H_F^0(BH^+)$ is calculated and used to obtain $PA(B)$ from the equation:

$$PA(B) = \Delta H_F^0(B) + \Delta H_F^0(H^+) - \Delta H_F^0(BH^+) \tag{42}$$

One criticism which can be made for this method is that the BH^+ ion is formed with a kinetic energy (translational and vibrational) which is far in excess of its thermal energy. The values of $\Delta H_F^0(BH^+)$ obtained are therefore too high and the $PA(B)$ values too low. Haney and Franklin[96] have shown that the excess energy of ions formed in the source of a mass spectrometer can be calculated from their translational energies. They measured these translational energies in a time-of-flight mass spectrometer and corrected their AP values for excess energy to obtain $PA(H_2O) = 165$, $PA(H_2S) = 170$, $PA(NH_3) = 207$. These corrected values agree more closely with the values obtained from crystal lattice energies than those where no corrections have been made.[93][95]

b. *Relative Proton Affinities.* Relative PA values can be obtained by studying proton transfer reactions occurring in mass spectrometers:

$$B_1H^+ + B_2 \rightleftharpoons B_2H^+ + B_1 \tag{43}$$

If B_1H^+ is observed to transfer a proton to B_2, it is deduced that the reaction (Eq. 43) is thermoneutral or exothermic and that $PA(B_2) \geq PA(B_1)$. By observing several proton-transfer reactions involving B_2 and B_2H^+ it is possible to set limits for the value of $PA(B_2)$. However, more precise PA values are obtained if the equilibrium constant corresponding to Eq. 43 can be determined. If $PA(B_1)$ is known, $PA(B_2)$ is then calculated from the equation:

$$-RT \ln K = \Delta G^0 \approx \Delta H^0 = PA(B_1) - PA(B_2) \tag{44}$$

It is generally assumed that the entropy change for the symmetrical reaction (Eq. 43) is small, although corrections have been made in some cases.[97] For the results obtained either by the bracketing technique or the equilibrium technique to be meaningful and self-consistent two conditions should be verified: (1) The reaction given by Eq. 43 does indeed occur in the reaction chamber. Furthermore, if equilibrium constants K are to be calculated, the reaction must reach equilibrium. (2) The ions involved in the reaction (B_1H^+ and B_2H^+) must attain thermalization before they react.

Although these two factors have not always been verified by the authors who have reported PA values, the close agreement between values determined by different methods might be said to confirm the validity of these methods.

2. EXPERIMENTAL METHODS

We will now give a short description of the experimental mass spectrometric methods used to obtain relative PA values.

a. *Time of Flight Spectrometry.* This method has been used mainly by Haney and Franklin,[96,98] and Long and Munson.[99] The experimental procedure consists in introducing a pair of bases (B_1 and B_2), at a given mole ratio and a usually low (0.03–0.05 torr) total pressure, into the source of the mass spectrometer. Ionization is produced by electron pulses of 0.1 to 1.5 μsec duration. Each electron pulse is followed by an electrical pulse (-250 V, 1.5–5 μsec duration) which draws the positive ions formed out of the source into an accelerating region. The ions are accelerated into a field-free drift tube from 1–2 m long. Since all the ions present have about the same kinetic energy, mass separation takes place. The lighter ions travel faster and reach an ion collector first. Mass spectra, obtained in the form of ion current versus arrival time, are recorded for different residence times of the ions in the spectrometer. Residence times can be varied by delaying the ion-drawout pulses. The intensities of ion current, I, for B_1H^+ and B_2H^+ are plotted against residence time, t. If the mass spectra recorded consist mostly of the two protonated species B_1H^+ and B_2H^+, then a decrease in $I(B_1H^+)$ and an increase in $I(B_2H^+)$ with t means that $PA(B_1) < PA(B_2)$. The interpretation of the results is more difficult, when large percentages of other ions are present, because B_1H^+ and B_2H^+ are then formed by several reactions. A disadvantage of this method is that conditions of pressure and residence times are such that it is impossible to insure that B_1H^+ and B_2H^+ have undergone enough collisions to attain thermalization before reacting.

b. *High Pressure Mass Spectrometry.* This method has been used mainly by Bone et al.,[100,101] Franklin et al.,[102–104] and Kebarle et al.[105,106] The experimental procedure calls for the introduction of a mixture of two bases B_1 and B_2 into the spectrometer source at fixed mole fraction and high

(0.2–3 torr) pressure. This high pressure is achieved by making the ion exit hole very small so that there is a high-pressure differential between the source and the highly evacuated analyzer region. Ionization is produced either by electron impact or by photon irradiation. The ions formed react in the field-free source. Those which come close to the exit hole drift out into the highly evacuated region, are accelerated, mass filtered (usually by a quadrupole filter), and detected. The ion current intensities of B_1H^+ and B_2H^+ are recorded. In the work of Bone et al.[100,101] the experiment is repeated at different total pressures, P, and the relationship between I and P is analyzed. Because of the high pressures at which these experiments are carried out there is a high probability that the ions involved have attained thermalization before reaction. However, because of the short residence times, equilibrium conditions may not prevail. Equilibrium constant calculations were nevertheless carried out by assuming steady-state kinetics for the reaction given by Eq. 43. In the work of Franklin et al.,[102–104] ion current intensities were measured for a fixed pressure (~ 0.5 torr) of the first base B_1 and for small increments of the second base B_2, 0–0.01 torr. If B_2H^+ was found to increase at the expense of B_1H^+, $PA(B_2)$ was assumed greater than $PA(B_1)$. If no B_2H^+ could be detected, it was concluded that $PA(B_1) > PA(B_2)$.

Kebarle et al.[105,106] kept constant pressure (3 torr) and mole ratio of B_1 to B_2. The ionizing electron beam was pulsed, being on for 10 μsec and off for 8 μsec. The ion current intensities of B_1H^+ and B_2H^+ were monitored during these 8 μsec. Equilibrium conditions, i.e., constant ratio for B_2H^+/B_1H^+, were usually observed for $t < 100$ μsec and equilibrium constants calculations could be carried out. Assuming a rate constant of approximately 10^{-9} cm^2 molecule^{-1} sec^{-1} for the reaction (Eq. 43) it was calculated that, under the experimental conditions, an ion underwent about 100 collisions before reacting, so that thermalization could be assumed. The validity of the results were verified by performing similar experiments with low partial pressures of B_1 and B_2 and a high pressure of a third gas such as methane. Under these conditions, the ions undergo some 10^5 collisions before reaction, and thermalization is assured. These experiments gave identical results for the equilibrium constants.

c. *Ion Cyclotron Resonance Spectrometry.* Ion cyclotron resonance (ICR) spectrometry is the most recent mass spectrometric method and its development is by no means complete.[107,108] It has the advantage of allowing the specific reactions occurring in the mass spectrometer to be identified. Furthermore, the residence times of the ions in the reaction chamber are long so that thermalization and equilibrium are ensured.[109] The method has been used by different research groups: Beauchamp et al.,[110–112] Aue et al.,[113,114] Kriemler and Buttrill,[115] Brauman *et al.*,[116] and McDaniell *et al.*[117] Several techniques have been used:

i. Single resonance mode. A mixture of two bases at a given mole ratio and pressure, 10^{-6}–10^{-3} torr, is introduced into the source of the spectrometer. Ionization is produced by electron impact and the positive ions formed are drawn out of the source into a resonance region by appropriate electric fields. A uniform magnetic field B is applied in this region. The angular frequency of an ion circular motion in a plane perpendicular to B is given by $\omega_c = (eB/m)\,\omega_c$, which is called the cycloton frequency of a given ion, is therefore a function of the mass to charge ratio of the ion. An alternating electric field of variable frequency ω_i supplied by a radiofrequency (rf) marginal oscillator is applied in a direction perpendicular to B. When ω_i becomes equal to ω_c, the ion absorbs energy from the rf field, the Q factor of the oscillator is changed, and a signal is produced. The intensity of the signal is proportional to the concentration of the ion of cyclotron frequency ω_c. The mass spectrum is obtained as energy absorption versus frequency.

Most experiments performed with the single resonance mode were carried out at different pressures to yield intensity–pressure plots for the protonated species B_1H^+ and B_2H^+. Such plots allow the calculation of equilibrium constants and subsequent determination of PA values to ± 0.2 kcal/mole. Although the pressure values are small, the residence times (≈ 1 msec) of the ions in the spectrometer are sufficiently long to ensure thermalization before reaction.

ii. Double resonance mode. This working mode of the ICR spectrometer differs from the previous single resonance mode in that two rf alternating fields are applied instead of one. The frequency of the first oscillator, the observing oscillator, is kept constant at $\omega_c(B_1H^+)$ while the frequency of the second oscillator ω_i is varied. When ω_i becomes equal to $\omega_c(B_2H^+)$, the B_2H^+ ions go into resonance and gain kinetic energy. If B_1H^+ and B_2H^+ are related by reaction, this results in a change in the energy absorption of B_1H^+. If, on the other hand, B_1H^+ and B_2H^+ are not related by reaction, the fact that B_2H^+ goes into resonance does not change the energy absorption of B_1H^+.

The double resonance technique not only serves to establish a relationship between B_1H^+ and B_2H^+, but also to fix limits to PA values. For example, if the energy absorption of B_1H^+ drops off when B_2H^+ goes into resonance, it is then concluded that $PA(B_2) > PA(B_1)$.

iii. Pulse mode. Ionization is produced by a short electron pulse. The ions formed are kept trapped in the source region for a given time by applying appropriate voltages. They are then allowed to drift into the resonance region and are detected by an rf oscillator. This experimental procedure yields intensity-ion transit time plots for B_1H^+ and B_2H^+. Though the experiment is usually performed at very low pressures, 10^{-6}–10^{-5} torr, the long residence times, up to 500 msec, ensures that both thermalization and equilibrium are attained. This technique allows the calculation of precise equilibrium constants and is very useful for studying systems where formation of $(B)_nH^+$ competes with proton transfer (Eq. 43). Such reactions are eliminated by working at very low pressures, while the long reaction times still allow for thermalization and equilibrium.

3. GAS PHASE PROTONATION DATA

Proton affinity values are reported in Table XIII for a number of bases commonly used as solvents. These values are not yet as numerous as we would wish. When a choice had to be made between several values, those determined by ICR spectroscopy were preferred for the reasons given above. Because PA spans some 120 kcal/mole, different reference values have often been used for different classes of bases. For example, a PA of 207 (± 2) has been taken for NH_3 to obtain PA for most amines, while PA for a number of oxygen bases has been referred to a value of 179 (± 2) for isopropylene. However, the PA differences between NH_3 and isopropylene is not yet firmly established. Comparisons for various classes of bases should be made with such limitations in mind. Before closing this section we take this opportunity to point out that the correlation between PA and core-electron ionization potentials for double-bonded oxygen has been nicely used to conclude that the site of protonation of esters is the keto oxygen.[118]

B. Boron Compounds as Reference Acids

Gas-phase thermodynamic data for reactions between bases and Brønsted or Lewis acids are rather scanty and obviously restricted to volatile bases and acids. There are a few isolated values for Brønsted acids, such as that for HCl with dimethylether.[119] Data have been reported mostly by Tamres et al.[120] for the reaction of BF_3 with a few bases. Brown and Gerstein[121] have studied the thermodynamics of adduct formation between trimethylboron and a few nitrogen bases. In both cases, ΔG^0 values were obtained by a manometric method, and the ΔH^0 values were calculated from the variation of the equilibrium constant with temperature. Some relevant enthalpy data are given in Table XIV.

It has been noted that the affinity for $B(CH_3)_3$ increases from NH_3 to CH_3NH_2 and $(CH_3)_2NH$ but decreases for $(CH_3)_3N$, thus following surprisingly the trend observed for the affinity of these amines for the proton in

TABLE XIII

PROTON AFFINITIES OF SOLVENT
MOLECULES IN THE GAS PHASE[q]

Solvent	PA^a (kcal/mole)	Solvent	PA^a (kcal/mole)
Ammonia	207[b]	Acetaldehyde	183[i]
n-Butylamine	223[c]	Acetone	190[i]
Piperidine	230[d]	Cyclohexanone	204[k]
Pyrrolidine	229[e]	Formic acid	175[j]
Morpholine	223[d]	Acetic acid	188[j]
Ethylenediamine	232[d]	Trifluoroacetic acid	167[j]
Pyridine	225[f]	Ethyl acetate	205[j]
Quinuclidine	234[g]	Acetamide	210[f]
Acetonitrile	186[h]	Nitromethane	180[l]
Water	165[b]	Nitroethane	185[l]
Methanol	180[i]	Benzene	178[m]
Ethanol	186[i]	Toluene	187[m]
1-Propanol	189[j]	Xylene (o-, p-, or m-)	188[m]
2-Propanol	195[i]	Ethyl chloride	167[n]
2-Butanol	197[i]	Ethyl bromide	170[n]
tert-Butyl alcohol	206[i]	Hydrogen chloride	141[n]
Diethyl ether	205[j]	Hydrogen fluoride	112[o]
Tetrahydrofuran	199[k]	Nitric acid	168[p]
Diethyl sulfide	197[k]		

[a] Proton affinity: enthalpy change for the reaction $SH^+(g) \rightarrow H^+(g) + S(g)$.

[b] M. A. Haney and J. L. Franklin, *J. Chem. Phys.* **48**, 4093 (1968).

[c] D. H. Aue, H. M. Webb, and M. T. Bowers, *J. Am. Chem. Soc.* **94**, 4726 (1972).

[d] D. H. Aue, H. M. Webb, and M. T. Bowers, *J. Am. Chem. Soc.* **95**, 2699 (1973).

[e] M. T. Bowers, D. H. Aue, H. M. Webb, and R. T. McIver, Jr., *J. Am. Chem. Soc.* **93**, 4314 (1971).

[f] R. Yamdagni and P. Kebarle, *J. Am. Chem. Soc.* **95**, 3504 (1973).

[g] R. H. Staley and J. L. Beauchamp, *J. Am. Chem. Soc.* **96**, 1604 (1974).

[h] M. A. Haney and J. L. Franklin, *J. Phys. Chem.* **73**, 4328 (1969).

[i] J. L. Beauchamp and M. C. Caserio, *J. Am. Chem. Soc.* **94**, 2638 (1972).

[j] J. Long and M. S. B. Munson, *J. Am. Chem. Soc.* **95**, 2427 (1973).

[k] E. M. Arnett, E. J. Mitchell, and T. S. S. R. Murty, *J. Am. Chem. Soc.* **96**, 3875 (1974).

[l] P. E. Kriemler and S. E. Buttrill, Jr., *J. Am. Chem. Soc.* **95**, 1365 (1973).

[m] S. Chong and J. L. Franklin, *J. Am. Chem. Soc.* **94**, 6630 (1972).

[n] J. L. Beauchamp, D. Holtz, S. D. Woodgate, and S. L. Patt, *J. Am. Chem. Soc.* **94**, 2798 (1972).

[o] M. S. Foster and J. L. Beauchamp, *Inorg. Chem.* **14**, 1229 (1975).

[p] P. E. Kriemler and S. E. Buttrill, Jr., *J. Am. Chem. Soc.* **92**, 1123 (1970).

[q] See Note added in proof.

aqueous solution, but not the expected trend set by gas-phase PA values. However, this time in agreement with PA values, the $B(CH_3)_3$ scale indicates that NH_3 is a weaker base than pyridine but this is opposite to the aqueous pK_a order. Data in Table XIV also indicate that BF_3 and $B(CH_3)_3$ do not

TABLE XIV

ENTHALPIES OF ADDUCT FORMATION BETWEEN
BF_3 OR $B(CH_3)_3$ AND SOLVENT MOLECULES
IN THE GAS PHASE

Solvent	$\Delta H^0(BF_3)$ (kcal/mole)	$\Delta H^0(B(CH_3)_3)$ (kcal/mole)
Ammonia	—	−13.8[f]
Methylamine	—	−17.6[f]
Dimethylamine	—	−19.3[g]
Trimethylamine	−26.6[a]	−17.6[f]
Piperidine	—	−19.7[g]
Pyrrolidine	—	−20.4[g]
Pyridine	—	−17.0[h]
Diethyl ether	−11.9[b]	—
Tetrahydrofuran	−16.8[c]	—
Tetrahydropyran	−15.4[c]	—
Diethyl sulfide	−2.9[d]	—
Trimethylphosphine	—	−16.5[i]
Triethylphosphine	−9.5[e]	—

[a] W. A. G. Graham and F. G. A. Stone, *J. Inorg. Nucl. Chem.* 3, 164 (1956).
[b] D. E. McLaughlin and M. Tamres, *J. Am. Chem. Soc.* 82, 5618 (1960).
[c] D. E. McLaughlin and M. Tamres, *J. Am. Chem. Soc.* 82, 5621 (1960).
[d] H. L. Morris, N. I. Kulevsky, M. Tamres, and S. Searles, *Inorg. Chem.* 5, 124 (1966).
[e] H. L. Morris, M. Tamres, and S. Searles, *Inorg. Chem.* 5, 2156 (1966).
[f] H. C. Brown, H. Bartholomey, Jr., and M. D. Taylor, *J. Am. Chem. Soc.* 66, 435 (1944).
[g] H. C. Brown and M. Gerstein, *J. Am. Chem. Soc.* 72, 2926 (1950).
[h] H. C. Brown and G. K. Barbaras, *J. Am. Chem. Soc.* 69, 1137 (1947).
[i] H. D. Kaesz and F. G. A. Stone, *J. Am. Chem. Soc.* 82, 6213 (1960).

correlate very well. On the BF_3 scale, the ΔH^0 values for $(CH_3)_3N$ and $(C_2H_5)_3P$ differ by some 17 kcal/mole, while on the $B(CH_3)_3$ scale, the difference between $(CH_3)_3N$ and $(CH_3)_3P$ [$(C_2H_5)_3P$ was not studied] is only 1 kcal/mole. It is unlikely that the basicity of the two alkyl phosphines differ by as much as 16 kcal/mole. It may be that $B(CH_3)_3$ gives some specific interactions with $(CH_3)_3N$. However, it would appear that, at the present time, gas-phase data for reactions with these boron compounds are not very useful as indicators of solvent basicity.

VI. CORRELATION OF SOLVENT BASICITY PARAMETERS

We have now reviewed the main operational definitions of solvent basicity. We have also gathered some corresponding basicity parameters. These parameters were expressed as free energy and enthalpy changes for the reaction of the solvents with reference acids, the solvents being in the bulk state or as isolated molecules in various media. Bulk solvent basicity is the property of prime interest in nonaqueous solvent chemistry. However, the corresponding ΔH^0 or ΔG^0 values referring to reactions of common bulk solvents with acids, particularly with H^+, are still relatively scarce in the literature. It is therefore worthwhile looking for possible correlations with other more numerous basicity data to supplement results for bulk solvents. Furthermore, useful reaction patterns between bases and reference acids are more likely to appear in studies dealing with inert media rather than with bulk solvents where solute–solvent interactions may add to the complexity of the interpretation of results.

A. Basicity as Related to Protonation Reactions

The role of the proton in chemistry is such that the definition of basicity with reference to the proton is clearly the most important one. We will, therefore, look for relationships between parameters of protonation reactions in the bulk solvent and in other media. The first comparison which comes to mind is between the enthalpy of solvation of the gaseous proton and gas phase proton affinity of the solvent molecule. One would expect $-\Delta H^0_{g \to s}(H^+)$ to increase with PA but this is not always observed. For example, the PA values for water and acetonitrile are 165 and 186 kcal/mole, respectively, while the corresponding values for $-\Delta H^0_{g \to s}(H^+)$ are 270 and 257 kcal/mole. On the strength of the PA values, one concludes that gaseous acetonitrile is more basic than water, whereas the solvation enthalpies $\Delta H^0_{g \to s}$ point to the opposite for the bulk solvents. Reasons for this reversal[122] can be advanced from the work of Kebarle[123] on the thermodynamics of gas phase formation of species H^+B_n. It appears that when B is a hydrogen-bond donor, such as water and alcohols, reaction with H^+ leads, through cooperative hydrogen bonding, to a series of H^+B_n species of decreasing stability. However, when B is not a hydrogen-bond donor, the stability of species H^+B_n where $n > 2$ drops drastically because of blocking of hydrogen bonding past the structure H^+B_2. It then becomes clear why the difference between $-\Delta H^0_{g \to s}(H^+)$ and PA, which accounts for the reversal of basicities, is much higher for protic water (105 kcal/mole) than for aprotic acetonitrile (71 kcal/mole), since this difference can be equated to the formation of H^+B_n followed by solvation of this ion to lead to $H^+(s)$. Another example of reversal of basicities which could be similarly explained is given by the protic ammonia aprotic pyridine pair. While their PA values, 207 and 225 kcal/mole respectively, indicate that gaseous ammonia is less basic than pyridine, the $\Delta G^0_{g \to s}(H^+)$ values in Table I show the basicity of bulk ammonia to be at least 10 kcal/mole larger than that of bulk pyridine. It is also remarkable that, just as for water,[9] the basicity of bulk HF is much higher than one would expect from its very low PA.[97]

We have nevertheless plotted in Fig. 3 some $\Delta G^0_{g \rightarrow s}(H^+)$ values against PA. $\Delta G^0_{g \rightarrow s}(H^+)$ was used instead of $\Delta H^0_{g \rightarrow s}(H^+)$ because there are not enough $\Delta H^0_{g \rightarrow s}(H^+)$ values available. Although a trend is apparent, in that the lower the proton affinity, the lower $-\Delta G^0_{g \rightarrow s}(H^+)$, the free energy of solvation of the proton, the correlation is poor. Entropy factors may partly

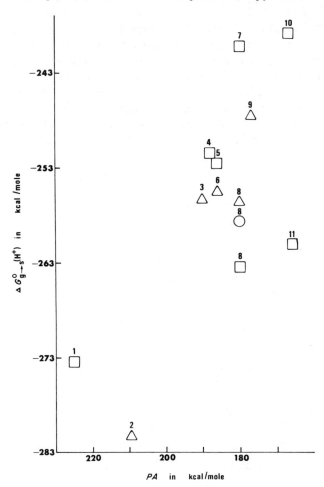

FIG. 3. Free energy of solvation of H^+ as a function of proton affinity of solvent molecules. 1, pyridine; 2, ammonia; 3, acetone; 4, acetic acid; 5, acetonitrile; 6, ethanol; 7, nitromethane; 8, methanol; 9, formic acid; 10, trifluoroacetic acid; 11, water; □, $\Delta G^0_{g \rightarrow s}$ obtained using the ferrocene assumption; ○, tetraphenylborate assumption; △, Izmailov's extrapolation method.

account for this poor correlation. The fact that different extrathermo-dynamic assumptions lead at times to widely different $\Delta G^0_{g \to s}(H^+)$ values, methanol being a case in point, makes matters difficult. That the points for protic NH_3, CH_3OH, and H_2O fall below a line which connects the points for aprotic pyridine, acetonitrile, and nitromethane, can now at least be understood.

There is a fair correlation in Fig. 4 between the few values of $\Delta G^0_{g \to s}(H^+)$ and ΔH^0_i, the enthalpy of protonation in HSO_3F, if both points for protic solvents H_2O and CH_3OH are ignored. A similar correlation would hold between pK_{SH^+} and $\Delta G^0_{g \to s}(H^+)$ since Arnett et al.[1] established that ΔH^0_i and pK_{SH^+} are linearly related. The available $\Delta H^0_{g \to s}(H^+)$ values for five dipolar aprotic solvents also correlate reasonably well with ΔH^0_i in Fig. 5, the point for H_2O being off the line. At this stage, in view of the fair correla-tion between bulk basicity data for aprotic solvents and corresponding ΔH^0_i values, we reexamined the potential relationship between PA and ΔH^0_i. More data are available for such a comparison than for the previously attempted correlation between PA and $\Delta H^0_{g \to s}(H^+)$, or $\Delta G^0_{g \to s}(H^+)$. If, on

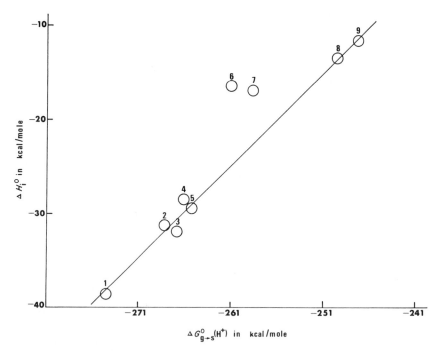

FIG. 4. Correlation between free energies of solvation of H^+ and enthalpies of protonation of solvent molecules in HSO_3F. 1, pyridine; 2, N-methyl-2-pyrrolidone; 3, dimethylacetamide; 4, dimethyl sulfoxide; 5, dimethylformamide; 6, water; 7, methanol; 8, acetonitrile; 9, sulfolane.

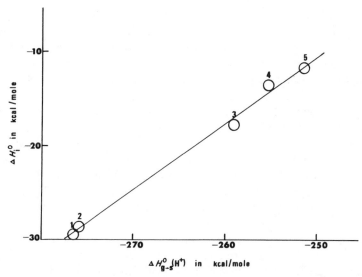

FIG. 5. Correlation between enthalpies of solvation of H^+ and enthalpies of protonation of solvent molecules in HSO_3F. 1, dimethylformamide; 2, dimethyl sulfoxide; 3, propylene carbonate; 4, acetonitrile; 5, sulfolane.

the basis of existing data (see Table V), we assume for acetamide and nitromethane ΔH_i^0 values of -30 and -6 kcal/mole respectively, we get with pyridine and acetonitrile a reasonably straight line for PA against ΔH_i^0. The line passes close to a group of points corresponding to esters, ethers, and ketones. Some of these points are nevertheless some 8 kcal/mole off the PA value given by the line, and 8 kcal/mole is higher than the ± 2 kcal/mole usually quoted as margin of error. There is obviously a need for more numerous and accurate PA values, particularly for commonly used solvents, before it can be ascertained to what extent such gas phase data are useful to predict basicity of aprotic solvents.

B. Basicity as Related to Adduct Formation Reactions

Solvent basicity defined with respect to acids other than H^+ is also of interest, considering the importance of all acid–base reactions. We will examine correlations between parameters of adduct formation reactions with Brønsted and Lewis acids in the bulk solvent, and the corresponding parameters in other media. Arnett et al.[1] have carefully compared enthalpy changes, ΔH_h^0 (see Table III), for the reaction of HA (p-fluorophenol and phenol) with a series of pure bases, after correction for non-hydrogen-bonding interactions, and ΔH_f^0 for the reaction of HA with the same bases at high dilution in an inert medium. For the reaction of

HA = p-fluorophenol with 15 bases, the average difference between ΔH_h^0 and ΔH_f^0 was found to be 0.4 kcal/mole. For HA = phenol the same average difference was 0.5 kcal/mole for 29 compounds. It would be worthwhile to extend such comparisons to other acids HA in order to determine how successfully basicity parameters of bulk solvents toward HA can be predicted from reaction parameters in inert media. However, it may be that protic compounds do not follow the close correspondence generally observed by Arnett et al.[1] for their series of compounds. The large difference noted by these authors between ΔH_h^0 and ΔH_f^0 for N-methylformamide, the one solvent studied possessing some protic character may be significant. When HA interacts with protic solvents, several solvent molecules may be involved through cooperative hydrogen bonding, just as in the case of H^+. Such cooperative hydrogen bonding would not occur between HA and aprotic solvents.

Predictions as to the basicity of bulk solvents with respect to HA, at least where phenol-type acids are concerned, can be made from the enthalpy of reaction of the solvent with the Lewis acid $SbCl_5$ in inert media, i.e., from Gutman's donor number, DN. These predictions result both from the good linear correlation observed between DN and ΔH_f^0 for p-fluorophenol[1] and from the previously quoted close correspondence between ΔH_f^0 values and ΔH_h^0 for bulk solvents. Protic solvents are likely, however, not to follow this correlation, as evidenced by the addition of another DN value (~ 33) for bulk water.[80a,124] It is interesting to note that $SbCl_5$ gives the same basicity order for five dipolar aprotic solvents regardless of whether the reaction is with the bulk solvent (see Table IV) or with the dilute solvent in $C_2H_4Cl_2$ (see Table XI). Furthermore, the spread between the corresponding values, given in Tables IV and XI, is seen to increase with the basicity of the solvent in a manner not unlike that observed for HA = p-fluorophenol. There are not enough data for other acids such as $CHCl_3$ or I_2 to make some worthwhile comparisons between reaction parameters for bulk solvents and isolated solvent molecules.

It is interesting to note that an NMR parameter has been used as an indicator of solvent basicity. Popov et al.[124] have reported that, for $NaBPh_4$ solutions in several aprotic solvents, the plot of the chemical shifts of ^{23}Na nucleus against Gutman's donor numbers gives a reasonable straight line. These authors took advantage of this correlation to estimate some DN values for protic basic solvents (see Table II). Such values would likely refer to the basicity of the bulk solvents.

C. Protonation and Adduct Formation Reactions in Bulk Solvents

Finally, we will consider relationships between basicity parameters for bulk solvents defined with respect to both H^+ and molecular acids. Due to the small number of enthalpy or free energy values for the solvation of the proton and the solvation (or transfer from a reference phase) of Brønsted and Lewis acids, comparisons are clearly limited. This is why we will focus our attention only on what we consider to be typical enthalpy data gathered in Table XV for the reaction of $SbCl_5$, several Brønsted acids and the proton with five dipolar aprotic solvents possessing different functional group donors. Although there is a rough correlation between the different basicity parameters, some differences remain apparent. Thus, dimethylformamide and dimethyl sulfoxide have the same basicity towards H^+ (on the ΔG^0 scale DMSO is more basic by a not very significant 1 kcal/mole). However, dimethyl sulfoxide is slightly more basic than dimethylformamide when the reference acid is $SbCl_5$ or HA, although with p-fluorophenol this is only so when ΔH^0 is "corrected." On the other hand, while tetramethylene sulfone and acetonitrile have nearly the same basicity toward $SbCl_5$ and HA, with respect to H^+ tetramethylene sulfone is the lesser basic solvent. Larger differences would probably have appeared if data on protic solvents had been available for our comparison. On the basis of determinations, for a wide series of bases, of the enthalpies of protonation in HSO_3F, ΔH_i^0, and of

TABLE XV

ENTHALPY DATA (kcal/mole) FOR PROTONATION AND ADDUCT FORMATION REACTIONS WITH FIVE DIPOLAR APROTIC SOLVENTS

	Solvent				
	Tetra-methylene sulfone	Acetonitrile	Propylene carbonate	Dimethyl-formamide	Dimethyl-sulfoxide
$-\Delta H_{g \to s}^0(H^+)^a$	252.6	256.6	259.5	276.7	276.1
$-\Delta H_i^{0b}$	11.8	14.6	17.8	29.5	28.6
$-\Delta H_{g \to s}^0(SbCl_5)^c$	26.9	26.8	29.9	44.6	46.5
$-\Delta H^0(SbCl_5\text{-B})^d$	14.3	14.1	15.1	26.6	29.8
$-\Delta H_{g \to s}^0(HCl)^e$	5.8	5.7	6.3	17.8	18.3
$-\Delta H_b^0(p\text{-FP})^e$	4.25^f, 4.9^g	—	4.53^f, 4.99^g	6.97^f, 8.13^g	7.21^f, 7.66^g
$-\Delta H_{g \to s}^0(CH_3CO_2H)^e$	3.4	3.1	—	6.4	6.7
$-\Delta H_{g \to s}^0(H_2O)^e$	8.9	8.5	8.6	11.4	11.8
$-\Delta H_{g \to s}^0(H \cdots OH)^e$	6.15	6.13	6.20	7.85	8.13

[a] Based on data in Table II and $-\Delta H_{g \to w}^0(H^+) = 270.0$.

[b] Enthalpy of protonation in HSO_3F; see Table V.

[c] See Table IV.

[d] Enthalpy of reaction in $C_2H_4Cl_2$; see Table XI.

[e] See Table III.

[f] Enthalpy of solution of p-fluorophenol from CCl_4 corrected for nonspecific interactions; see Table III.

[g] As in f but uncorrected.

the enthalpies of hydrogen bonding with p-fluorophenol, ΔH_f^0, Arnett et al.[1] concluded that there was no general correlation between ΔH_i^0 and ΔH_f^0. Although we are considering bulk solvents here, where solvation effects may be different, we might well draw a similar conclusion and suspect that there is no definite correlation between protonation and adduct formation.

REFERENCES

1. E. M. Arnett, E. J. Mitchell, and T. S. S. R. Murty, J. Am. Chem. Soc. **96**, 3875 (1974).
2. T. M. Krygowski and W. R. Fawcett, J. Am. Chem. Soc. **97**, 2143 (1975).
3. R. Domain, M. Rinfret, and R. L. Benoit, Can. J. Chem. **54**, 2101 (1976); R. Domain, Ph.D. Dissertation, University of Montreal, Montreal (1975).
4. G. Charlot and B. Trémillon, "Les réactions chimiques dans les solvants et les sels fondus." Gauthier-Villars, Paris, 1963.
5. I. M. Kolthoff, Anal. Chem. **46**, 1992 (1974).
6. E. S. Amis and J. F. Hinton, "Solvent Effects on Chemical Phenomena," Vol. 1. Academic Press, New York, 1973.
7. E. M. Arnett, L. Joris, E. Mitchell, T. S. S. R. Murty, T. M. Gorrie, and P. v. R. Schleyer, J. Am. Chem. Soc. **92**, 2365 (1970).
8. M. Herlem and A. I. Popov, J. Am. Chem. Soc. **94**, 1431 (1972); J. J. Dechter and J. I. Zink, ibid. **97**, 2937 (1975); C. N. R. Rao, K. G. Rao, and N. V. R. Reddy, ibid. p. 2918.
9. E. M. Arnett, Acc. Chem. Res. **6**, 404 (1973).
10. R. L. Martin and D. A. Shirley, J. Am. Chem. Soc. **96**, 5299 (1974).
11. M. M. Davis, in "The Chemistry of Nonaqueous Solvents" (J. J. Lagowski, ed.), Vol. 3, p. 1. Academic Press, New York, 1970.
12. J. J. Lagowski, Anal. Chem. **46**, 460R (1974).
13. D. Bauer and M. Bréant, Electroanal. Chem. **9**, 281 (1975).
14. H. L. Friedman and C. V. Krishnan, in "Water: A Comprehensive Treatise" (F. Franks, ed.), Vol. 3, p. 1. Plenum, New York, 1973.
15. H. F. Halliwell and S. C. Nyburg, Trans. Faraday Soc. **59**, 1126 (1963), and references therein.
16. J. E. Desnoyers and C. Jolicoeur, Mod. Aspects Electrochem. **5**, 1 (1969).
17. W. A. Millen and D. W. Watts, J. Am. Chem. Soc. **89**, 6051 (1967).
18. H. J. Nedermeijer-Denessen and C. L. De Ligny, J. Electroanal. Chem. **57**, 265 (1974).
19. L. G. Hepler and E. M. Wooley, in "Water: A Comprehensive Treatise" (F. Franks, ed.), Vol. 3, p. 145. Plenum, New York, 1973.
20. A. J. Parker, Chem. Rev. **69**, 1 (1969); R. Alexander, A. J. Parker, J. H. Sharp, and W. E. Waghorne, J. Am. Chem. Soc. **94**, 1148 (1972).
21. I. M. Kolthoff, Pure Appl. Chem. **25**, 305 (1971).
22. O. Popovych, Crit. Rev. Anal. Chem. **7**, 73 (1970).
23. H. Strehlow, in "The Chemistry of Non-Aqueous Solvents" (J. J. Lagowski, ed.), Vol. 1, p. 129. Academic Press, New York, 1966.
24. N. A. Izmailov, Zh. Fiz. Khim. **34**, 2414 (1960).
25. D. Feakins and P. Watson, J. Chem. Soc. p. 4734 (1963).
26. M. Alfenaar and C. L. De Ligny, Recl. Trav. Chim. Pays-Bas **86**, 929 (1967).
27. N. A. Izmailov, Dokl. Akad. Nauk. SSSR **149**, 884, 1103, and 1364 (1963).
28. N. Bjerrum and E. Larsson, Z. Phys. Chem. (Leipzig) **127**, 358 (1927).
29. A. J. Parker and R. Alexander, J. Am. Chem. Soc. **90**, 3313 (1968).

30. I. T. Oiwa, *Sci. Rep. Tohoku Univ., Ser. 1* **41**, 129 (1957), quoted in Popovych.[22]
31. O. Popovych and A. J. Dill, *Anal. Chem.* **41**, 456 (1969).
32. V. A. Pleskov, *Usp. Khim.* **16**, 254 (1947).
33. H. M. Koepp, H. Wendt, and H. Strehlow, *Z. Elektrochem.* **64**, 483 (1960).
34. D. Bauer and J. P. Beck, *Bull. Soc. Chim. Fr.* p. 1252 (1973); J. Verastegin, G. Durand, and B. Trémillon, *J. Electroanal. Chem.* **54**, 269 (1974).
35. E. Grunwald, G. Baughman, and G. Kohnstam, *J. Am. Chem. Soc.* **82**, 5801 (1960).
36. R. Alexander and A. J. Parker, *J. Am. Chem. Soc.* **89**, 5539 (1967).
37. O. Popovych, *Anal. Chem.* **38**, 558 (1966).
38. I. M. Kolthoff and M. K. Chantooni, Jr., *J. Phys. Chem.* **76**, 2024 (1972).
39. J. W. Diggle and A. J. Parker, *Electrochim. Acta* **18**, 975 (1973).
40. M. Herlem, *Bull. Soc. Chim. Fr.* p. 1687 (1967); G. Petit and J. Bessière, *J. Electroanal. Chem.* **34**, 489 (1972).
41. B. G. Cox and A. J. Parker, *J. Am. Chem. Soc.* **95**, 402 (1973).
42. M. H. Abraham, *J. Chem. Soc., Faraday Trans. 1* **69**, 1375 (1973).
43. C. M. Criss, *J. Phys. Chem.* **78**, 1000 (1974).
44. C. V. Krishnan and H. L. Friedman, *J. Phys. Chem.* **75**, 3598 (1971).
45. W. C. Duer and G. L. Bertrand, *J. Am. Chem. Soc.* **92**, 2587 (1970).
46. E. M. Arnett, T. S. S. R. Murty, P. v. R. Schleyer, and L. Joris, *J. Am. Chem. Soc.* **89**, 5955 (1967).
47. W. Gerrard and E. D. Macklen, *Chem. Rev.* **59**, 1105 (1959); W. Gerrard, A. M. A. Mincer, and P. L. Wyvill, *Chem. Ind. (London)* p. 894 (1958); J. Charalambous, M. J. Frazer, and W. Gerrard, *J. Chem. Soc.* p. 1520 (1964).
48. L. G. Hepler and D. V. Fenby, *J. Chem. Thermodyn.* **5**, 471 (1973); G. L. Bertrand, *J. Phys. Chem.* **79**, 48 (1975).
49. J. E. Gordon, *J. Org. Chem.* **26**, 738 (1961), and references therein.
50. C. Jambon and R. Philippe, *J. Chem. Thermodyn.* **7**, 479 (1975).
51. E. M. Arnett, *Prog. Phys. Org. Chem.* **1**, 223 (1963).
52. M. Liler, *in* "Reaction Mechanisms in Sulfuric Acid and Other Strong Acid Solutions." Academic Press, New York, 1971.
53. I. M. Kolthoff, M. K. Chantooni, Jr., and S. Bhowmik, *J. Am. Chem. Soc.* **90**, 23 (1968); I. M. Kolthoff and M. K. Chantooni, Jr., *ibid.* **95**, 8539 (1973); **90**, 3320 (1968).
54. E. M. Arnett and J. F. Wolf, *J. Am. Chem. Soc.* **95**, 978 (1973); E. M. Arnett, R. P. Quirk, and J. J. Burke, *ibid.* **92**, 1260 (1970).
55. G. Barrow, *J. Am. Chem. Soc.* **78**, 5802 (1956).
56. D. V. Fenby and L. G. Hepler, *Chem. Soc. Rev.* **3**, 193 (1974).
57. L. Lamberts, *Ind. Chim. Belge* **36**, 347 (1971).
58. J. J. Christensen, J. Ruckman, D. J. Eatough, and R. M. Izatt, *Thermochim. Acta* **3**, 203 (1972).
59. L. Lamberts and Th. Zeegers-Huyskens, *J. Chim. Phys. Phys.-Chim. Biol.* **60**, 435 (1963).
60. T. F. Bolles and R. S. Drago, *J. Am. Chem. Soc.* **87**, 5015 (1965).
61. S. Cabani and P. Gianni, *J. Chem. Soc. A* p. 547 (1968).
62. G. A. Aknes and T. Gramstad, *Acta Chem. Scand.* **14**, 1485 (1960); J. N. Spencer, R. A. Heckman, R. S. Harner, S. L. Shoop, and K. S. Robertson, *J. Phys. Chem.* **77**, 3103 (1973); J. N. Spencer, K. S. Robertson, and E. E. Quick, *ibid.* **78**, 2236 (1974); T. Gramstad, *Spectrochim. Acta* **19**, 829 (1963).
63. F. L. Slejko, R. S. Drago, and D. G. Brown, *J. Am. Chem. Soc.* **94**, 9210 (1972).
64. T. Gramstad, *Acta Chem. Scand.* **16**, 807 (1962); T. Kitao and C. H. Jarboe, *J. Org. Chem.* **32**, 407 (1967).
65. R. S. Drago and K. F. Purcell, *Prog. Inorg. Chem.* **6**, 271 (1964).

66. R. L. Middaugh, R. S. Drago, and R. Niedyielski, *J. Am. Chem. Soc.* **86**, 388 (1964).
67. N. J. Rose and R. S. Drago, *J. Am. Chem. Soc.* **81**, 6138 (1959).
68. M. D. Joesten and R. S. Drago, *J. Am. Chem. Soc.* **84**, 2037 and 2696 (1962).
69. S. D. Christian, E. H. Lane, and F. Garland, *J. Phys. Chem.* **78**, 557 (1974).
70. C. J. Creswell and A. L. Allred, *J. Phys. Chem.* **66**, 1469 (1962).
71. S. Nishimura, C. H. Ke, and N. C. Li, *J. Am. Chem. Soc.* **90**, 234 (1968).
72. G. R. Wiley and S. I. Miller, *J. Am. Chem. Soc.* **94**, 3287 (1972).
73. D. Gurka and R. W. Taft, *J. Am. Chem. Soc.* **91**, 4794 (1969).
74. S. D. Christian, K. O. Yeo, and E. E. Tucker, *J. Phys. Chem.* **75**, 2413 (1971).
75. S. D. Christian, R. Frech, and K. O. Yeo, *J. Phys. Chem.* **77**, 813 (1973).
76. R. S. Drago, M. S. Nozari, and G. C. Vogel, *J. Am. Chem. Soc.* **94**, 90 (1972).
77. M. S. Nozari and R. S. Drago, *J. Am. Chem. Soc.* **94**, 6877 (1972).
78. R. M. Guidry and R. S. Drago, *J. Phys. Chem.* **78**, 454 (1974).
79. A. Allerhand and P. v. R. Schleyer, *J. Am. Chem. Soc.* **85**, 371 (1963).
80. G. Olofsson and I. Olofsson, *J. Am. Chem. Soc.* **95**, 7231 (1973).
80a. V. Gutmann and R. Schmid, *Coord. Chem. Rev.* **12**, 263 (1973); H. Mayer and V. Gutmann, *Struct. Bonding (Berlin)* **12**, 113 (1972); V. Gutmann, "Coordination Chemistry in Non-Aqueous Solutions." Springer-Verlag, Berlin and New York, 1968.
81. R. G. Pearson, *J. Chem. Educ.* **45**, 581 (1968).
82. R. S. Drago, *Struct. Bonding (Berlin)* **15**, 73 (1973).
82a. M. L. Junker and W. R. Gilkerson, *J. Am. Chem. Soc.* **97**, 493 (1975); H. W. Aitken and W. R. Gilkerson, *ibid.* **95**, 8551 (1973).
83. S. D. Christian and B. M. Keenan, *J. Phys. Chem.* **78**, 432 (1974).
84. R. M. Badger and S. H. Bauer, *J. Chem. Phys.* **5**, 839 (1937).
85. T. D. Epley and R. S. Drago, *J. Am. Chem. Soc.* **89**, 5770 (1967).
86. M. D. Joesten and R. S. Drago, *J. Am. Chem. Soc.* **84**, 3817 (1962).
87. R. S. Drago and T. D. Epley, *J. Am. Chem. Soc.* **91**, 2883 (1969).
88. G. C. Pimentel and A. L. McClellan, "The Hydrogen Bond." Freeman, San Francisco, California, 1960.
89. G. Olofsson, *Acta Chem. Scand.* **22**, 377 (1968).
90. D. P. Eyman and R. S. Drago, *J. Am. Chem. Soc.* **88**, 1617 (1966).
91. J. Sherman, *Chem. Rev.* **11**, 164 (1932).
92. S. I. Vetchinkin, E. I. Pshenichnov, and N. Sokolov, *Zh. Fiz. Khim.* **33**, 1269 (1959).
93. D. Van Raalte and A. G. Harrison, *Can. J. Chem.* **41**, 3118 (1963).
94. M. S. B. Munson and J. L. Franklin, *J. Phys. Chem.* **68**, 3191 (1964).
95. A. G. Harrison, A. Irko, and D. Van Raalte, *Can. J. Chem.* **44**, 1625 (1966).
96. M. A. Haney and J. L. Franklin, *J. Chem. Phys.* **48**, 1527 (1968); **50**, 2028 (1969).
97. M. S. Foster and J. L. Beauchamp, *Inorg. Chem.* **14**, 1229 (1975).
98. M. A. Haney and J. L. Franklin, *J. Phys. Chem.* **73**, 4328 (1969).
99. J. Long and M. S. B. Munson, *J. Am. Chem. Soc.* **95**, 2427 (1973).
100. L. Y. Wel and L. I. Bone, *J. Phys. Chem.* **78**, 2527 (1974).
101. J. M. Hopkins and L. I. Bone, *J. Chem. Phys.* **58**, 1473 (1973).
102. S. Chong, R. Myers, and J. L. Franklin, *J. Chem. Phys.* **56**, 2427 (1972).
103. S. Chong and J. L. Franklin, *J. Am. Chem. Soc.* **94**, 6347 (1972).
104. S. Chong and J. L. Franklin, *J. Am. Chem. Soc.* **94**, 6630 (1972).
105. J. P. Briggs, R. Yamdagni, and P. Kebarle, *J. Am. Chem. Soc.* **94**, 5128 (1972).
106. R. Yamdagni and P. Kebarle, *J. Am. Chem. Soc.* **95**, 3504 (1973).
107. J. D. Baldeschwieler and S. D. Woodgate, *Acc. Chem. Res.* **4**, 114 (1971).
108. G. A. Gray, *Adv. Chem. Phys.* **19**, 141 (1971).
109. J. M. S. Henis, *J. Am. Chem. Soc.* **90**, 844 (1968).

110. R. H. Staley and J. L. Beauchamp, *J. Am. Chem. Soc.* **96**, 1604 and 6252 (1974).
111. D. A. Dixon, D. Holtz, and J. L. Beauchamp, *Inorg. Chem.* **11**, 960 (1972).
112. J. L. Beauchamp and M. C. Caserio, *J. Am. Chem. Soc.* **94**, 2638 (1972).
113. D. H. Aue, H. M. Webb, and M. T. Bowers, *J. Am. Chem. Soc.* **94**, 4726 (1972).
114. D. H. Aue, H. W. Webb, and M. T. Bowers, *J. Am. Chem. Soc.* **95**, 2699 (1973).
115. P. E. Kriemler and S. E. Buttrill, Jr., *J. Am. Chem. Soc.* **95**, 1365 (1973).
116. J. I. Brauman, J. M. Riveros, and L. K. Blair, *J. Am. Chem. Soc.* **93**, 3914 (1971).
117. D. H. McDaniel, N. B. Coffman, and J. M. Strong, *J. Am. Chem. Soc.* **92**, 6697 (1970).
118. T. X. Carroll, S. R. Smith, and T. D. Thomas, *J. Am. Chem. Soc.* **97**, 659 (1975).
119. A. S. Gilbert and H. J. Bernstein, *Can. J. Chem.* **52**, 674 (1974).
120. H. L. Morris, M. Tamres, and S. Searles, *Inorg. Chem.* **5**, 2156 (1966).
121. H. C. Brown and M. Gerstein, *J. Am. Chem. Soc.* **72**, 2926 (1950).
122. R. L. Benoit and S. Y. Lam, *J. Am. Chem. Soc.* **96**, 7385 (1974).
123. P. Kebarle, *Mod. Aspects Electrochem.* **9**, 1 (1974); E. P. Grimsrud and P. Kebarle, *J. Am. Chem. Soc.* **95**, 7939 (1973).
124. R. H. Erlich, E. Roach, and A. I. Popov, *J. Am. Chem. Soc.* **92**, 4989 (1970).

NOTE ADDED IN PROOF

Since this chapter was submitted in August 1975, new values have been reported for solvent basicity parameters, particularly proton affinities. For example, PA(NH$_3$) is now taken as 202.3 kcal/mole so that our PA values for more basic molecules should be lowered by some 5 kcal/mole. Recent papers by J. L. Beauchamp, P. Kebarle, and R. W. Taft should be consulted for up to date PA values [e.g., *J. Am. Chem. Soc.* **99**, 5417 (1977)].

～ 3 ～

Nonaqueous Solvents in Organic Electroanalytical Chemistry

⚬

PETR ZUMAN*AND STANLEY WAWZONEK†

* Department of Chemistry, Clarkson College of Technology
Potsdam, New York

†Department of Chemistry, University of Iowa
Iowa City, Iowa

I. Introduction

The use of nonaqueous solvents in polarography and related electro-analytical techniques offers some indisputable advantages over analogous investigations carried out in systems containing water or other proton-donating solvents. Nevertheless, the introduction of such solvents is not a general cure of all the problems faced in electroanalytical studies of organic compounds and their applications. Decisions to use nonaqueous or aqueous (and water-similar) solvents for the solution of electroanalytical problems should be based on the realization and thorough understanding of the pros and cons of both types of such approaches.

II. Advantages of the Use of Nonaqueous Solvents

Advantages offered by the use of nonaqueous solvents for electro-analytical studies covers a variety of aspects. The increased solvating properties when compared with aqueous solutions is discussed first, followed by a discussion of the elimination of solvolytic reactions and covalent hydration. Specific effects of organic solvents on the reduction of halogen compounds and the extension of the accessible potential scale are discussed. A description of the advantages in the study of one-electron reversible processes follows together with a comparison of the different types of reactions of products of such reversible reactions, namely radical ions or carbenes. Further on, reasons are presented to explain why organic solvents are less useful in the study of chemical reactions preceding the electrode process proper and, finally, possibilities are indicated as to how electrochemical methods can contribute to classification and better understanding of solvent effects.

The majority of the work on which the following discussion is based has been carried out in N,N-dimethylformamide, acetonitrile, or dimethylsulfox-ide. Some of the information is based on studies using pyridine, hexamethyl-phosphoramide, methylene chloride, liquid sulfur dioxide, methylacetamide, tetramethylurea, nitromethane, sulfuric acid, methanesulfonic acid, and tri-fluoroacetic acid as solvents. Occasional use of other solvents like dioxane, tetrahydrofuran, glyme, diglyme, sulfolane, acetone, acetic anhydride, other amides and nitriles, nitro compounds, and halogen derivatives has also been reported. Sometimes mixtures such as methanol–benzene, ethanol–ethyl acetate, methanol–ethylene chloride, methanol–dioxane, and trifluoro-ethanol–dimethylformamide were used.

Less frequently used solvents and mixtures have some specific properties. The attempts to generalize which follow should be considered approximations; the exact limits of validity should be checked for each individual solvent.

A. Solvating Properties

The most important advantages nonaqueous solvents offer are their solvating properties. Even when the majority of organic compounds, which are important because of their physiological or biochemical properties or interesting because of their chemical behavior, can be dissolved in water-containing or water-similar solvent systems to reach concentrations $(10^{-6}-10^{-4}\ M)$ usually used in modern electroanalytical methods, the better solvent properties of nonaqueous solvents enable us to investigate biologically important, fat-soluble compounds, some polycyclic hydrocarbons including carcinogens, dialkylperoxides, and anthraquinone derivatives.

Those increased solvent properties were practically exploited in determinations of styrene in polystyrene,[1] and analyses of coal extracts,[2] liquefied coal samples,[3] and of various oils.[4,5] In more theoretical studies, oxidation of slightly soluble 3,4-dimethoxypropenylbenzene[6] and reduction of methylenequinones[7] was followed.

B. Hydrolysis and Hydration

Presence of nonaqueous solvents can prevent hydrolysis and hydration. Species undergoing cleavage in the presence of traces of water can be studied in nonaqueous systems. With such unstable systems, even when lower temperatures and special rapid sample handling is used, no useful information can be obtained in the presence of water, where only solvolysis products can be investigated, if they are electroactive.

Examples of the use of the inert character of nonaqueous solvents in the study of easily hydrolyzed compounds include reductions of the triphenylcyclopropenyl and triphenylcarbonium ions in acetonitrile[8] and trialkylchlorosilanes in dimethyl sulfoxide.[9] The latter reduction is ascribed to the product from the following reaction:

$$R_3SiCl + (CH_3)_2SO \rightarrow R_3SiOS(CH_3)_2{}^+ + Cl^- \tag{1}$$

Carbonium ions have also been studied in methanesulfonic acid.[10] Strongly solvated systems, particularly those where water or alcohol is covalently bound, often give small waves, the heights of which are governed by the rate of dehydration or the loss of alcohol. Such waves are less suitable for practical applications and theoretical studies, and large, diffusion-controlled waves in nonaqueous systems offer practical advantages. Thus, e.g., tri-fluoroacetophenone gives in aqueous solution a reduction wave only if a millimolar or more concentrated solution is used, whereas in dimethylform-amide or acetonitrile the well-developed diffusion controlled wave is ideally suitable for analysis or kinetic studies.[11] Among other areas of applications, the electrochemistry of carbohydrates in nonaqueous systems seems parti-cularly promising.

C. Halogen Reduction

In some instances the presence of a nonaqueous solvent enables electrode processes to take place which do not occur in the presence of water. Hence reduction of chlorobenzene[12,13] and p-chloroacetanilide,[14] for which no wave is observed in aqueous solutions, was reported in dimethylformamide in the presence of quaternary ammonium salts. It is not certain whether such a synergistic effect of the nonaqueous solvent reflects the change in solvation of the electroactive species or its orientation at the electrode surface, or whether it affects reducibility indirectly by changing the composition of the electrical double layer, but practical advantages of such phenomenon are obvious.

D. Extension of the Potential Scale

Another important advantage is the extension of the potential scale to more positive values. Whereas in aqueous solutions the range is usually limited to $+1.0$ V, and at more positive potentials decomposition of the solvent occurs, potentials up to $+2.0$ V (vs. Ag^+) have been reported[15][17] to be reached in acetonitrile. This particularly extends the range of oxidation processes which can be followed, making it possible to study and use current–voltage curves even of those compounds which are less readily oxidized.

E. Reversible Systems

Perhaps the most publicized and utilized advantage of nonaqueous sol-vents is the possibility to measure potentials of one-electron reversible systems[18-39] such as

$$R + e \rightleftharpoons R^{\cdot} \tag{2}$$

$$RH \rightleftharpoons RH^+ + e \tag{3}$$

In some instances the potentials characterizing these couples can be ob-tained in aqueous solutions at sufficiently high pH values, where preprotonation and protonation of the radical anion are too slow to interfere.[40] But nonaqueous media offer much wider possibilities for obtaining such data. The reason for this advantage of nonaqueous solvents is that, in the presence of water and other proton donors, acid-base processes affect the nature of the species before electrolysis or proton transfers involving the product of the one-electron transfer occur.[41] In the first case, the measured half-wave potential is a function of pH and the value obtained is often a function not only of the free energy of the electron transfer but also of the pK_a of the corresponding acid–base reaction (i.e., the free energy of the proton transfer reaction). In the second case, products formed by protonation of radical anions or dissociation of radical cations are highly reactive species which are able to undergo rapidly subsequent reactions, like dimerizations or interactions with the parent compound, solvent, supporting electrolyte, or other component of the electrolyzed solution.

Data obtained for potentials of one-electron systems in nonaqueous media are suitable for application in further theoretical treatments as they are thermodynamic values and have been successfully used, e.g., for correlations with molecular orbitals,[42-50] ionization potentials, and electron affinities.[46,51,52] Several words of caution are, nevertheless, necessary. First, it should be realized that the reference state involves the organic solvent. Second, the oxidation-reduction potential of a reversible system is a measure of the energy difference between the ground state and the reaction product, i.e., between the electroactive species and the radical ion. Any structural change which affects the energy of both of these states, by increasing or decreasing the energy by exactly the same amount, will show no change in the value of the half-wave potentials. There are few structural changes which would affect only the form R or the form R $^-$, hence any shift of the half-wave potential arising from a structural change is due to a greater increase in energy in R_1 than in $R_1 {}^-$ when compared to the increase in R_0 and $R_0 {}^-$ or vice versa. Hence quantum chemical calculations should estimate energies of R_0, R_1, R_2, and of $R_0 {}^-$, $R_1 {}^-$, and $R_2 {}^-$ and correlate the corresponding differences to half-wave potentials. All too frequently, only the energies of R_0, R_1, and R_2 are estimated and the chemical noncrossing rule is assumed (sometimes tacitly) to operate. This brings MO treatments to a level of semiempirical correlations. Third, it should be ascertained that all data really correspond to one-electron transfer on R. The danger is, e.g., in the comparison of potentials for various heterocyclic aromatic systems, where for some compounds a one-electron and for others a two-electron process predominates. Reporting of currents under well-defined conditions together with potentials is essential. Another example of unsuitable data are the values used for MO correlations by Dewar et al.[53] Values obtained in mixtures containing 80% dioxane and 20% water do not correspond to aprotic conditions. The fact that good correlations were reported does not add to a feeling of confidence in assumptions made in MO calculations.

Half-wave potentials measured in organic solvents, both for reversible and irreversible processes, can also be utilized in semiempirical correlations with structural parameters, as in the Hammett and Taft equations.[54]

F. Reactions of Radical Ions

The possibility of generating species, such as radical ions, enables studies of their reactions, both preparatively, when products of reactions of electro-generated radical ions are identified and their yields followed, or kinetically. In the latter case, two alternatives are possible. Either the electrochemically generated radical is followed by nonelectrochemical methods or cyclic methods are used. In the first case the combination with ESR is of particular importance. Because generation and detection of concentration changes of radical ions in the cavity of the ESR spectrometer is complicated by the nonhomogeneity of the solution, external generation and rapid transfer to the ESR cavity, introduced by Kastening[55] seems to be the most promising technique for kinetic study. The monitoring of the time change in radical concentration can be also followed spectrophotometrically or by electro-analytical methods. Combination with stop-flow methods seems promising.

ESR measurements are, naturally, also the most powerful methods for the detection of the presence of radicals and radical ions in the course of elec-trolysis and for the elucidation of the structure of such species. Alternatively, even when on a more limited scale, information about the existence or nonexistence of species with unpaired electrons can also be obtained from current–voltage curves, effects of radical trapping substances, formation of polymers, or even from the effect of glass wool added to the electrolysis cell.

For the study of the kinetics of radicals and radical ions, cyclic voltam-metry, current-reversal chronopotentiometry, and related methods have been widely used, particularly in the investigations of hydrodimerizations. Many studies in this area offer qualitative or semiquantitative information, but the work of Savéant, Baizer, and Bard offers some solid kinetic evidence.[56-60]

G. Cation Effects

Another area where the possibility of the electrogeneration of radical ions has extended our understanding of their chemical interactions involves the reactions with ions of the supporting electrolyte, yielding ion-pairs or adducts.

Shifts of half-wave potentials and formation of new waves was followed[27,61] as a function of the nature and concentration of the cation for processes in which the uptake of one electron led to formation of a radical anion.

Observed shifts made it possible[62] to distinguish between two mechanisms. When the first reduction wave with increasing concentration or decreasing radius of the cation shifted to more positive and the second wave to more negative potentials, only the solvated radical anion was reduced in the second wave and Scheme I (without the boxed portion) was operating. When both half-wave potentials of the more positive and negative wave shifted to more positive potentials with increasing salt concentration or decreasing cation radius, full Scheme I (including the boxed portion) operated; in the more negative wave both radical anion R_{solv}^- and the ion-pair $(R^{\bar{\ }} \cdots M^{n+})_{solv}$ was reduced.

For processes involving two one-electron steps, where the more negative wave was shifted to more positive values by increasing concentration of an added neutral salt, or when at constant salt concentration the ionic radius of the cation was decreased, it has been sometimes observed that the half-wave potential of the more positive wave remains practically unchanged. This has been interpreted as a proof of the reversibility of the first electron uptake. Such conclusions may not be correct, as it is possible that the shifts of $(E_{1/2})_1$ are simply so small that they are comparable with experimental errors, as was observed for 1,4-naphthoquinone.[61] An interaction of the metal cation with dianion formed also cannot be excluded.

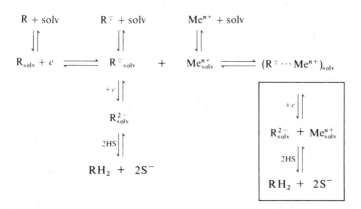

SCHEME I. If the boxed part of Scheme I can be neglected, $(E_{1/2})_1$, becomes more positive, $(E_{1/2})_2$ more negative with increasing metal concentration. If the complete scheme is operating, both $(E_{1/2})_1$ and $(E_{1/2})_2$ become more positive. When $(E_{1/2})_2$ is shifted to positive potentials and $(E_{1/2})_1$ is practically independent, ion pair formation of $R^{\bar{\ }}$ is not extensive.

Shifts of half-wave potentials caused by cation effects can be due to ion-pair formation or to change in the energy level of the radical anion resulting from the disturbance caused by the presence of a cation. The contributions of these two factors were considered and attempts have been made[27,61] to distinguish between them and to evaluate their individual contributions. Nevertheless, it should be pointed out that some part of the resulting half-wave potential shifts can be due to a change in the double-layer composition and its effect on the reduction potential.

H. Carbenes

Another type of reactive intermediates that can be generated electro-chemically are the carbenes. The study of carbene reactions should be done rapidly as the reactive carbene is easily further reduced. Hence, e.g., carbon tetrachloride and bromide are reduced in acetonitrile and dimethylform-amide through a dihalocarbene[63] in one step to the methylene compound

$$CCl_4 + 2e \rightarrow CCl_3^- + Cl^- \tag{4}$$

$$CCl_3^- \rightarrow CCl_2 + Cl^- \tag{5}$$

$$CCl_2 + 2e + \text{solvent} \rightarrow H_2CCl_2 \tag{6}$$

Another reactive intermediate, benzyne, is assumed[13] to be formed in reductions of 1,2-dibromobenzene and 1-bromo-2-chlorobenzene to benzene.

I. Preceding Reactions

The number of reactions that have been studied in nonaqueous media which involve electroactive species (rather than the electrolysis product) is more limited. There seem to be two main reasons for such more restricted scope of applications. First of all, electroanalytical methods can be applied straightforwardly for studies of equilibrium reactions only when the concentration of one component can be kept constant during the experiment by using excess or buffered media. Under such conditions, first-order kinetics operate and the mathematical treatment of the relationships between the current or another measured quantity and the reaction rate is simplified. The majority of precedent reactions studied in aqueous solutions are acid–base reactions. Only some of them can be studied in the presence of excess acid or base. Usually, investigations are carried out in aqueous buffers. As our knowledge about buffers in nonaqueous systems is minimal,[64] it is not surprising that the information available on the effect of solvents on the rate of preprotonation is predominantly qualitative.

Linear correlation has been found for the half-wave potentials of the reduction of hydrogen from the pyridinium ion in pyridine as a solvent and titrimetric half-neutralization potentials by Elving.[65] A similar relationship was also observed to be fulfilled for these half-wave and pK values of corresponding acids in water.

The second limitation stems from the fact that theoretical treatments of the kinetics of reactions preceding electron transfer proper are based on the assumption that the equilibrium constant for the reaction studied is known and that its numerical value is available, usually by an independent method. This is true for numerous acid–base and other reactions in aqueous systems, but few reliable values for such equilibrium constants are currently available or accessible for reactions in nonaqueous media.

Strong or, more frequently, weak acids are sometimes added,[27] in concentrations comparable to those of the electroactive species, to the investigated system in order to show an effect of preprotonation on the organic solvent. Such an addition results in the formation of a wave of the protonated form, usually at more positive potentials than does reduction of the unprotonated species. The increased current reaches a limiting value, frequently at a nonstoichiometric concentration of the added acid, usually higher than the equivalent concentration. Moreover, the concentration at which the limiting value is reached often depends on the nature of the added acid. Whether the reaction results in proton transfer, ion-pair formation, or adduct formation, the reaction involved is not necessarily slow, and care should be taken if the resulting more positive wave of the protonated form is governed by diffusion or by the rate of reaction. In practice, criteria for diffusion or kinetic currents are rarely applied to more positive waves resulting in the addition of a substoichiometric amount of acid. It should be pointed out here that criteria for the kinetic diffusion character of the current should be applied after additions of small amounts of the acid, when the more positive wave is less than 20% of the original wave height.

When the character of the more positive wave indicates a kinetic current, the limiting current can be used for obtaining kinetic data provided that treatment for a second-order process of the given type is available.

Only when the more positive wave has been proved to have diffusion character throughout is it possible to draw conclusions about how many protons are involved preceding the given electron transfer. To achieve conditions under which transport by diffusion is the governing step, the equilibrium between the electroactive species and the added acid must be established slowly. As information about slowly established acid–base equilibria in nonaqueous solution are not generally available, the character of the current should be tested in all cases when a substoichiometric amount of acid is added to a nonaqueous system.

Addition of strong acids or bases to polar aprotic solvents can result in the hydrolysis or cleavage of such solvents; changes of solvents and their properties with time have to be followed, therefore, if such studies are attempted. Solutions of sulfuric acid in glacial acetic acid proved useful.[66,67]

J. Acidity Functions

In strongly acidic nonaqueous solutions and strongly basic solutions, where either the strong acid has the function of the nonaqueous solvent or strong acid or base are added to an nonaqueous, usually organic, solvent, definition of acidity may become a problem.

Numerous acidity scales have been developed, but it is essential to keep in mind that, in order to be reliably applicable, the structure of the indicator used for characterization of acidity scale and the acid–base process it undergoes should resemble the acid–base process investigated as closely as possible. For example, when an acidity scale was developed for strongly alkaline media a very considerable difference (for a given concentration of base) between aqueous and dimethyl sulfoxide (DMSO) solutions (about 14 powers of ten) was found for the acidity function H^- when aromatic amines were used as indicators,[68] but much smaller (about three powers of ten) when benzaldehydes were used.[69] Hence, the first acidity scale is useful for the study of the oxidation of anilines in alkaline solutions, whereas for the investigation of reduction or oxidation[70] of benzaldehydes the second acidity scale must be used.

Alternatively, potentials of a glass electrode in strongly acidic or alkaline media, when measured polarographically against ferrocene or cobaltocene reference electrodes, have been recently recommended for the characterization of acidity.[71]

K. Other Equilibria

Reactions preceding the electrode process proper other than acid–base reactions have been rarely studied in nonaqueous media, but comparison of results in aqueous and nonaqueous systems in some instances offered information important for the better understanding of the given reaction in aqueous solutions. An example of such an application of organic solvents is the study of the reversible hydration of some carbonyl compounds. A decrease in the water content results in dramatic change both of UV spectra and polarographic curves, accompanied by a change in mechanism. Studies of systems containing small and varying amounts of water for α,α,α-trifluoroacetophenone,[72] chloral,[73] and, most recently, hexachloroacetone[74] offer information on the equilibria and kinetics of water addition in organic solvents. Unfortunately, this information cannot be directly applied to the interpretation of mechanisms of hydration in aqueous solutions.

L. Electrochemical Data and Solvent Effects

Relatively recently attempts have been made to use electrochemical data for the quantitative measurement, characterization, and interpretation of solvent effects.

Approaches used, for example, a comparison of solvent effects on electrochemical data with solvent effects on homogeneous kinetics and spectral data, are similar as those used by Gutmann,[75,76] but the complexity of the organic solute interactions indicates that there is little hope for such an elegant, general, and basically simple approach, as has been used in the development of the principle of donicity and its use for reactions of ionic species.

When the effect of the nature of nonaqueous solvent on the reactions of various organic compounds is compared, it is obvious that variation in the solvent changes not only relative effectiveness, but also the sequence of individual structure effects. This indicates that solvent–solute interactions depend not only on the properties of the solvent but also on the nature of the solute; solvent effects are not generally additive when solutes of different structures are compared. It can be also concluded that several types of solvent–solute interactions are operating. Undoubtedly, the dielectric constant is not the only and frequently not even the predominant factor in the observed solvent effect.

One of the possible approaches used to rule out some of the operating solute–solvent interactions is based on comparison with known solvent polarity parameters. One of such parameters is Dimroth's parameter E_t based on transition energies for the intramolecular charge-transfer band of a pyridinium phenyl betaine which is related to, and a linear function of, Kosower's Z value. Merocyanine dyes of the type

have been studied,[77] the electronic spectra of which are very sensitive to solvent effects; changes from water to acetonitrile, dimethylformamide (DMF), and DMSO result in shifts by 35, 75, and 80 nm, respectively. In view of the Franck–Condon principle it was of interest to prove whether the observed solvent effect on spectra reflected predominantly the change in interaction of the solvent with the ground state or with the transition state. As for merocyanine dyes, very small effects have been observed on polarographic half-wave potentials,[78] so it can be deduced that the decisive interaction takes placed in the excited state. The changes in the ground state energies with variation of solvent would equally affect λ_{max} and $E_{1/2}$. Nevertheless, it cannot be excluded that the presence of the electrode imposes on the stilbazolium ion an orientation or conformation in which either the solvation is minimized or the intramolecular charge–transfer interaction is prevented.

It is thus not possible to use E parameters or Z values for predicting solvent effects on electrochemical data. Because of the specific nature of solvent–solute interactions it seems hardly possible to find or even expect the existence of a series of constants quantitatively characterizing solvent effects for a wide variety of processes and systems, allowing predictions in the way as the Hammett equation made it possible for predictions of substituent effects on homogeneous rates and equilibria. Each solvent parameter will be valid only for processes closely related to the process from which the parameter was obtained. Development of solvent parameters characterizing solvent effects on the most common types of electrochemically reducible and oxidizable compounds is therefore desirable.

Alternatively, it is possible to study and correlate solvent effects on chemical reactions preceding or following electron transfer. One of the types of reactions investigated in this context was the interaction between cations and radical anions. Comparison of solvent effects on the shifts of half-wave potentials of the first wave of 1,2-naphthoquinones due to cations has shown[79] that these shifts can be correlated with the values of donicities (DN).[75,76] This strongly indicates that the observed solvent effects are predominantly due to an interaction between the cation and the solvent rather than to an interaction between the solvent and the radical anion. Similar correlation applies also[62] to half-wave potentials of the second wave of 1,2-naphthoquinones (Fig. 1). Deviations shown in acetonitrile in the presence of $N(C_2H_5)_4^+$, K^+, and Na^+ can be due to the competition of ion pairing with proton transfer, which is more strongly preferred in acetonitrile than in the other solvents used.

The values of the reaction constant χ, characterizing the susceptibility of ion-pair formation to the solvent effect for a given cation in equation $\Delta E = \chi DN$, vary strongly in magnitude and sign depending upon the process followed,[62] namely, the nature of the organic species reduced, as demonstrated in the Table I. Attempts to interpret these differences should be made only when more data of this type is available.

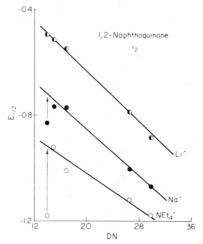

FIG. 1. Dependence of half-wave potentials (in volts) of the second reduction wave of 1,2-naphthoquinone on donicities. Supporting electrolyte contains 0.05 M $N(C_2H_5)_4$ with the addition of 0.005 M Li^+ (Halved circles), Na^+ (full circles), and without addition (open circles).

III. PROBLEMS INVOLVED IN THE USE OF NONAQUEOUS SOLVENTS

As is frequently observed in chemical research, the replacement of aqueous by nonaqueous media offers not only various advantages, but also brings some problems not usually encountered in the study of aqueous solutions. These problems can be either theoretical, like complications in

TABLE I

VALUES OF REACTION CONSTANT χ FOR $\Delta E_{1,2} = \chi DN$
FOR SHIFTS OF HALF-WAVE POTENTIALS DUE TO CATION EFFECTS
IN DIFFERENT SOLVENTS

	Values of χ			
	1,2-Naphthoquinone		1,4-Naphthoquinone	
	i_1	i_2	i_1	i_2
Et_4N^+	6.16	-17.5	—	-18.0
K^+	4.27	-18.0	≈ 7.3	-22.5
Na^+	-2.16	-22.0	—	-25.0
Li^+	-11.4	-23.0	—	≈ -27.0

mechanisms caused by the high reactivity of the primary electrolysis product, or practical. Among the practical problems there are some, e.g., the resistance of nonaqueous solutions, the toxicity of nonaqueous solvents, and the presence of water and oxygen in the medium studied, which can be experimentally demanding but which eventually can be solved. There is, nevertheless, another group of practical problems for which no final solution exists, which must be accepted as existing and their role minimized by compromise. To this latter group belong the limited solubility of strong electrolytes in the majority of solvents used and particularly the problems involved in the choice of a reference electrode, the possibilities of comparing data obtained in various solvents, and problems related to the choice and preparation of liquid junctions. Finally, it is important to realize what kind of information for general and applied chemistry can be obtained by studying organic compounds in nonaqueous solutions.

A. Reactivity of Primary Electrolysis Products

As long as we are interested in the ease with which the first electron uptake occurs, investigations of nonaqueous, aprotic media can offer such information more readily and are less complicated by other chemical reactions than studies of aqueous and water-similar solvents. The situation, nevertheless, can be different when elucidation of the mechanism of the complete process, involving second electrode uptake and other reactions following the first electron uptake is sought. The understanding of the overall electrode process can be sometimes more difficult in nonaqueous than in aqueous media. The main reason for the complex processes occurring in the absence of proton donors lies in the nature of the product of the first electron uptake. Radical anions formed in one-electron reductions are relatively strong bases, and radical cations produced in one-electron oxidations of neutral molecules are relatively strong acids. Due to their acid–base properties these ions frequently do not persist in the solution. In the absence of other proton donors or acceptors, they interact with solvents, even those considered aprotic, with tetraalkylammonium salts used as supporting electrolytes or even with other molecules of the starting material.[80] Because ways of keeping the concentration of proton donors or acceptors constant during the course of electrolysis, by buffering or with an excess of a strong acid or base, are usually not available in nonaqueous media, the resulting systems involve reactions of second and higher orders. The exact treatment of such systems is complicated and accessible to a rigorous evaluation in a few cases. Dimerization reactions of primary electrolysis products and interactions of cations of the supporting electrolyte with radical anions formed are reasonably well understood and have been mentioned before, but these two processes do not cover the whole spectrum of possible consecutive or interposed processes.

It is useful to realize that the impossibility of controlling the acidity at the electrode surface deprives us of a useful diagnostic tool that is available in the study of aqueous solutions. Further studies concerning the nature and kinetics of reactions following the first electron uptake are needed, but real progress will be possible only when the general chemistry of each solvent involved is better understood, in particular with respect to its acid–base properties, solvent–solute, and solute–solute (including ion-pair formation) interactions.

B. Toxicity

All nonaqueous solvents used in electroanalytical should be dealt with carefully because solvents are usually used in considerably larger amounts and for longer periods of time than other chemicals studied. Moreover, numerous solvents have relatively low boiling points and considerable vapor pressures at normal laboratory temperatures. Particular care should be exerted when dealing with hexamethylphosphoroamide, reported to be carcinogenic, and acetonitrile, reported to cause, on excess exposure, reversible nephritis. Reports on the toxicity of dimethylsulfoxide vary, but it rapidly penetrates skin and can carry with it into body fluids toxic substances dissolved in it. Working with gloves in well ventilated hoods or closed systems is therefore recommended.

C. Water Content

If work in nonaqueous solvents is to correspond to conditions in the absence of proton donors, then water should not be present in the investigated solvent in concentrations comparable with those of the electroactive species, and if possible, it should be one-order of magnitude lower. Taking into consideration that the electroactive species should be (for theoretical studies) present in the solution in concentrations of the order of 10^{-4} M or lower (the study of millimolar solutions can offer additional difficulties), means that the water content in the given solvent should be 10^{-4} M or less. This is not an easy goal and for a number of solvents specific procedures for purification have been described.[81,82] If a general principle can be outlined, it would be that in many cases combination of rectification and the use of molecular sieves proved useful. (Linde AW-500 molecular sieves are superior to commonly used Linde 4A sieves.) Such treatment might introduce alkali metal ions into the solvent and distillation of the solvent prior to use may be necessary. Sometimes traces of water can be scavenged by alumina or trifluoroacetic anhydride. In any case, the life of a chemist working in organic electroanalytical chemistry in nonaqueous solvents remains a constant battle against water. Moreover, after purification has been carried out, the water content should be determined. The statement that a specific procedure was used for purification does not offer information as to how effective it was in the hand of the particular chemist. The choice of analytical method for water determination is also important. Numerous reports state that the Karl Fischer method is not reliable at low water concentrations. Gas chromatography seem to be more generally applicable and additional information can be obtained from the UV spectra[83] (particularly the position of the cutoff at short wavelengths) and from the available voltage range, with Pt electrode measuring the limit toward positive and DME toward negative potentials.[81]

It is not only important to know how much water is present in the organic solvent, but also if it is readily available as a proton donor; whereas, e.g., in acetonitrile, traces of water react with radical anions, the same amount of water in DMF or DMSO has almost no effect on the waves of some organic compounds. Water is thus present in DMF and DMSO solutions in a form that does not act as a proton donor; it is assumed that this may be caused by hydrogen bonding. Thus following the shapes of current-voltage curves of species yielding radical anions of limited stability might be the most useful evaluation of water content for electroanalytical studies.

D. Oxygen

Another problem which is of particular importance in nonaqueous electrochemistry is the reactivity of radical ions with oxygen. No data seem to be available on the relative reactivities of radical cations, but radical anions are more reactive towards oxygen than the radicals which themselves are easily trapped by oxygen. Moreover, removal of oxygen from solutions in organic solvents is more time consuming than removing it from aqueous solutions and presents particular problems for organic compounds which are volatile with the given organic solvent. Use of vacuum lines has been recommended for such systems.

E. Supporting Electrolyte

The choice of supporting electrolyte for use in nonaqueous media is limited by the solubilities of the salts of strong electrolytes in such solutions. Only lithium and tetraalkylammonium salts are soluble enough to allow preparation of 0.1 M or more concentrated supporting electrolytes. It is necessary to take into account specific phenomena accompanying the use of such salts. Whereas in aqueous solutions hydrated lithium cations show little tendency to interact with radical anions,[84] in nonaqueous solvents strong interactions[85] with some radical anions have been postulated.

Tetraalkylammonium salts can exert several types of effects, depending on the nature of the alkyl group, on the solvent used and the electroactive species studied. The size of the alkyl group will affect the tendency to form ion pairs or adducts with radical anions, the composition of the double layer, and the adsorptivity of the cation. With large (e.g., n-butyl) or branched (e.g., isopropyl) groups the adsorption of the hydrocarbon chains (e.g., of an added surfactant) can complicate the electrode process.

Tetraalkylammonium salts, such as $(C_2H_5)_4NBr$ have been found[86-88] to affect the availability of added proton donors like water, phenol, or acetic acid at the electrode surface, probably due to preferential adsorption of the quaternary ammonium ions.

Whenever tetraalkylammonium salts are used, it is strongly recommended that the behavior of the organic compound be studied in a number of supporting electrolytes, composed of tetraalkylammonium salts with varying alkyl groups.

For reductions, tetraalkylammonium halides can be used. There is, nevertheless, the possibility that the halide may coordinate or act as a nucleophile and hence the use of these salts, particularly of bromides and iodides, is not favored, even when they can be more easily purified by recrystallization than other salts. Perchlorates, hexafluorophosphates, and tetrafluoroborates seem to be advantageous from all points of view, with the exception of price. Tetraalkylammonium perchlorates and tetrafluoroborates have to be used for anodic oxidations as halides are too readily oxidized. Tetrafluoroborate salts are somewhat easier to purify and dry than perchlorates and they are safer, even when tetraethylammonium perchlorates have been frequently prepared in kilogram batches. Nevertheless, the initial reaction mixture which contains perchloric acid must never be allowed to become overheated or evaporated to dryness. Explosions are possible and do occur.

F. Reference Electrodes and Liquid Junctions

A group of problems where a complete and definite solution seems currently impossible and compromises are necessary involves the choice of a reference electrode[82] and the definition of its potential, and the use of salt bridges and the definition of liquid junction potentials involved.[89] There are two aspects of the problem, two requirements which a reliable reference electrode must fulfill.[90] First, the potential of the electrode must remain practically constant in the course of recording of current-voltage curves when small currents flow through the system. It is also desirable, if a separated reference electrode is used, that as little mixing as possible occurs on the interface. Second, the potential of the reference electrode should be known and it should be possible to express its numerical value relative to some chosen standard.

Liquid junction potentials between aqueous and organic phases can often be as large as several hundred millivolts.[91] Moreover, any boundary or separation which is not a source of excessive resistance is a possible source of water. As investigations in organic solvents are usually carried out in the absence of water, it is obvious that aqueous reference electrodes cannot be recommended for work in nonaqueous systems. By making this decision we lose, nevertheless, one advantage of aqueous reference electrodes, namely, that conventions regarding their potentials are reasonably well established.

Hence reference electrodes in the same solvent as the studied species in cells with or without junction are recommended. Reference electrodes such as calomel, silver–silver chloride, or mercurous sulfate can be used in hydroxylic and some other, e.g., dichloromethane, organic solvents but not in the most frequently used aprotic polar solvents. The solubility of the metal salts and formation of anionic complexes in solvents like DMF, DMSO, acetonitrile, or propylene carbonate make these electrodes unsuitable in organic solvents. Hydrogen and glass electrodes can be used in a wide range of solvents but are not convenient for practical use. Wide use has been made of silver–silver ion electrodes,[92] which consist of a silver wire in a solution of silver nitrate or perchlorate in the particular organic solvent. A special salt bridge or the use of a three-compartment cell is recommended for this reference electrode. As the electrode is polarizable,[93] it should be used as a reference electrode in a three-electrode arrangement. Similar arrangements are recommended for other electrodes of the kind involving metal ions, e.g., Li^+ or Tl^+, and for electrodes consisting of Cd^{2+} solution in dimethylformamide and cadmium amalgam.[94] Alternatively, reference electrodes consisting of a platinum electrode immersed in a solution containing both oxidized and reduced forms (e.g., ferrocene–ferricinium ion) or I_3^- and I^- can be used.

Each of these electrodes makes it possible to compare half-wave potentials in the given solvent. It is also possible to measure in a given solvent potentials relative to the potential of one compound chosen as the internal standard.

When values of the half-wave potentials of a given organic compound are to be compared, direct measurement of the potential differences between the reference electrodes in different solvents is practically impossible because of unknown values of liquid junction potentials. Therefore, another approach is needed. A system is chosen which is considered to possess the same potential in all solvents compared. It is possible to define the potential of the hydrogen electrode as zero in all solvents at all temperatures. The handling of the hydrogen electrode is nevertheless inconvenient and the unit activity of hydrogen ions not well defined. Attempts were thus made to find ions whose solvations would not change with the solvent. The first attempt made by Pleskov,[95] who chose the large rubidium ion, was an approximation which showed some limitations. In the search of a univalent ion where the central atom would be screened off by large organic ligands, Strehlow[96] first recommended ferrocene and cobaltocene, which were later followed by the use of an Fe (o-phenanthroline) complex[97] and most recently by the use of bisbiphenylchromium(I) proposed by Gutmann and co-workers.[98]

The difference between half-wave potentials of ferrocene and bisbiphenyl-chromium(I) is practically independent (1.12–1.13 V) of the solvent used indicating suitability of both these systems. The chromium complex offers the additional advantage of being able to be used in mixed solvents containing water[98] and so, by extrapolation, potentials in water can be compared with potentials in nonaqueous solvents.

Finally, the reference electrode used should be always described. A frequently quoted paper by Fujinaga et al.[61] states that a three-electrode system was used, but not what reference electrode was used. Potential data are quoted against this unknown electrode and ferrocene and show systematic differences between these two scales. It is fortunate that only relative values of potentials are often needed.

At this stage a general plea to authors of research papers seems appropriate. As published data become part of data collections[99] their presentation in journals should be such as to not leave any doubts about their meaning and to make them comparable with other data. A detailed discussion of the presentation of electroanalytical data is published elsewhere[100] and can only be briefly summarized here; the structure, name, and concentration of the compound studied should be given together with a detailed description of the solvent composition (particularly with respect to water content and analytical method used), of the nature and concentration of the supporting electrolyte and the possible addition of surfactants, of the temperature and of the indicator and reference electrodes used, including the cell type and the nature of the separation of the two electrodes. Properties of the indicator electrode and mode of measurement (technique, conditions such as stirring or light) should be described in sufficient detail, as well as the means of measurement and evaluation of potentials, current, curve shapes, and other characteristic quantities. Information on products and proposed mechanisms should be included.

G. Applicability

The interest of organic chemists in the use of electrochemical methods is reflected by the recent upsurge in publication of monographs and compendia on the subject.[57,101 107] It can be assumed that organic chemists want to obtain an information about the reactivity of an organic compound toward a nucleophilic attack or other properties of organic materials, or about optimum conditions for electrosynthetic work. The reactivities at the DME and in homogeneous reactions can be compared only when the processes are followed under identical conditions. A majority of mechanisms of homogeneous chemical reactions has been studied in solutions containing water and proton donors and hence electrochemical studies in such media seem to be more pertinent. Also, biochemically-important reactions (including oxidation–reduction processes), which are currently the center of great interest, take place in aqueous solutions. On the other hand, workers in the area of radical and radical ion reactions will evidently be able to apply the knowledge obtained by electrochemical studies and there are assumptions that some biochemically-important reactions take place in enzymatic systems under virtually aprotic conditions.

Large-scale synthetic work in the absence of traces of water may be possible in some exceptional cases, but the costs of water-free solvents might be prohibitive in others. On the other hand, in laboratory-scale synthesis and in the manufacture of small batches (as in the production of pharmaceuticals and fine chemicals), the use of nonaqueous media is promising and for such purposes electroanalytical methods can offer contributions in the elucidation of mechanisms of electrode processes.[107]

IV. CONCLUSIONS

The preceding paragraphs have indicated possibilities which the extension of electroanalytical methods to nonaqueous solvents offer, together with some areas which seem to be particularly promising for future research. Considerable progress can be expected when more information on the fundamental chemistry of such solvents, particularly on acid–base properties and nonaqueous electroinactive buffers, on ion-pair formation, solvent–solute interactions, and on the effects of the ions of supporting electrolytes on the solvent structure become available. Such developments will also contribute to our better understanding of aqueous organic electrochemistry.

REFERENCES

1. J. Pasciak, *Chem. Anal.* **5**, 477 (1960).
2. P. H. Given and M. E. Peover, *J. Chem. Soc.* p. 394 (1960).
3. D. D. Whitehurst, M. Farcasiu, and T. O. Mitchell, Mobil Research and Development Corp., "The Nature and Origin of Asphaltenes in Processed Coals," EPRI AF-252, Proj. 410-1, Annu. Rep. Mobil Res. Dev. Corp., 1976. Princeton, N.J.
4. W. R. Lewis, F. W. Quackenbush, and T. de Vries, *Anal. Chem.* **21**, 762 (1949).
5. M. L. Whisman and B. H. Eccleston, *Anal. Chem.* **30**, 1638 (1958).
6. J. J. O'Connor and I. A. Pearl, *J. Electrochem. Soc.* **111**, 335 (1964).

7. A. S. Lindsey, M. E. Peover, and N. G. Savill, *J. Chem. Soc.* p. 4558 (1962).
8. R. Breslow, W. Bahary, and W. H. Reinmuth, *J. Am. Chem. Soc.* **83**, 1763 (1961).
9. V. Gutmann, P. Heilmayer, and G. Schöber, *Monatsh. Chem.* **92**, 240 (1961).
10. S. Wawzonek, R. Berkey, and D. Thomson, *J. Electrochem. Soc.* **103**, 513 (1956).
11. W. J. Scott and P. Zuman, *Electrochim. Acta* (in press).
12. F. L. Lambert and K. Kobayashi, *J. Org. Chem.* **23**, 773 (1958).
13. S. Wawzonek and J. H. Wagenknecht, *J. Electrochem. Soc.* **110**, 420 (1963).
14. R. Jones and B. C. Page, *Anal. Chem.* **36**, 35 (1964).
15. S. Wawzonek, *Talanta* **12**, 1229 (1965).
16. B. Kratochvil, *Crit. Rev. Anal. Chem.* **1**, 415 (1971).
17. D. T. Sawyer and J. L. Roberts, Jr., " Experimental Electrochemistry for Chemists," p. 170. Wiley, New York, 1974.
18. S. Wawzonek, E. W. Blaha, R. Berkey, and M. E. Runner, *J. Electrochem. Soc.* **102**, 235 (1955).
19. S. Wawzonek and D. Wearring, *J. Am. Chem. Soc.* **81**, 2067 (1959).
20. A. C. Aten, C. Buthker, and G. J. Hoijtink, *Trans. Faraday Soc.* **55**, 324 (1959).
21. A. C. Aten and G. J. Hoijtink, *Z. Phys. Chem. (Frankfurt am Main)* **21**, 192 (1959).
22. P. H. Given, *J. Chem. Soc.* p. 2684 (1958).
23. P. H. Given and M. E. Peover, *J. Chem. Soc.* p. 385 (1960).
24. W. Kemula and J. Kornacki, *Rocz. Chem.* **36**, 1835, 1849, and 1852 (1962).
25. P. G. Grodzka and P. J. Elving, *J. Electrochem. Soc.* **110**, 225 and 231 (1963).
26. P. H. Given and M. E. Peover, *Collect. Czech. Chem. Commun.* **25**, 3195 (1960).
27. M. E. Peover, *Electroanal. Chem.* **2**, 1 (1967).
28. S. Wawzonek, R. Berkey, E. W. Blaha, and M. E. Runner, *J. Electrochem. Soc.* **103**, 456 (1956).
29. P. H. Given, M. E. Peover, and J. Schoen, *J. Chem. Soc.* p. 2674 (1958).
30. I. M. Kolthoff and T. B. Reddy, *J. Electrochem. Soc.* **108**, 980 (1961).
31. S. Wawzonek and A. Gundersen, *J. Electrochem. Soc.* **111**, 324 (1964).
32. S. Wawzonek and A. Gundersen, *J. Electrochem. Soc.* **107**, 237 (1960).
33. R. H. Philip, Jr., R. L. Flurry, and R. A. Day, Jr., *J. Electrochem. Soc.* **111**, 328 (1964).
34. R. H. Philip, Jr., T. Layloff, and R. N. Adams, *J. Electrochem. Soc.* **111**, 1189 (1964).
35. P. H. Rieger, I. Bernal, W. H. Reinmuth, and G. K. Fraenkel, *J. Am. Chem. Soc.* **85**, 683 (1963).
36. A. Zweig, J. E. Lehnsen, W. G. Hodgson, and W. H. Jura, *J. Am. Chem. Soc.* **85**, 3937 (1963).
37. A. K. Hoffmann, W. G. Hodgson, and W. H. Jura, *J. Am. Chem. Soc.* **83**, 4675 (1961).
38. D. H. Geske and A. H. Maki, *J. Am. Chem. Soc.* **82**, 2671 (1960).
39. W. Kemula and R. Sioda, *Bull. Acad. Pol. Sci. Ser. Sci. Chim.* **10**, 107 (1962).
40. P. Zuman, *Collect. Czech. Chem. Commun.* **33**, 2548 (1968).
41. P. Zuman, *Prog. Polarogr.* **3**, 73 (1972).
42. G. J. Hoijtink, *Recl. Trav. Chim. Pays-Bas* **74**, 1525 (1955).
43. G. J. Hoijtink, *Recl. Trav. Chim. Pays-Bas* **77**, 555 (1958).
44. G. J. Hoijtink, *Adv. Electrochem. Electrochem. Eng.* **7**, 221 (1970).
45. A. Streitwieser, Jr. and I. Schwager, *J. Phys. Chem.* **66**, 2316 (1962).
46. E. S. Pysh and N. C. Yang, *J. Am. Chem. Soc.* **85**, 2124 (1963).
47. W. C. Neikam and M. M. Desmond, *J. Am. Chem. Soc.* **88**, 4811 (1964).
48. C. Parkányi and R. Zahradnik, *Collect. Czech. Chem. Commun.* **30**, 4287 (1965).
49. G. J. Gleicher and M. K. Gleicher, *J. Phys. Chem.* **71**, 3693 (1967).
50. R. Zahradnik and C. Parkányi, *Talanta* **12**, 1289 (1965).
51. G. Briegleb, *Angew. Chem.* **76**, 326 (1964).

52. L. L. Miller, G. D. Nordblom, and E. A. Mayeda, *J. Org. Chem.* **37**, 916 (1972).
53. M. J. S. Dewar, J. A. Hashmall, and N. Trnajstić, *J. Am. Chem. Soc.* **92**, 5555 (1970).
54. P. Zuman, "Substituent Effects in Organic Polarography." Plenum, New York, 1967.
55. B. Kastening, *Collect. Czech. Chem. Commun.* **30**, 4033 (1965).
56. M. M. Baizer and J. P. Petrovich, *Prog. Phys. Org. Chem.* **7**, 189 (1970).
57. M. M. Baizer, ed., "Organic Electrochemistry," p. 679ff. Dekker, New York, 1973.
58. L. Nadjo and J. M. Savéant, *J. Electroanal. Chem.* **33**, 419 (1971).
59. V. J. Puglisi and A. J. Bard, *J. Electrochem. Soc.* **119**, 829 and 833 (1972); **120**, 748 (1973).
60. E. Lamy, L. Nadjo, and J. M. Savéant, *J. Electroanal. Chem.* **42**, 189 (1973).
61. T. Fujinaga, K. Izutsu, and T. Nomura, *J. Electroanal. Chem.* **29**, 203 (1971).
62. P. Zuman, *Electrochim. Acta* **21**, 687 (1976).
63. S. Wawzonek and R. C. Duty, *J. Electrochem. Soc.* **108**, 1135 (1961).
64. J. C. Halle, R. Gaboriaud, and R. Schaal, *Bull. Soc. Chim. Fr.* p. 2047 (1970).
65. P. J. Elving, *J. Electroanal. Chem.* **29**, 55 (1971).
66. L. Stárka and A. Vystrčil, *Collect. Czech. Chem. Commun.* **23**, 216 (1958).
67. L. Stárka, A. Vystrčil, and B. Stárková, *Collect. Czech. Chem. Commun.* **23**, 206 (1958).
68. D. Dolman and R. Stewart, *Can. J. Chem.* **45**, 911 (1967).
69. T. J. M. Pouw and P. Zuman, *J. Org. Chem.* **41**, 1614 (1976).
70. W. J. Bover and P. Zuman, *J. Electrochem. Soc.* **122**, 368 (1975).
71. J. Janata and R. D. Holtby-Brown, *J. Electroanal. Chem.* **44**, 137 (1973).
72. W. J. Scott and P. Zuman, *Electrochimica Acta* (in press).
73. W. Szafranski and P. Zuman, unpublished results.
74. M. Romer, N. Sleszynski, and P. Zuman, unpublished results.
75. V. Gutmann, "Chemische Funktionslehre." Springer-Verlag, Berlin and New York, 1971.
76. V. Gutmann and R. Schmid, *Coord. Chem. Rev.* **12**, 263 (1974).
77. P. Zuman and M. Szyper, *J. Chem. Soc., Faraday Trans. I* **77**, 1017 (1977).
78. M. Szyper, P. Zuman, and H. W. Gibson, *J. Chem. Soc., Faraday Trans. I* **73**, 1032 (1977).
79. T. M. Krygowski, *J. Electroanal. Chem.* **35**, 436 (1972).
80. J. Janata, J. Gendell, R. G. Lawton, and H. B. Mark, *J. Am. Chem. Soc.* **90**, 5226 (1968).
81. D. T. Sawyer and J. L. Roberts, Jr., "Experimental Electrochemistry for Chemists," p. 203ff. Wiley, New York, 1974.
82. J. N. Butler, *Adv. Electrochem. Electrochem. Eng.* **7**, 77 (1970), and references therein.
83. D. T. Sawyer and J. L. Roberts, Jr., "Experimental Electrochemistry for Chemists," p. 183. Wiley, New York, 1974.
84. D. Barnes and P. Zuman, *Trans. Faraday Soc.* **65**, 1681 (1969).
85. L. Holleck and D. Becher, *J. Electroanal. Chem.* **4**, 321 (1962).
86. M. M. Baizer, *J. Electrochem. Soc.* **111**, 215 (1964).
87. A. J. Fry and R. G. Reed, *J. Am. Chem. Soc.* **93**, 553 (1971).
88. A. J. Fry and R. G. Reed, *J. Am. Chem. Soc.* **94**, 8475 (1972).
89. D. T. Sawyer and J. L. Roberts, Jr., "Experimental Electrochemistry for Chemists," p. 14ff. Wiley, New York, 1974.
90. M. J. Barbier and J. J. Rameau, *Bull. Soc. Chim. Fr.* p. 1268 (1973).
91. R. C. Larson, R. T. Iwamoto, and R. N. Adams, *Anal. Chim. Acta* **25**, 371 (1961).
92. V. A. Pleskov, *Zh. Fiz. Khim.* **22**, 351 (1948).
93. B. Kratochvil, E. Lorah, and C. Garber, *Anal. Chem.* **41**, 1793 (1973).
94. J. L. Hall and P. W. Jennings, *Anal. Chem.* **48**, 2026 (1976).
95. V. A. Pleskov, *Usp. Khim.* **16**, 254 (1947).
96. H. Strehlow, *in* "The Chemistry of Non-Aqueous Solvents" (J. J. Lagowski, ed.), Vol. 1, p. 129, and references therein. Academic Press, New York, 1966.
97. I. M. Kolthoff and F. G. Thomas, *J. Phys. Chem.* **69**, 3049 (1965).

98. O. Duschek and V. Gutmann, *Monatsh. Chem.* **104**, 990 (1973), and references therein.
99. L. Meites, P. Zuman, W. J. Scott, B. H. Campbell, and A. M. Kardos, "Electrochemical Data," Vol. A, Part I. Wiley, New York, 1974.
100. L. Meites, B. H. Campbell, and P. Zuman, *Talanta* (in press).
101. N. L. Weinberg, ed., "Techniques of Chemistry," Vol. 5, Parts I and II. Wiley, New York, 1974.
102. A. J. Fry, "Synthetic Organic Electrochemistry." Harper, New York, 1972.
103. M. R. Rifi and F. H. Covitz, "Introduction to Organic Electrochemistry." Dekker, New York, 1974.
104. L. Eberson and H. Schafer, *Fortschr. Chem. Forsch.* **21**, 1 (1971).
105. M. Fleischmann and D. Pletcher, *Adv. Phys. Org. Chem.* **10**, 155 (1973).
106. S. D. Ross, M. Finkelstein, and E. J. Rudd, "Anodic Oxidation." Academic Press, New York, 1975.
107. L. Eberson and K. Nyberg, *Adv. Phys. Org. Chem.* **12**, 1 (1976).

~ 4 ~

Ion-Selective Electrodes in Nonaqueous Solvents

∞

E. Pungor and K. Tóth

Institute for General and Analytical Chemistry
Technical University, Budapest, Hungary

I. Introduction

After the discovery of pH and alkaline ion-sensitive glass electrodes in 1906, a renewed interest in the field of ion-selective electrodes took place in the 1960's. As a result, electrodes selective to various ions were developed and significant results were achieved in interpreting the behavior of these newly developed electrodes.

Ion-selective electrodes can be classified from different points of view according to the structure of the electrode's membrane, the ionic species detected by the appropriate ion-selective electrodes, or the mechanism of the electrode response.

The membrane structure of the electrode determines whether the ion-selective electrode is homogeneous or heterogeneous. Homogeneous ion-selective electrodes have a membrane layer which contains only the material responsible for the electrochemical behavior of the electrode. They can be pressed from amorphous particles or polycrystals, but they can also be made of single crystals. The membrane layer of heterogeneous ion-selective electrodes consists of an electrochemically active material and a supporting material.

According to the primary ion sensed by the ion-selective electrode, there are cation- and anion-sensitive electrodes, the most important of which are listed in Table I.

From an electrochemical point of view, the ion-selective electrodes can be divided into the following main groups:

1. Fundamental ion-selective electrodes

   A. Electrodes operating on electron exchange reactions

      Electrodes of the first kind (metal/metal ion electrodes)
      Electrodes of the second kind (e.g. Ag/AgCl, KCl)
      Electrodes of the third kind [e.g. $Pb/Pb(COO)_2$, $Cu(COO)_2/Cu^{2+}$]
      Electrodes involving organic or inorganic redox couples

   B. Electrodes operating on chemisorption (on precipitate-exchange reactions)

      Homogeneous electrodes
      Heterogeneous electrodes

   C. Electrodes operating on ion-exchange reactions

      Glass electrodes
      Organic ion-exchanger type electrodes

   D. Electrodes operating on complex forming reactions (carrier electrodes)

      Electrodes with charged matrix
      Electrodes with uncharged matrix

2. Sensitized ion-selective electrodes (covered surface electrodes)

   A. Coverage with inactive membrane (gas electrodes)

      Gas-permeable electrodes
      Air-gap electrodes

TABLE I

FUNDAMENTAL ION-SELECTIVE SYSTEMS[a]

Ion	Material/Form
H^+	Glass/bulb
Li^+	Glass/bulb
Na^+	Glass/bulb, glass/disk
K^+	Glass/bulb, valinomycin/L or PVC or SR, potassium tetrachlorophenylborate/L
NH_4^+	Glass/bulb, monactin-nonactin/L
Ca^{2+}	Calcium didecylphosphate/L or PVC, thenoyltrifluoracetone/PVC, $CaF_2 - LaF_3$/disk
Ca^{2+}/Mg^{2+}	Calcium didecylphosphate/L(decanol)
Ba^{2+}	Nonylphenoxypolyoxyethyleneethanol/L
Cu^{2+}	$Ag_2S + CuS$/disk or SR, $Cu_{1.79}S$/disk, CuS/C, chalcogenide glass/disk, $(R—S—CH_2COO)_2Cu$/L
Fe^{3+}	Chalcogenide glass/disk
Ag^+	Glass/bulb, Ag_2S/disk
Zn^{2+}	$ZnSe + Ag_2S$/disk
Mn^{2+}	$MnSe + Ag_2S$/disk
Cd^{2+}	$CdS + Ag_2S$/disk
Ni^{2+}	$NiSe + Ag_2S$/disk
Pb^{2+}	$PbS + Ag_2S$/disk or SR, $(R—S—CH_2COO)_2Pb$/L
F^-	LaF_3/single crystal or SR
Cl^-	$AgCl + Ag_2S$/disk, AgCl/SR, $AgCl + Ag_2S$/G, dimethyldistearylammonium chloride/L
Br^-	$AgBr + Ag_2S$/disk, AgBr/SR, $AgBr + Ag_2S$/G
I^-	$AgI + Ag_2S$/disk, AgI/SR, I_2/G/CCl_4/
S^{2-}	Ag_2S/disk, Ag_2S/SR, Ag_2/S/G
ClO_4^-	Fe/II/1,10-phenanthroline perchlorate/L, tetra-n-heptylammonium perchlorate/L
NO_3^-	Ni/II/1,10-phenanthroline nitrate/L or PVC, tridodecylhexadecylammonium nitrate/L or PVC
BF_4^-	Ni/II/1,10-phenanthroline fluoroborate/L
SCN^-	$AgSCN + Ag_2S$/disk, AgI/SR
SO_4^{2-}	$PbSO_4 + PbS + Ag_2S + Cu_2S$/disk

[a] L, Liquid ion-exchange form; G, graphite ("selectrode"); SR, silicone rubber heterogeneous membrane; PVC, polyvinyl chloride membrane.

From A. K. Covington, *Crit. Rev. in Anal. Chem.*, **3** (4), 357 (1974). CRC Press, Inc. 1974. Used by permission of CRC Press, Inc.

B. Coverage with active membrane (enzyme electrodes)

Enzyme activity measuring electrodes

Substrate activity measuring electrodes

Group 1A electrodes have a potential response which is interpreted on the basis of electron exchange equilibria. These electrochemical sensors have been well known for a long time and they really cannot be considered as ion-selective electrodes. The application of such sensors in nonaqueous media has been summarized by Buck.[1,2]

For Group 1B electrodes chemisorption on the electrode membrane's surface is the most important potential-determining step. These electrodes are classified in the literature as precipitate-based electrodes, single-crystal electrodes, or pressed-crystal electrodes, the sensing part of which can be either a homogeneous or heterogeneous structure. It is characteristic that the active component is a slightly soluble salt—a precipitate—in aqueous solutions.

The papers on precipitate-based ion-selective membrane electrodes that have so far appeared in the literature often discuss the electrochemical behavior of heterogeneous and homogeneous electrodes separately.[1-6] On the basis of experimental results this differentiation is not necessary because they differ only in mechanical and not in electrochemical respects. In nonaqueous media, however, this differentiation may be of importance because of the swelling effects caused by the appropriate organic solvents used for the measurements.

Group 1C contains those electrodes which operate on ion-exchange principles, e.g., glass electrodes and the organic ion-exchanger or charged-matrix electrodes. Of the organic ion-exchanger electrodes, the liquid ion-exchanger electrodes are of significant importance; their development can be attributed to the fundamental research of Eisenman.[7,8] This type of electrode can be prepared either as a liquid-type or a polyvinyl chloride $(PVC)_1$[9,10] incorporated electrode. The latter was introduced by Thomas et al.[11] in order to increase the lifetime of the ion-exchanger type electrodes and to produce a less fragile type of electrode.

The ion-exchanger electrode with an electrically charged matrix suffers from various interferences and its selectivity to a series of ions is rather poor.

The so-called neutral carrier type electrodes introduced by Simon and co-workers[12,13] belong to Group 1D. These electrodes can be prepared as liquid electrodes or as PVC-incorporated electrodes. In the former, the electroactive components, such as antibiotics or crown compounds, are dissolved in a nonpolar organic solvent, e.g., diphenyl ether in the case of the valinomycin-based potassium electrode. The membrane layer of the potassium-selective electrode can also be prepared by incorporating the valinomycin into an inert nonpolar matrix such as PVC or silicone rubber. This carrier-type of electrode is one of the most useful ion-selective electrodes since its sensitivity and selectivity are much better than those of electrodes prepared for the same purposes on a different base.

The applications of fundamental ion-selective electrodes have been extended with the introduction of the so-called sensitized ion-selective electrodes. A sensitized ion-selective electrode is made up of a fundamental ion-selective electrode whose surface is covered with either an inactive or an active membrane layer. Accordingly this group can be further subdivided into gas electrodes and enzyme electrodes.

In the case of gas electrodes, a pH-sensitive glass electrode is most commonly used as a fundamental electrode and it is separated from the sample solution by an appropriate gas-permeable membrane. The surface of the pH-sensitive glass electrode is wetted with a buffer solution of a given pH which is usually in contact with a reference electrode. The membrane separating the fundamental electrode from the sample solution permits only the diffusion of gases and hinders the transport of ions and solvent molecules. The gas concentration of the sample solution is detected by the fundamental ion-selective electrode through a pH change in the buffer solution, caused by the shift of appropriate chemical equilibria.

A new type of the gas electrode is the so-called air-gap electrode in which diffusion of the appropriate gas to be measured occurs through an air layer of well-defined thickness.[14-16]

In the case of enzyme electrodes designed for the measurement of either enzyme or substrate activities, the separating membrane layer is chemically active.[17] For the latter, the separatory layer, i.e., the reaction layer, contains the specific enzyme in an immobilized form, which catalyzes the decomposition of the substrate; the alteration of the activity of the products in the reaction layer is measured with the fundamental ion-selective electrode. The potential established is a result of steady-state equilibrium reactions.

The urea electrode[17] is an example of a substrate activity measuring electrode. It consists of an ammonium ion-selective electrode and an immobilized urease-containing reaction layer. The urea hydrolysis is catalyzed by the enzyme in the reaction layer and the ammonia produced is detected by the fundamental electrode.

II. FUNDAMENTAL THEORETICAL CONSIDERATIONS

A. Potential Response

The mechanism of the potential response of ion-selective electrodes, including glass electrodes, is far from being completely understood. Being aware of this problem we deal only in those aspects of the theory of ion-selective electrodes that are necessary to demonstrate the possible applications and limitations of ion-selective electrodes in nonaqueous solvents.

The potential response of ion-selective electrodes (Fig. 1) can uniformly be

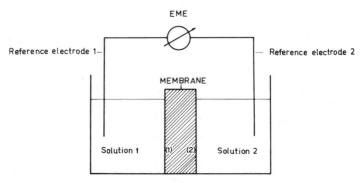

FIG. 1. Schematic diagram of an ion-selective membrane electrode assembly. EME is the cell potential measured at zero current.

described on a thermodynamic basis considering the phase-boundary potentials E_B and the diffusion potential E_D within the membrane.[18] Accordingly,

$$E = \psi_{2(2)} - \psi_{1(1)} = \underbrace{[(\psi_2 - \psi_{(2)}) - (\psi_1 - \psi_{(1)})]}_{E_B} + \underbrace{[\psi_{(2)} - \psi_{(1)}]}_{E_D} \qquad (1)$$

where ψ is the Galvani potential; and 1, 2, (1), and (2) are subscripts denoting the solution and the membrane phase, respectively (see Fig. 1).

1. PHASE-BOUNDARY POTENTIAL

The phase-boundary potential of ion-selective membranes can be described thermodynamically by considering the electrochemical reactions occurring at the solution–membrane interphase. At zero current in a solution containing the ion to which the electrode is reversible, the equilibrium between the solution and membrane phases is attained when the difference between the electrochemical potentials of the solvated ion and the ion bonded to the membrane phase equals zero. If the electrochemical potential of the appropriate ith ion in the solution is

$$\eta_1 = \mu_1 + z_i \mathscr{F} \psi_1 \qquad (2)$$

while in the membrane

$$\eta_{(1)} = \mu_{(1)} + z_i \mathscr{F} \psi_{(1)} \qquad (3)$$

where η is the electrochemical potential, μ is the chemical potential, z is the valency, $\mathscr{F}$ is the Faraday constant.

In equilibrium η_1 is equal to $\eta_{(1)}$.

$$\mu_1 + z_i \mathscr{F} \psi_1 = \mu_{(1)} + z_i \mathscr{F} \psi_{(1)} \tag{4}$$

$$z_i \mathscr{F} (\psi_{(1)} - \psi_1) = \mu_1 - \mu_{(1)} \tag{5}$$

$$\underbrace{z_i \mathscr{F} (\psi_{(1)} - \psi_1)}_{E_{B_{1(1)}}} = \underbrace{\mu_1^0 - \mu_{(1)}^0}_{E_0} + RT \ln \frac{(a_i)_1}{(a_i)_{(1)}} \tag{6}$$

$$E_{B_{1(1)}} = E_0 + \frac{RT}{z_i \mathscr{F}} \ln \frac{(a_i)_1}{(a_i)_{(1)}} \tag{7}$$

where E_B is the phase-boundary potential, E_0 is a constant and independent of activities, μ^0 is the standard chemical potential, z_i is the valency of the ith ion, a_i is the activity of the ith ion.

Naturally, the difference between the phase-boundary potentials can also be expressed:

$$E_B = E_{B_{1(1)}} - E_{B_{2(2)}} = E_0 + \frac{RT}{z_i \mathscr{F}} \ln \frac{(a_i)_1 (a_i)_{(2)}}{(a_i)_2 (a_i)_{(1)}} \tag{8}$$

Equations 7 and 8 are valid for every positively or negatively charged ion able to pass through the phase boundaries.

Boksay and Csákváry[19] have found some contradictions when employing Eq. 7 to describe the potential of an interphase between the glass electrode and the aqueous solution—the electrode process was found to be different than described by the equation. In fact, the real electrode process should be described by taking into account that when an ion leaving the glass enters the solution a vacancy is produced in the glass. In the reverse process, the ion entering the glass from the solution can occupy a vacancy only. Accordingly, if this real electrode process is considered

$$\text{Vacancy} + H_1^+ \rightleftharpoons H_{(1)}^+ \tag{9}$$

then the phase-boundary potential can be described by the following equation:

$$E_{1(1)} = E_0 + \frac{RT}{\mathscr{F}} \ln \left[c_v \frac{a_{H,1}}{a_{H,(1)}} \right] \tag{10}$$

Thus, the potential of the glass electrode depends, in addition to the activity of the hydrogen ion, on the concentration of the vacancies (c_v). The concentration of the vacancies may change if the ion-exchange reaction is related to volume changes in the glass or if the vacancies in the glass phase are occupied by the neutral molecules in the solution. These possibilities must be taken into account, for example, when the solvent containing the ion tested is replaced by another one.

It has been found experimentally that when a glass electrode is transferred from an aqueous solution into a nonaqueous one, the potential of the glass electrode changes significantly (20–100 mV) as a function of time, and the final potential value is reached within 30–120 min. even when the oxonium ion activity is almost equal in the two appropriate solutions. This can be attributed either to the change in concentrations of the vacancies or to the variations in proton activity at the surface layer of the glass. The change in proton activity does not necessarily involve a change in proton concentration, since the proton is bonded in the surface layer to different proton acceptors such as solvent molecules or certain groups in the silicate lattice. Consequently, if the glass electrode is intended to be used in nonaqueous media, then the electrode must be soaked for a few hours in the appropriate solvent in order to obtain an electrode surface layer having a constant proton activity as well as a constant vacancy concentration characteristic of the nonaqueous solvent. In this case the phase-boundary potential can be described as follows:

$$E_{1(1)} = E_0' + \frac{RT}{\mathscr{F}} \ln \frac{a_{H1}}{a_{H(1)}} \tag{11}$$

where E_0' is a constant also involving the concentration of vacancies.

2. THE DIFFUSION POTENTIAL

A general equation describing the diffusion potential of all types of selective membranes is given by

$$E_D = \frac{RT}{\mathscr{F}} \int_{(1)}^{\cdot(2)} \frac{\partial}{\partial x} \frac{[\sum u_c a_c(x) - \sum u_c a_a(x)]}{\sum u_c a_c(x) + \sum u_a a_a(x)} dx \tag{12}$$

where u is the mobility of a particle in the membrane, $a(x)$ is the activity of a particle in the membrane, c and a are subscripts denoting cations and anions, respectively. This rather complicated equation, however, can be simplified[20] under the following conditions.

a. The membrane is perm selective which means that the mobilities of either the cations or the anions in the membrane can be neglected. This condition is fulfilled if either the mobilities or the activities of ions of identical charge are negligible. When only cations of charge z_i can move across the membrane, then the diffusion potential is described by Eq. 13.

$$E_D = \frac{RT}{z_i \mathscr{F}} \ln \frac{\sum u_c a_c(1)}{\sum u_c a_c(2)} \tag{13}$$

b. The activities of all ions in the membrane are constant, except for ions with charges similar to z_i. In this case the diffusion potential can be given as

$$E_D = \frac{RT}{z_i \mathscr{F}} \ln \frac{\sum z_c^2 u_c a_c(1) + \sum z_a^2 u_a a_a(1)}{\sum z_c^2 u_c a_c(2) + \sum z_a^2 u_a a_a(2)} \tag{14}$$

c. The activity of all stationary ions is constant in the membrane with the exception of ions of extremely small mobilities. Under this condition, the value of the diffusion potential can be neglected.

$$E = 0 \tag{15}$$

If one of the above-mentioned conditions is fulfilled in the case of different types of ion-selective membrane electrodes, then the overall membrane potential, i.e., the potential of the ion-selective electrode, can be described by considering the diffusion potential appropriate for the membrane type. At the same time the potential of all types of ion-selective electrodes can be described with the modified Nikolsky equation if the effect of the diffusion potential is considered by the selectivity coefficient of the appropriate ion-selective membrane electrode:

$$E = E_0' + \frac{RT}{z_i \mathscr{F}} \ln \left[a_{i,1} + \sum_{j \neq i} K_{ij}^{pot} (a_{j,1})^{z_i/z_j} \right] \tag{16}$$

where E_0' is a constant and independent of ion activities, $a_{i,1}$ and $a_{j,1}$ are the activities of the ith and jth ion in solution 1 (see Fig. 1), and K_{ij} is the selectivity coefficient.

B. Selectivities of Different Types of Ion-Selective Electrodes

1. PRECIPITATE-BASED ION-SELECTIVE ELECTRODES

This type of electrode consists of a water-insoluble precipitate either in homogeneous or in heterogeneous form. The electrochemical behavior of the electrodes is interpreted[21] on the basis of a precipitate–exchange equilibrium. Accordingly, a general equation has been deduced[22] for the calculation of the selectivity coefficients by considering the precipitate–exchange reaction for ions having a valency of more than one. The basic reaction is

$$(1/a)[(I_k)_a(I_i)_b]_{(1)} + (m/n)(I_j)_1 \rightleftharpoons (1/n)[(I_k)_n(I_j)_m]_{(1)} + (b/a)(I_i)_1 \tag{17}$$

where I_k represents the cation of the precipitate built into the membrane, e.g., Ag; I_i and I_j are two anions taking part in the precipitate exchange reaction; and a, b, n, m are the stoichiometric constants of the precipitates built into the membrane or formed during the exchange reaction, respectively.

At coprecipitation the component (ions) of the solution taking part in the exchange reaction is in equilibrium with those in the solid membrane phase and the equilibrium can be expressed simply as follows:

$$K_{ij} = (I_i)_{(1)}^{b/a}/(I_j)_{(1)}^{m/m} \tag{18}$$

If the activities of the anions in the boundary phases and in the bulk solution are equal to each other, then

$$K_{ij} = (I_i)_1^{b/a}/(I_j)_1^{m/n} \tag{19}$$

The selectivity coefficient of the ion-selective electrode to various precipitate forming ions other than the primary ion can be expressed as a function of the solubility products of the precipitates originally in the membrane and formed during the exchange reaction. If it is assumed that the composition of the two precipitates is $(I_k)_a(I_i)_b$ and $(I_k)_n(I_j)_m$, where S_{ki} and S_{kj} are the appropriate solubility products, then

$$S_{ki} = (I_k)^a(I_i)^b \tag{20}$$

and

$$S_{kj} = (I_k)^n(I_j)^m \tag{21}$$

If the same solubility products are valid in the membrane phase as in the solution phase, then

$$(I_k)_{(1)} = S_{ki}^{1/a}/(I_i)_{(1)}^{b/a} \tag{22}$$

or

$$(I_k)_{(1)} = S_{kj}^{1/n}/(I_j)_{(1)}^{m/n} \tag{23}$$

At equilibrium the I_k concentration is the same for both precipitates so

$$K_{ij} = S_{ki}^{1/a}/S_{kj}^{1/n} \tag{24}$$

When monovalent ions form the precipitate, then Eq. 24 reduces to

$$K_{ij} = S_{ki}/S_{kj} \tag{25}$$

Equations 24 and 25 can be used for the theoretical calculation of the selectivity coefficients of precipitate-based ion-selective electrodes. The validity of the equations has been proved potentiometrically.[22]

Using a different principle, Simon and co-workers[23] have independently come to the same conclusion concerning the theoretical calculation of the selectivity coefficients of precipitate-based ion-selective electrodes.

2. GLASS AND OTHER ANALOGOUS ION-EXCHANGE MEMBRANES

The selectivity coefficient of the glass membrane electrodes can be determined with the following equation which was derived by considering the activities of the anionic groups in the membrane to be constant

$$K_{ij}^{pot} = (u_j/u_i)K_{ij} \tag{26}$$

where u_i and u_j are the mobilities of the ith and jth ion in the membrane, and K_{ij} is the ion-exchange equilibrium constant.

The liquid ion-exchanger and carrier electrodes consist of a specific lipophilic ligand dissolved in an organic solvent immiscible with the solution containing the ion to be studied. According to the structure of the specific ligands the following differentiations can be made.

a. *Liquid membranes with electrically charged ligands.* If the activity of the anions in the membrane can be neglected with respect to that of cations, then the potentiometric selectivity coefficient of this type of electrode can be given as

$$K_{ij}^{pot} = (u_j/u_i)K_{ij} \tag{27}$$

where K_{ij} is the exchange equilibrium constant, and u_j and u_i are the mobilities of the ith and jth ion in the membrane.

In this case the selectivity of the membrane electrode is solely defined by the solvent of the membrane phase.

However, if the activity of all ions, except for those of charge z_i, is constant in the membrane, then the potentiometric selectivity coefficient of the appropriate selective membrane electrode is described as

$$K_{ij}^{pot} = \frac{u_j + u_s}{u_i + u_s} K_{ij} \tag{28}$$

where u_s is the mobility of the ligand in the membrane.

If the ith and jth ions are trapped in the membrane with association or complex formation and the degree of their dissociation is almost constant, then the selectivity coefficient of the appropriate electrode can be expressed by

$$K_{ij}^{pot} = \frac{K_{js}}{K_{is}} K_{ij} \tag{29}$$

where K_{js} and K_{is} are the equilibrium constants of the complex-formation reactions of the ligand with the ith and jth ion.

Sandblum and Eisenman[8,24] have studied the selectivity coefficients of ion-exchanger membrane electrodes in the limiting case of strong association of the appropriate ions in the membrane and derived the following equation:

$$K_{ij}^{pot} = [1 - \tau]\frac{u_j + u_s}{u_i + u_s} K_{ij} + \tau \frac{u_{js}K_{js}}{u_{is}K_{is}} K_{ij} \tag{30}$$

$$\tau = \frac{u_s/u_{is}K_{is} - u_{js}K_{js}}{(u_j + u_s)u_{is}K_{is} - (u_i + u_s)u_{js}K_{js}} \quad (0 \leq \tau \leq 1) \tag{31}$$

where τ is the transference member and u_{js} and u_{is} are the mobilities of the appropriate complexes formed.

Unfortunately, the behavior of this type of liquid membrane electrode does not match any of the conditions discussed, so they do not exhibit outstanding selectivities.

b. *Membranes with electrically neutral ligands.* This type of membrane measures ions in a complex form while the activities of the anions are limited to a relatively small value. In an ideal case, the activity of the ligand in the membrane is constant and the selectivity coefficient is given as follows:[20,23]

$$K_{ij}^{pot} = \frac{u_{js} K_{js}}{u_{is} K_{is}} K_{ij} \qquad (32)$$

where K_{js} is the stability constant of the complex in the membrane. However, in other limiting cases the potentiometric selectivity coefficients can be expressed by different equations.[23]

Eisenman et al.[25] have assumed that in the case of very thin membranes— if the mobilities of anions and cations in comparison with that of the complex are negligible—the selectivity coefficient can be described by the following equation

$$K_{ij}^{pot} = \frac{u_{js} K_{js}^*}{u_{is} K_{is}^*} K_{is,js} \qquad (33)$$

where K_{is}^*, K_{js}^* are the stability constants of the complexes in the solution phase and $K_{is,js}$ is the exchange equilibrium constants of the complexes.

Simon and co-workers[23] have pointed out that, if the ion to be tested is completely trapped with the ligand in the membrane, then the selectivity coefficient of the appropriate membrane does not depend on the nature of the solvent for the membrane. In addition, they have shown that if the ligand is a synthetic "crown" compound which does not completely cover the central ion, the solvent plays an important role in the selectivity of the appropriate membrane electrode.

III. Practical Applications

A. Potentiometric Measuring Cells

In general, two types of measuring cells are used for potentiometric investigations, i.e., potentiometric cells with and without liquid junction (hereafter referred to as electrodes of the first and second kinds, respectively). In the former, the reference electrode is joined to the solution to be studied

through a salt-bridge electrolyte, and the reference electrode used is generally
an electrode of second kind. The construction of the cell is as follows:

| Indicator electrode | Solution to be tested | Salt-bridge electrolyte | Reference electrode | (34) |

| Pt, H$_2$ | Solution to be tested | KCl (sat) | 0.1 M KCl | Hg$_2$Cl$_2$, Hg | (35) |

where | indicates a solution-electrode interphase, ‖ means a liquid junction
potential. The cell potential

$$E = (E_{ref} + E_j) - E_{ind}$$

where E_{ref} is the potential of the reference electrode, E_{ind} is the potential of
the indicator electrode, and E_j is the liquid junction potential.

A similar cell construction can be given for the measurement of other pX
values also

| pX selective electrode | Test solution containing ion X^{z+} | KCl (sat) | Hg$_2$Cl$_2$, Hg | (36) |

or

| pX selective electrode | Test solution containing ion X^{z-} | KCl (sat) | Hg$_2$Cl$_2$, Hg | (37) |

and the corresponding cell potential

$$E = (E_{ref} + E_j) - E_X^{0'} - (RT/z\mathscr{F}) \ln a_{X^{z+}}$$ (38)

or

$$E = (E_{ref} + E_j) - E_X^{0'} + (RT/z\mathscr{F}) \ln a_{X^{z-}}$$ (39)

This type of potentiometric cell has been employed most widely for pH
and other pX measurements. However, difficulties may arise when the individual potential values are to be measured. In fact, the greatest problem is to
determine the liquid junction potential value even in aqueous solutions
which significantly increases by the introduction of aqueous and nonaqueous interphases. Data concerning the liquid junction potential values
formed at aqueous and nonaqueous interphase are shown in Table II.

In addition it must be emphasized that when an aqueous electrode of
the second kind is employed as a reference electrode for the measurements
of pH or pX values in nonaqueous media, then the final potential value is
always influenced by the liquid-junction potential formed at the boundaries of
aqueous and nonaqueous solutions. A nonaqueous reference electrode
prepared with the same nonaqueous solvent as the sample should be used.

When using potentiometric cells without transference, the reference and

TABLE II

LIQUID JUNCTION POTENTIALS E_j
AT NONAQUEOUS–AQUEOUS BOUNDARIES[a]

Boundary	E_j (mV)
KCl (sat) ‖ 1 C_2H_5OH, 1 H_2O	25
KCl (sat) ‖ 4 C_2H_5OH, 1 H_2O	75
KCl (sat) ‖ C_2H_5OH	140

[a] Estimated by Gutbezahl and Grunwald.[26]

the indicator electrodes are directly connected to the sample solution. The construction of the cell is as follows:

$$\begin{array}{c|c|c} \text{Indicator} & \text{Solution to be} & \text{Reference} \\ \text{electrode} & \text{tested} & \text{electrode} \end{array} \qquad (40)$$

i.e.,

$$\text{Pt, H}_2 \mid \text{HCl} \mid \text{Hg}_2\text{Cl}_2, \text{Hg} \qquad (41)$$

or

$$\text{pX} \mid \text{XCl} \mid \text{Hg}_2\text{Cl}_2, \text{Hg} \qquad (42)$$

Accordingly, the cell potential

$$E = E_{ref} - E_{ind}$$

$$= [E^0_{Hg,Hg_2Cl_2} - (RT/\mathscr{F}) \ln a_{Cl^-}] - [E^0_{Pt,H_2} + (RT/\mathscr{F}) \ln a_{H^+}] \quad (43)$$

or

$$E = E_{ref} - E_{ind}$$

$$= [E^0_{Hg,Hg_2Cl_2} - (RT/\mathscr{F}) \ln a_{Cl^-}] - [E^{0'}_X + (RT/\mathscr{F}) \ln a_{X^+}] \quad (44)$$

where a_{H^+}, a_{Cl^-}, and a_{X^+} are the activities of the hydrogen, chloride, and X ion, respectively.

In this case the cell potential is not influenced by the liquid-junction potential. However, the inaccuracy of the determination of the chloride ion activity may cause errors in the determinations.

In addition to the electrodes of second kind, other electrodes may also be used as reference electrodes.

This type of potentiometric cell is recommended for potentiometric studies in nonaqueous media using reference electrodes having a known behavior in the appropriate nonaqueous solvent.

B. pH or pX Measurements

In practice the pH or pX measurements are carried out in an appropriate potentiometric cell by reference to a standard system. The potentiometric cells most commonly used for this purpose involve a pH indicator electrode and an electrode of second kind as a reference. The standard system consists of a series of solutions whose pH and pX values are defined. With the help of these standard solutions a conventional or operational pH or pX scale can be established. The accuracy of the pH and pX measurements in the unknown system depends to a great extent on the real pH and pX values of these standard solutions. The establishment of a conventional pH scale was carried out by Bates and Guggenheim.[27,28]

The standardization of the hydrogen electrode and other pH electrodes can be carried out in potentiometric cells with transference using one pH standard solution if the slope of the electrode calibration graph, i.e., E vs. pH is linear in a relatively wide activity range. Accordingly,

$$pH(x) = pH(s) + (E_x - E_s)/RT \ln 10 \qquad (45)$$

where $pH(x)$ is the pH of the unknown sample solution, $pH(s)$ is the pH of the standard solution, E_x is the potential measured in the unknown sample solution, and E_s is the potential measured in the standard solution.

In practice it is advisable to use at least two standard solutions (s_1 and s_2) for the standardization pH electrodes and these should preferably surround the pH of the test solution (x). In this case potentiometric measurements are generally carried out in a potentiometric cell with transference and the $pH(x)$ values can be calculated from the following equation

$$pH(x) - pH(s_1) = \frac{E_x - E_{s_1}}{E_{s_2} - E_{s_1}} [pH(s_2) - pH(s_1)] \qquad (46)$$

Unfortunately, the standardization of pX electrodes other than pH electrodes has not been studied as thoroughly as that of pH electrodes. The development of conventional pX (activity) scales being consistent among themselves and with the pH convention currently is required and in progress.[29]

The standardization of a cation-selective electrode is carried out in solutions containing a well-dissociated salt of the appropriate cation, while that of an anion-selective electrode is done in solutions of the well-dissociated salt of the corresponding anion.[29] The pX values or pa_x values of a few standard solutions are given in Table III. For the calculation of the pX values the mean activity coefficients have been taken from the literature. The pX values, determined with the help of the relating pX standards, can be calculated from the following equation:

$$pX(x) = pX(s) \pm \frac{z(E_x - E_s)\mathscr{F}}{RT \ln 10} \qquad (47)$$

TABLE III

STANDARD REFERENCE VALUES FOR UNIVALENT AND DIVALENT IONS, pX(s)^a

Molality (mole kg^{-1})	LiCl pLi	NaCl pNa	NaCl pCl	KCl pK	RbCl pRb	CsCl pCs	NH$_4$Cl pNH$_4$	KF pF
					Reference material			
0.01	2.043	2.044	2.045	2.043	2.044	2.045	2.043	2.044
0.1	1.097	1.106	1.112	1.112	1.116	1.121	1.112	1.111
0.2	0.810	0.827	0.838	0.840	0.846	0.858	0.840	0.837
0.5	0.406	0.455	0.481	0.482	0.495	0.519	0.483	0.475
1.0	0.054	0.157	0.208	0.206	0.266	0.264	0.208	0.190
2.0	-0.392	-0.180	-0.072	-0.086	-0.055	0.003	-0.080	-0.119
3.0	-0.754	-0.417	-0.245	-0.274	-0.232	-0.157	-0.261	-0.325
4.0	—	-0.618	-0.374	-0.421	-0.368	-0.278	-0.397	-0.494
5.0	—	-0.804	-0.477	—	-0.481	-0.376	-0.509	—
6.0	—	-0.981	-0.563	—	—	—	-0.602	—

Molality (mole kg^{-1})	MgCl$_2$ pMg	CaCl$_2$ pCa	SrCl$_2$ pSr	BaCl$_2$ pBa	CaCl$_2$ pCl
			Reference material		
0.0333	1.884	1.901	1.901	1.905	1.282
0.1	1.554	1.570	1.575	1.587	0.842
0.2	1.321	1.349	1.360	1.389	0.562
0.333	1.123	1.165	1.184	1.232	0.349
0.5	0.932	0.991	1.022	1.083	0.177
1.0	0.463	0.580	0.646	0.777	-0.140
2.0	-0.459	-0.186	-0.035	0.385	-0.507

^{a} At 25°C based on the Bates, Stapes, and Robinson[30] Hydration Theory Approach. From A. K. Covington, *Crit. Rev. in Anal. Chem.* **3** (4), 396 (1974). Used by permission of CRC Press, Inc.

where $pX(x)$ is the pX value of the unknown solution containing ion X, $pX(s)$ is the pX value of the pX standard solution, z is the valency of ion X, E_x is the emf measured in the unknown solution containing ion X, E_s is the emf measured in the standard solution, and X can be either a cation or an anion.

Bates and co-workers[29,30] have carried out fundamental research on the conventional pX activity scales.

Recently, efforts have been made by Havas and his co-workers[31] to develop anion buffer systems based on heterogeneous precipitate equilibria. These systems, however, are open to the criticism that, because they involve heterogeneous equilibria at low concentrations, the rate at which the electrode systems approach equilibrium is necessarily low and they are really not useful as buffers.

Metal–ion buffer systems based on complex equilibria have been discussed by Růžička and co-workers.[32] These buffer systems are very suitable for studying the capability of the appropriate cation selective electrodes, but for evaluating analytical results they are recommended only if appropriate care is taken.

C. Nonaqueous pH Standards

The applications of aqueous pH or pX standards to measurements carried out in nonaqueous solvents presents several practical and theoretical difficulties. From the practical point of view the change in the liquid-junction potential when passing from aqueous solutions into nonaqueous ones often introduces problems, as was shown earlier. This may result in a decrease in the reproducibility of the measurements. Of course, the problem is not as serious if the studies are carried out in solvents of relatively high dielectric constants such as methanol, hydrogen peroxide, and acetic acid since they are very similar to water in electrochemical behavior.

Another practical problem is the reproducibility of the appropriate pX indicator electrode in nonaqueous media compared to those in aqueous solutions. It is often pointed out in the literature that, for example, the response of the glass electrode in polar solvents is sluggish and the results obtained are rather poor.

From a theoretical point of view the pX determinations carried out in nonaqueous media employing aqueous pX standards have little significance. The fundamental reason for this is in the so-called medium effect introduced by the nonaqueous solvent and as a result

$$f_i = f_m f_i' \tag{48}$$

where f_i is the activity coefficient of the aqueous solvent, f_i' is the activity coefficient defined as approaching unity for the solvent concerned, and f_m is the primary medium effect.

Two approaches have been made to establish nonaqueous pX scales.

1. OPERATIONAL pH SCALES

Operational pH scales can be obtained in the same way as described for aqueous operational pH scales. The potentiometric cells usually involve the appropriate pH glass electrode and an electrode of second kind as a reference electrode. The reference electrode may be aqueous if the liquid-junction potential in the appropriate nonaqueous media is reproducible.

Wynne-Jones[33] has established a standard scale in hydrogen peroxide–water mixtures for a wide range of solvent compositions. For these measurements an aqueous potassium chloride calomel electrode was employed. However, this scale can only be recommended for use if the solutions tested have the same ionic strength and background composition as those used in the establishment of the scale. Since the total ionic strength of the standard solutions is kept constant, a concentration pH scale is defined.

Glasoe and Long[34] as well as Mikkulsen and Nielsen[35] have employed a similar approach to the establishment of a pD scale. Gary et al.[36] obtained similar results for the deuterium scale and showed that the pD values are higher than the pH values by 0.45 units on the molarity scale and by 0.40 on the molality scale. Bruckenstein and Kolthoff,[37,38] have developed a concentration pH scale in acetic acid medium. In their work a nonaqueous saturated calomel reference electrode was used which contained glacial acetic acid instead of the normal aqueous internal filling.

2. ACTIVITY pH SCALES

Activity pH scales have been worked out by De Ligny and co-workers[39–41] for anhydrous methanol and methanol–water systems. For the study, a hydrogen electrode and a silver–silver halide reference electrode were employed. Attempts have been made to establish a pD scale for a deuterium oxide solvent by Gary and co-workers.[42]

The pH activity scales worked out for nonaqueous media have the same theoretical importance as the pH scale determined for aqueous media. The disadvantage of nonaqueous pH scales is that they are only suitable for pH studies in identical solvent compositions.

D. pX Standards

The same principles discussed in connection with the establishment of pH scales may be employed for the establishment of nonaqueous pX scales. A series of solutions of known X ion concentration can be prepared in a background medium of constant ionic strength and composition. The pX values of these standard solutions can be determined with the help of an appropriate pX indicator electrode and a suitable reference electrode either with or without a liquid junction. Naturally, if the pX values of these standard solutions are defined as standard pX values, then the composition of the standard solution must be as similar to the unknown one as possible. Kazarjan and Pungor[43] have employed potassium halide salts for the calibration of the appropriate halide selective electrodes in mixtures of water and nonaqueous solvents such as ethanol, methanol, isopropanol, and acetonitrile.

E. Reference Electrodes

Two main approaches have been used for the application of the reference electrodes in nonaqueous media. The common practice of using the aqueous saturated calomel electrode as a reference electrode is very convenient, but in some cases it introduces uncertainties as, for example, that caused by the liquid-junction potential. A better approach is to use the aqueous saturated calomel electrode joined to the unknown solution through a salt bridge made with the appropriate nonaqueous solvent.

Another approach employs the appropriate reference electrode prepared with the same nonaqueous solvent as tested. For this purpose the silver–silver chloride and the calomel electrodes are widely used; tetramethylammonium salts are employed as electrolytes. Naturally, among other factors, the solubility of the reference electrolyte in the appropriate nonaqueous medium may restrict the successful generalization of this type of application of the reference electrodes.

In some cases electrodes of the first kind and ion-selective electrodes may also be used in nonaqueous media as reference electrodes if their electrochemical behavior is known in the nonaqueous solvent studied. Excellent reviews on the application of reference electrodes in nonaqueous media were made by Ives and Janz[44] and Mann and Barnes.[45]

F. Application of Ion-Selective Electrodes in Nonaqueous Solvents

The application of ion-selective electrodes in nonaqueous media as well as in aqueous and nonaqueous solvent mixtures is rather limited. This may be attributed to the fact that the theoretical interpretation of the behavior of ion-selective electrodes even in aqueous media is far from being complete. Furthermore, only a few attempts to employ these sensors in nonaqueous media have been made. When employing a number of electrodes in nonaqueous solvents, difficulties may arise due to the swelling of the electrode membrane or the electrode body in the appropriate organic solvent. In other cases the electroactive material of the electrode may dissolve in the non-aqueous solvent, which means that the partition coefficient of the component between the electrode and the solution phase will be influenced so that a proper electrochemical function cannot be obtained. The former occurs mainly in the case of heterogeneous electrodes, while the latter occurs with organic ion-exchanger and carrier-type electrodes.

As the literature shows, glass electrodes have been most widely used in nonaqueous solvents up to this point. This may be explained by the fact that ion-selective electrodes other than the glass electrodes are rather new tools since they have been developed in the last two decades. Moreover, in some cases, the mechanism of the electrode response restricts the application of the newly developed ion-selective electrodes in nonaqueous solvents.

1. GLASS ELECTRODES

a. *Nernstian response.* The ideal pH function of the glass electrodes in nonaqueous solvents generally extends only to a few pH units. Consequently, in nonaqueous media the glass electrodes can only be used for direct pH measurements in a narrow pH range. This is the reason that glass electrodes are used mainly as indicator electrodes for potentiometric acid–base titrations in nonaqueous media.

The first papers dealing with the application of pH-sensitive glass electrodes in alcohol–water and acetone–water solvent mixtures were published by Izmailov et al.[46,47] An ideal Nernstian pH response was found in the pH range 3 to 9.5 in 40 : 60 ethanol–water mixture, while in 70 : 30 ethanol–water mixture the upper limit of the ideal pH response was at pH 7.

Izmailov and Alexandrovna[48,49] first showed experimentally that the pH-sensitive glass electrodes can also be used for the measurement of the oxonium ion activity in water-free nonaqueous media, such as ethanol and methanol.

Mattock[50] has employed pH-sensitive glass electrodes in dioxane–water mixtures after conditioning them for 24 hours in the appropriate solvent mixtures. Similarly, Mitshell and Wynne-Jones[51] have suggested that glass electrodes be conditioned for 24 hours before using them in peroxide–water mixtures. These findings are in good correlation with the theoretical considerations described earlier.

In addition, the pH-sensitive glass electrodes also have been used as indicator electrodes for the direct or indirect measurement of oxonium ion concentration in ether, acetone,[52–54] pyridine,[55,56] quinoline, alcohols,[56,57] perchloric acid,[58] dioxane,[59–62] acetic acid,[63] and hydrogen peroxide.[64,65]

Recently Glasoe and Long[66] as well as Mikkulsen and Nielsen[67] used the glass electrode in the determination of deuterium ion activity.

The cation-sensitive glass electrodes have been used first of all for the determination of the activities of cations of valency one in mixtures of aqueous and nonaqueous solvents. Surprisingly, these electrodes show an almost ideal potential response even in solvent mixtures containing only a few percent water.

Rechnitz and Zamochnick[68] studied the alkaline ion function of cation-sensitive glass electrodes in amphiprotonic (alcohols), aprotonic (ketones), and protophilic (dimethylformamide) solvent mixtures in potentiometric cells with transference employing an aqueous reference electrode. The glass electrodes were preconditioned in the appropriate solvent for approximately 18–24 hours before use. After each measurement the electrodes were washed with the appropriate solvent and stored in the conditioning solution between measurements. Stable reproducible responses to alkali metal ions were found in solvent mixtures containing 0–90%(vol) organic solvent. The slopes of the electrode calibration graphs obtained in different solvent mixtures are summarized in Table IV. In the interpretation of the results it must be taken into account that the slope of the calibration graph of the glass electrode measured in aqueous solutions under otherwise identical conditions was only 56 mV per decade concentration change.

Concerning the application of the cation-sensitive glass electrodes in aqueous and nonaqueous solvent mixtures, Eisenman[69] and Rechnitz[68] independently found that the emf values of potentiometric cells employing aqueous electrodes of second kind as reference increase with the increasing fractions of the organic solvent in the solvent mixtures at identical cation concentrations. This effect may be attributed to changes in the liquid junction potential at the reference electrode rather than to a change in the response of the cation-sensitive electrode.

Lanier[70] studied the activity coefficients of sodium chloride with cation-sensitive glass electrodes in aqueous and nonaqueous solvent mixtures containing methanol, ethylene glycol, ethylene glycol diacetate, diethylene glycol monomethyl ether, dioxane, urea, and dimethylformamide as the organic component; potentiometric cells without liquid-junction were employed. The activity coefficient data determined with the cation-sensitive glass electrode were in agreement with those determined by other methods.

The application of a cation-sensitive glass electrode for the measurement of the ammonium ion activity in liquid ammonia solutions has been studied by Shiurba and Jolly.[71] A silver electrode was used as a reference electrode. The slope of the electrode calibration graph, i.e., emf values versus log $c_{NH_4^+}$, was 50 mV per decade change in concentration. Besides direct measurements, the cation-selective glass electrode was found suitable for acid–base titrations in ammonia.

Baumann and Simon[72] have successfully used a cation-sensitive glass electrode in liquid ammonia and obtained an electrode calibration curve

TABLE IV

RESPONSE OF CATION-SENSITIVE GLASS ELECTRODES
TO ALKALI METAL IONS IN MIXED SOLVENTS[a]

	Volume (%)				
	10	30	50	70	90
Ethanol					
KCl	56.0	55.0	55.0	56.0	53.0
KCl[b]	56.0	55.0	55.0	56.0	53.0
RbCl	56.0	55.0	55.5	56.0	53.0
NaCl	55.5	53.0	55.0	54.0	52.0
Acetone					
KCl	57.0	54.5	55.0	53.0	44.0
KCl[c]	57.0	54.5	55.0	53.0	44.0
RbCl	55.0	55.0	55.5	50.0	44.0
NaCl	55.0	54.0	54.5	49.0	43.7
Ethylene glycol					
KCl	56.0	57.0	57.0	56.0	56.0
RbCl	57.5	56.0	56.0	55.8	55.7
NaCl	56.0	56.0	55.0	55.0	55.0
	8.3	25.0	41.7	58.3	75
DMF					
KCl	52.5	53.3	54.0	51.0	50.0
RbCl	53.5	50.0	55.7	50.0	48.0
NaCl	51.0	51.0	55.0	51.0	52.5

[a] Slopes of plots of mV vs. $-\log [M^+]$. From Rechnitz and Zamochnick.[68]
[b] Vs. ESCE.
[c] Vs. ASCE.

with a slope of 47 mV per decade change in ammonium ion concentration when an ammonia-saturated cadmium–cadmium nitrate reference electrode was employed. In addition, titrations of the potassium salts of very weak acids (e.g., alcohols, phenols) were also carried out in this solvent using the same detector cell.

McClure and Reddy[73] investigated the alkaline ion (Li^+, Na^+, and K^+) response of the cation-sensitive glass electrode in propylene carbonate, acetonitrile, and dimethylformamide over a concentration range of 10^{-6}–10^{-2} M. The slopes of the mV versus log $[M^+]$ plots were less than the Nernstian value (Table V). In the measurements the reference electrode was a saturated calomel electrode which was separated from the test solution by a nonaqueous salt bridge.

Wilcox[74] reported on the measurement of various alkaline nitrate salts in

TABLE V

SLOPES OF mV vs. log $[M^+]$ PLOTS[a]

Solvent	Li^+	Na^+	K^+
Acetonitrile	56	56	58
Propylene carbonate	53	52	53
Dimethylformamide	b	b	59

[a] After McClure and Reddy.[73]
[b] Li^+ and Na^+ apparently react with an impurity in this solvent.

fused ammonium nitrate salts with a Pyrex glass membrane electrode involving a ternary nitrate eutectic for lithium, potassium, and sodium as well as silver nitrate as inner electrolyte. The electrical contact was made with a silver wire. The reference electrode was the asbestos junction type filled with the same silver nitrate containing eutectic as was the bulb. Both the reference electrode and Pyrex bulb electrode were stored in fused ammonium nitrate between measurements.

The potentiometric results were interpreted by the following equation if the system contained no more than two electroactive species:

$$E = E' + (RT/\mathscr{F}) \ln (C^0 + k_1 X_1 + k_2 X_2) \qquad (49)$$

where C^0 is the contribution due to the solvent (i.e., the solvent itself, and the impurities), k is the selectivity coefficient, and X is the mole fraction of the electroactive species. The results obtained for sodium were found to be in best agreement with those of Notz and Keenan.[75]

b. *Selectivities.* The successful application of glass electrodes in any solvent greatly depends on their selectivity data. In spite of this only a few results are available on this subject in the literature, especially for nonaqueous solvents.

Rechnitz[68] reported on the selectivity of a cation-sensitive glass electrode in various mixed solvents and made the following conclusions.

1. The order of the selectivity of the particular cation-sensitive glass electrode studied in the solvent mixtures is as follows: $K^+ > Rb^+ > Na^+ > Cs^+ > Li^+$.

2. By increasing the ethanol concentration in ethanol–water mixtures, the selectivity coefficients of the cation-sensitive glass electrode to cations other than the hydrogen ion remain constant.

3. The selectivity coefficient of the cation-sensitive glass electrode to hydrogen ions decreases with the increase in ethanol concentration in

ethanol–water mixtures, whereby the useful measuring range of the electrode may be shifted toward lower pH values.

McClure and Reddy[73] have studied the selectivity ratios of the cationic glass electrode for lithium, potassium, and sodium in acetonitrile and propylene carbonate, employing the so-called separate solution method. The selectivity data calculated at $[M^+] = 10^{-3}\ M$ are collected in Table VI. The effect of solvent on glass-electrode selectivity was tested by the addition of methanol to the potentiometric cells containing the appropriate alkaline salts and $0.1\ M\ Bu_4NClO_4$ in propylene carbonate. It was found that the selectivity ratio of sodium to lithium apparently increases as the percentage of methanol in propylene carbonate is increased. The increase in the selectivity ratio is explained by the change in the nature of the swollen layer on the glass surface.

TABLE VI

SELECTIVITY RATIOS[a]

	Na^+/Li^+	K^+/Li^+	Na^+/K^+
Acetonitrile	1.9	1.6	1.2
Propylene carbonate	1.5	1.1	1.4

[a] After McClure and Reddy.[73]

Wilcox[74] studied the selectivity coefficients of the Pyrex glass membrane electrode to lithium, potassium, and silver over sodium in fused ammonium nitrate: sodium ions, 1; silver ion, 0.48; potassium ion, 0.09; and lithium ion, 0.046. In addition, it must be emphasized that in this solvent the mole fraction of the ion to be measured or the product of the appropriate (corresponding) selectivity coefficient and the mole fraction of interfering ion cannot be lower than the constant value (0.064) characteristic of the solvent (i.e., the solvent itself and the impurities).

c. *Response time.* Karlberg[76] carried out a detailed study on the response-time characteristics of some commercial glass electrodes in isopropanol solutions with the aim of clarifying some fundamental questions (e.g., storage, equilibration, and reconditioning) concerning the application of glass electrodes in nonaqueous solvents, as well as the processes in electrode surface layers. For the investigations, hydrated, etched, and dried electrodes were used and it was found that the response time of the electrodes significantly depend on their pretreatment; the dried electrode responded more slowly and the etched electrode more quickly than the hydrated electrode. Besides this, the response time of all electrodes was found to be shorter when going from the acidic to the basic solution. Moreover, it was interesting to learn that the response characteristics of the dried electrodes can be improved if these are immersed in water for only 1 min. Consequently, it was concluded that the response time of an electrode in an appropriate solvent significantly depends on the rate of movement of hydrogen ions in the hydrated gel layer of the appropriate glass electrode, which depends partly on the water content and partly on the thickness of this layer. During the etching treatment the thickness of the gel layer is reduced, while the drying process removes water from the surface layer. On the basis of these experimental results conclusions were drawn for the application of pH-sensitive glass electrodes in nonaqueous solvents. In addition, from the fact that rapid response can be restored by immersing the dried electrodes in water for a short time it may follow that only a very thin region of the hydrated layer is responsible for the response behavior.

McClure and Reddy[73] have investigated the response time of cation-sensitive glass electrodes, which were found in the range of 5–10 sec. These relatively long response time values were supposed to be due to the slow mixing process ensured by the experimental conditions used for the study.

d. *Titration in Nonaqueous Media.* Baumann and Simon[72] carried out the titration of glycerine, 2,6-diternary-butyl-4-methylphenol, and ammonium nitrate, respectively, with potassium amide in liquid ammonia at −34°C. The titrations were performed by adding potassium amide in known amount to the acids, whereupon the excess was back-titrated with ammonium chloride. The end point of the titration was determined potentiometrically using a pH-sensitive glass indicator electrode.

Shiurba and Jolly[71] have described a method for the titration of potassium hydroxide suspensions as well as the mixture of potassium hydroxide and potassium amide in liquid ammonia. In the course of the titration, electrode equilibrium was usually established within 20 sec. This method has been suggested for the determination of the end point of other acid–base titrations in ammonia.

Alessi and co-workers[77] have suggested a new class of "ionic associates" as a primary standard in the field of nonaqueous titrimetry. A comparison has been made of one of these pyridones, bis(1-butyl-2,6-dimethyl-4-pyridone)perchlorate(II), with bis(1,5-dimethyl-2-phenyl-3-pyrazolone)perchlorate(I) developed earlier by Busey et al. [78,79] as a nonaqueous primary standard. Since both compounds are "ionic associates" they can be used for the standardization of acids as well as bases. Moreover, along with antipyrine perchlorate, bis(1-butyl-2,6-dimethyl-4-pyridone)perchlorate is recommended as a nonaqueous primary standard because it can be obtained in high purity; it has a high equivalent weight; it can be used for titration in many nonaqueous solvents such as acetonitrile, acetone, methanol, ethanol, methyl ethyl ketone, dimethylformamide, dimethyl sulfoxide, and pyridine; and it displays a sharp inflection in the equivalence point region as both an acid and a base.

Some additional results concerning nonaqueous acid–base titrations and materials available at present for use in standardizing nonaqueous acids and bases are included in the excellent books on this subject.[80–82]

Akimoto and Hozumi[83,84] have carried out the microtitration of sulfate ion with barium ions at pH 5.5 in nonaqueous solvents using a sodium-selective glass electrode. The determination is based on the transient response of the glass electrode to barium ions, which is developed when the electrode is exposed to a sudden change of barium ion concentration. Rechnitz and Kugler[85–87] found a similar transient response to calcium and strontium ions, for which the glass electrode shows no appreciable selectivity. The phenomenon was explained by Rechnitz et al. through a calcium ion-exchange process occurring at the boundary of a glass electrode, which does not result in a steady-state potential since the mobilities of the calcium ions in the hydrated gel layer are extremely small. Furthermore, the transient signal gradually drops off to the initial potential value as the rate of the ion-exchange process decreases as a function of time.

The titration of sulfate ions has been carried out in various organic solvents such as methanol, isopropanol, and acetone first in 70%(vol) and later in 90%(vol) concentrations; the methanol yielded a smaller potential jump than any of the other solvents. Nevertheless, acetone is still recommended as the best solvent to use because it shows a slightly sharper end point for the titration than either ethanol or isopropanol. It was also stated that the potential jump becomes less evident with decreasing nonaqueous solvent concentration.

2. GRAPHITE ELECTRODES

Berčik[88] first suggested the use of graphite electrodes as indicator electrodes in potentiometric acid–base titrations. In the course of the investigations carried out with wax-impregnated graphite electrodes[89–91] it became clear that the electrode potential change per unit pH can be significantly increased if the electrodes are "activated," i.e., pretreated with a solution of an oxidizing agent. For example, the potential of the wax-impregnated graphite electrode, activated with permanganate dissolved in sulfuric acid, changed linearly with the pH in a certain pH range, i.e., about 70 mV per pH unit.

According to Berčik, in the course of the activation process a quinone–hydroquinone redox system is formed, the redox potential of which is influenced by the pH.

Berčik and Hladky,[92,93] as well as Berčik, Čakrt, and Derzsiova,[94,95] found the wax-impregnated activated graphite electrodes to be suitable for acid–base titrations, both in aqueous and nonaqueous solutions.

Miller[96,97] used pyrolytic graphite, while Doležal and Štulik[98] employed glassy carbon electrodes as indicator electrodes for acid–base titrations. According to their results, the potential of the pyrolytic graphite electrode changed by 50 mV, whereas that of the glassy carbon electrode changed by 20 mV per pH unit. Furthermore, Doležal and Štulik observed that the presence of anionic oxidants increased the magnitude of the potential change of the appropriate graphite electrodes.

In a detailed study Pungor and Szepesváry[99–101] found that activated and nonactivated silicone rubber-based graphite electrodes as well as silicone rubber-based graphite electrodes containing manganese dioxide are also pH sensitive and can be used as indicator electrodes for direct pH measurements in a given range of pH or potentiometric titrations in both aqueous and nonaqueous solvents such as methanol, ethanol, and dimethylformamide. The potential of a nonactivated graphite electrode changes at an average of 30 mV per pH unit, while that of an activated graphite electrode is at 70–90 mV. However, for electrodes containing a mixture of manganese dioxide and graphite, the electrode potential change is 60 mV per pH unit. The potential change of graphite electrodes containing manganese dioxide was also found to be smaller than that of activated graphite electrodes. The potential jump at the equivalence point, however, did not change even in the course of repeated titrations and, due to the rapid potential reset, this type of electrode could be applied to the end-point detection of automatic titrations, too.

In conclusion it can be stated that, in contrast to the pH-sensitive glass electrode, the various types of graphite electrodes do not require any conditioning in the appropriate solvents before measurements and, furthermore, their impedances are significantly lower than those of glass electrodes. Thus, they can be used advantageously as indicator electrodes especially for acid–base titrations in nonaqueous solvents.

3. HALIDE ION-SELECTIVE ELECTRODES

Kazarjan and Pungor[43,102–105] have thoroughly investigated the behavior of silicone–rubber halide ion-selective electrodes in mixed solvents; alcohols, acetone, dimethylformamide, acetonitrile, and a mixture of benzene and methanol were used. In every case the maximum ratio of the nonaqueous constituents was established and listed in Table VII.

It was found experimentally that when the ratio of the appropriate nonaqueous solvent exceeded that shown in Table VII, the silicone rubber

TABLE VII

THE MAXIMUM RATIO OF THE ORGANIC
CONSTITUENT IN MIXED SOLVENTS

Solvent mixture	Ratio of organic solvent v/v(%)
$CH_3OH : H_2O$	90
$C_2H_5OH : H_2O$	90
n-$C_3H_7OH : H_2O$	40
i-$C_3H_7OH : H_2O$	40–50
$Me_2CO : H_2O$	60
$DMF : H_2O$	60

matrix swelled continuously and at the same time, the color of the solvent mixture changed. The color change may indicate that a part of the precipitate in the boundary layer dissolves colloidally. Furthermore, it was also stated that silicone rubber membranes cannot be used in benzene–methanol solvent mixture because this solvent makes the membrane layer fragile.

In the solvent mixtures listed in Table VII the silicone rubber-based halide ion-selective electrodes exhibited a Nernstian response to the primary ion* in the range of 10^{-1}–10^{-5} M concentration. However, the calibration curves of the electrodes are shifted parallel to each other depending on the water content of the solvent mixture. This may be due to the liquid-junction potential.

Besides practical applications, the aim of these investigations was to study the validity of the correlations derived through solubility equilibria for precipitate-based ion-selective electrodes in aqueous media. Accordingly, the solubility products of the silver halide salts were determined in the solvent mixtures studied. On the basis of the solubility products it can be seen clearly that the application of nonaqueous solvents offers possibilities for extending the lower detection limits of the electrodes.

In accordance with this it was found that the lower detection limit of the iodide-selective electrode in 90 : 10 methanol–water mixture is at 10^{-9} M iodide, which can be explained by the decrease in the solubility of the silver iodide precipitate in the boundary layer of the electrode membrane.

The selectivity coefficients of halide ion-selective electrodes to various other ions determined potentiometrically in mixed solvents and calculated from Eq. 25 are compared in Table VIII. These results clearly show that the general theoretical equation derived for the calculation of the selectivity data in aqueous solutions is also valid in mixtures of aqueous and nonaqueous

---

* An ion to which the ion-selective electrode is primarily selective.

TABLE VIII

SELECTIVITY COEFFICIENTS OF AN
IODIDE-SELECTIVE ELECTRODE TO BROMIDE
ION IN MIXED SOLVENTS[a]

Solvent mixture	Ratio of organic solvent v/v (%)	$pK_{I,Br}$	
		Calculated	Experimental
$CH_3OH : H_2O$	10	3.64	3.78
	40	3.48	3.60
	60	3.34	3.38
	90	3.02	3.02
	100	2.93	—
$C_2H_5OH : H_2O$	10	3.62	3.70
	40	3.50	3.55
	60	3.34	3.30
	90	3.00	2.76
	100	—	—
$n\text{-}C_3H_7OH : H_2O$	10	3.70	3.74
	20	3.58	3.54
	30	3.58	3.46
	40	3.58	3.42
$iso\text{-}C_3H_7OH : H_2O$	10	3.74	3.78
	20	3.64	3.60
	30	3.56	3.49
	40	3.50	3.16
$Me_2CO : H_2O$	10	3.82	3.75
	20	3.68	3.60
	30	3.62	3.56
	40	3.56	3.55
$DMF : H_2O$	10	3.74	3.75
	40	3.26	3.20
	60	2.64	2.60

[a] After Kazarjan and Pungor.[43,105]

solvents. Naturally in this case the solubility products determined in the appropriate solvent mixture should be considered.

On the basis of the results of Kazarjan and Pungor it can be concluded that, for the determination of the halide ions in very small concentrations, the application of nonaqueous solvents is favored in analytical investigations. At the same time, the use of aqueous and nonaqueous solvent mixtures is unfavorable when the aim of the study is the determination of halides in the presence of each other because the selectivity coefficients of a halide-ion-selective electrode to another halide ion is higher than in aqueous solution.

More recently Kreskov, Kazarjan, and Syrykh[106,107] have investigated the behavior of a homogeneous chloride-selective electrode in methanol, ethanol, n-propanol, iso-propanol, and dimethylformamide, and in their 10–90% aqueous mixtures. The calibration of electrodes was done in tetraethylammonium chloride solutions. The slope of the electrode calibration curve was found to be 38–50 mV/decade in mixed solvents, whereas for pure organic solvents it was 35–40 mV/decade concentration change. In the study, a calomel electrode containing tetraethylammonium chloride in suitable solvents with a potassium nitrate bridge was used as a reference electrode. The chloride content of samples was determined to 10^{-6}–10^{-8} M with the help of a calibration graph.

The chloride ion-selective electrode was used in acetonitrile by Lemahieu, Lemahieu-Hodé, and Résibois[108] for the potentiometric titration of chloride ion with silver nitrate in the presence of other halides.

The behavior of the homogeneous bromide and iodide ion-selective electrodes was studied by Ficklin and Gotschall[109] in nonaqueous media. The electrode potential response was measured in methanol, ethanol, butanol, pentanol, and hexanol for bromide, and in methanol and ethanol for iodide. From the results it can be seen clearly that more negative potentials are observed in solvents with decreasing dielectric constant, while the electrode response was almost a Nernstian response in all solvents studied.

The organically bonded bromine content of pharmaceuticals such as bromocamphor, 4-bromoacetophenone, 5-bromosalicylic acid, and 4-methoxy-3,5-dibromobenzoyl chloride were determined by Gyenge and Lipták[110] with argentometric titration after decomposition by metal sodium in iso-butanol. The relative error of the determination was 1%.

Szarvas and co-workers[111] used thiocyanate-selective membrane electrodes for the determination of thiocyanate ion concentrations to 10^{-5} M concentrations in aqueous and nonaqueous solvent mixtures. Nonaqueous solvents included dioxane, ethanol, dimethylformamide, and dimethyl sulfoxide. In agreement with the results of Pungor and Kazarjan,[43] they also found that an increase in the concentration of the organic component in the solvent mixture increased the sensitivity of the electrode but decreased its selectivity.

Lingane[112] has found that the fluoride ion-selective electrode also exhibits a Nernstian response to fluoride ions in 60% aqueous ethanol at ionic strengths of 0.1 and 0.01 M. The detection limit of the electrode has also been the same as in aqueous solutions; however, the electrode potential values were approximately 100 mV more negative. The latter was supposed to be attributed to either a liquid junction or an asymmetry potential effect or to both.

Lithium has been determined by potentiometric titration with fluoride using a fluoride ion-selective electrode in a solvent containing 95% ethanol and 5% water.[112] Lithium fluoride was used as the titrant.

The lanthanum fluoride-based fluoride ion-selective electrode has also been used in an aprotic organic solvent, dimethyl sulfoxide.[113] The organic solvent appeared to cause no visible harm to the sensing mechanism or the electrode body. Attempts have been made to potentiometrically determine tetraethylammonium fluoride with lithium chloride in dimethyl sulfoxide using a fluoride-sensitive electrode. Unfortunately, no reproducible quantitative results could be obtained.

The microdetermination of fluorine in some inorganic and organic compounds[114] has been carried out by potentiometric titration in 50%(vol) dioxane. The titration was carried out at pH 5–6 with thorium nitrate using a fluoride ion-selective electrode. Prior to the titration, the organic bound fluorine compounds were decomposed by combustion. The simultaneous determination of fluoride and chloride has also been solved with a two-step potentiometric titration. In this study the fluoride ion-selective electrode exhibited an almost Nernstian response for 10^{-1}–10^{-6} M fluoride.

An indirect potentiometric method[115] has been reported for the determination of the fluoride complex-forming ions in 30–70% ethanol-containing aqueous solution with the help of a fluoride ion-selective electrode. The method is based on the measurement of the decrease in fluoride ion concentration caused by complex formation.

The chloride as well as the cadmium ion-selective electrode has been employed[116] for the study of cadmium(II) and zinc(II) chloro complexes in water–hexamethylphosphotriamide solution. In the course of the investigations it was found that both ion-selective electrodes can be used for the determination of β_n values.

The rapid determination of sulfite[117] was carried out by the argentometric titration of iodide released in the interreaction of sulfite and iodine. A mixture of methanol and water was used as solvent. The interfering effect of bromide, chloride, carbonate, and other sulfur-containing ions has been investigated. The method was applied to the determination of sulfur dioxide in chimney gases.

4. HEAVY METAL ION-SELECTIVE ELECTRODES

Selig and Salamon[118] suggested the lead ion-selective electrode as an indicator electrode for the potentiometric titration of sulfate and oxalate ions with lead ions in 50% methanol–water mixtures. They also investigated p-dioxane which was suggested earlier by Ross and Frant[119] as a solvent for the determination of sulfate ions. In the course of their investigations it was found that even a small amount of peroxide present in dioxane poisons the lead ion-selective electrode and therefore the application of methanol is favored.

Rechnitz and Kenny[120] carried out complexometric titrations of copper in methanol, acetone, and acetonitrile using a homogeneous copper ion-selective indicator electrode. Ethylenediamine, tetraethylenepentamine, and 5,6-dimethyl-1,10-phenanthroline were used as titrants. On the basis of the results it can be stated that the copper ion-selective electrode can be used advantageously for complexometric titrations carried out in nonaqueous media.

REFERENCES

1. R. P. Buck, *Anal. Chem.* **44**, 270R (1972).
2. R. P. Buck, *Anal. Chem.* **46**, 28R (1974).
3. E. Pungor, *Anal. Chem.* **39**, 28A (1967).
4. E. Pungor and K. Tóth, *Analyst* **95**, 625 (1970).
5. J. W. Ross, Jr., *Natl. Bur. Stand. (U.S.), Spec. Publ.* **314** (1969).
6. A. K. Covington, *Crit. Rev. Anal. Chem.* **3**, 355 (1974).
7. F. Conti and G. Eisenman, *Biophys. J.* **5**, 247 (1965).
8. J. Sandblom, G. Eisenman, and J. L. Walker, Jr., *J. Phys. Chem.* **71**, 3862 (1967).
9. J. W. Ross, Jr., *Science* **156**, 1378 (1967).
10. R. Bloch, A. Shatkay, and A. Saroff, *Biophys. J.* **7**, 865 (1967).
11. G. J. Moody, R. B. Oke, and J. D. R. Thomas, *Analyst* **95**, 910 (1970).
12. L. A. R. Pioda, N. Stankova, and W. Simon, *Anal. Lett.* **2**, 665 (1965).
13. W. Simon, W. E. Morf, and P. C. Meier, *Struct. Bonding (Berlin)* **16**, (1973).
14. J. Růžička and E. H. Hansen, *Anal. Chim. Acta* **69**, 129 (1974).
15. U. Fiedler, E. H. Hansen, and J. Růžička, *Anal. Chim. Acta* **74**, 423 (1975).
16. E. H. Hansen, J. Růžička, and N. R. Larsen, *Anal. Chim. Acta* **79**, 1 (1975).
17. G. Nagy and E. Pungor, *Hung. Sci. Instrum.* **32**, 1 (1975); G. G. Guilbault and J. G. Montalvo, *Anal. Lett.* **2**, 283 (1969).
18. K. J. Vetter, "Electrochemical Kinetics," pp. 10–73. Academic Press, New York, 1967.
19. Z. Boksay and B. Csákváry, *Acta Chim. Acad. Sci. Hung.* **67**, 157 (1971).
20. H. R. Wuhrmann, Theory of ion-selective sensors. Diss. No. 4805. Eidgenössische Technische Hochschule, Zürich (1972).
21. K. Tóth, Candidate's Thesis, University for Chemical Engineering, Veszprém (1969).
22. E. Pungor and K. Tóth, *Anal. Chim. Acta* **47**, 291 (1969).
23. H. R. Wuhrmann, W. E. Morf, and W. Simon, *Helv. Chim. Acta* **56**, 1011 (1973).
24. G. Eisenman, *Anal. Chem.* **40**, 310 (1968); J. Sandblom, *J. Phys. Chem.* **73**, 249 (1969).
25. S. Ciani, G. Eisenman, and G. Szabó, *J. Membr. Biol.* **1**, 1 (1969); G. Eisenman, S. Ciani, and G. Szabó, *ibid.* p. 294; G. Szabó, G. Eisenman, and S. Ciani, *ibid.* p. 346.
26. B. Gutbezahl and E. Grunwald, *J. Am. Chem. Soc.* **75**, 565 (1953).
27. R. G. Bates and E. A. Guggenheim, *Pure Appl. Chem.* **1**, 163 (1960).
28. R. G. Bates, "Determination of pH." Wiley, New York, 1964.
29. R. G. Bates and M. Alfenaar, *Natl. Bur. Stand. (U.S.), Spec. Publ.* **314** (1969).
30. R. G. Bates, B. R. Staples, and R. A. Robinson, *Anal. Chem.* **42**, 867 (1970).
31. J. Havas, M. Kaszás, and M. Varsányi, *Hung. Sci. Instrum.* **25**, 23 (1972).
32. E. H. Hansen, C. G. Lamm, and J. Růžička, *Anal. Chim. Acta* **59**, 403 (1972); J. Růžička and E. H. Hansen, *ibid.* **63**, 115 (1973); E. H. Hansen and J. Růžička, *ibid.* **72**, 365 (1974).
33. A. G. Mitchell and W. F. K. Wynne-Jones, *Trans. Faraday Soc.* **51**, 1690 (1955).
34. P. K. Glasoe and F. R. Long, *J. Phys. Chem.* **64**, 188 (1960).
35. K. Mikkulsen and S. O. Nielsen, *J. Phys. Chem.* **64**, 632 (1960).
36. R. Gary, R. G. Bates, and R. A. Robinson, *J. Phys. Chem.* **68**, 3806 (1964).

37. I. M. Kolthoff and S. Bruckenstein, *J. Am. Chem. Soc.* **78**, 1 (1956).
38. S. Bruckenstein and I. M. Kolthoff, *J. Am. Chem. Soc.* **78**, 2974 (1956); **79**, 5915 (1957).
39. C. L. De Ligny and P. F. M. Luykx, *Recl. Trav. Chim. Pays-Bas* **77**, 154 (1958).
40. C. L. De Ligny and A. A. Wieneke, *Recl. Trav. Chim. Pays-Bas* **79**, 268 (1960).
41. C. L. De Ligny, P. F. M. Luykx, M. Rehbach, and A. A. Wieneke, *Recl. Trav. Chim. Pays-Bas* **79**, 713 (1960).
42. R. Gary, R. G. Bates, and R. A. Robinson, *J. Phys. Chem.* **68**, 1186 (1964).
43. N. A. Kazarjan and E. Pungor, *Anal. Chim. Acta* **51**, 213 (1970).
44. D. J. G. Ives and G. J. Janz, "Reference Electrodes." Academic Press, New York, 1961.
45. C. K. Mann and K. K. Barnes, "Electrochemical Reactions in Nonaqueous Systems," p. 20. Dekker, New York, 1970.
46. N. A. Izmailov and M. A. Belgova, *Zh. Obshch. Khim.* **8**, 1873 (1938).
47. N. A. Izmailov and T. F. Frantsevich-Zabludovskaya, *Zh. Obshch. Khim.* **15**, 283 (1945).
48. N. A. Izmailov and A. M. Alexandrova, *Zh. Obshch. Khim.* **19**, 1404 (1949).
49. N. A. Izmailov and A. M. Alexandrova, *Zh. Obshch.* **20**, 2127 (1950).
50. G. Mattock, "pH Measurement and Titration." Heywood, London, 1961.
51. A. G. Mitshell and W. F. K. Wynne-Jones, *Trans. Faraday Soc.* **51**, 1690 (1955).
52. N. A. Izmailov, "Electrochemistry of solvents." Publication of the Kharkóv University, Kharkóv, 1959 (in Russian).
53. A. M. Shkodin, N. A. Izmailov, and N. P. Dzyuba, *Zh. Obshch. Khim.* **20**, 1999 (1950).
54. A. M. Shkodin, N. A. Izmailov, and N. P. Dzyuba, *Zh. Obshch. Khim.* **23**, 27 (1953).
55. P. Tutundzic and P. Putanov, *Glas. Hem. Drus., Beograd* **20**, 157 (1955).
56. F. N. Kozlenko, *Zh. Fiz. Khim.* **33**, 1866 (1959).
57. F. N. Kozlenko and S. P. Miskidzhyan, *Zh. Fiz. Khim.* **35**, 26 (1961).
58. J. C. Sullivan, J. C. Hindman, and A. J. Zielen, *J. Am. Chem. Soc.* **83**, 3373 (1961).
59. A. L. Bacarella, E. Grunwald, H. P. Marchall, and E. L. Purlee, *J. Phys. Chem.* **62**, 856 (1958).
60. E. Grunwald and B. Berkovitz, *J. Am. Chem. Soc.* **73**, 4939 (1951).
61. B. Gutbezahl and E. Grunwald, *J. Am. Chem. Soc.* **75**, 559 (1953).
62. E. L. Purlee and E. Grunwald, *J. Am. Chem. Soc.* **79**, 1366 (1957).
63. E. Warburg, *Ann. Phys. (Leipzig)* [1] **21**, 622 (1884).
64. A. G. Mitshell and W. F. K. Wynne-Jones, *Trans. Faraday Soc.* **51**, 1690 (1955).
65. J. R. Koloczynski, E. M. Roth, and E. S. Shanley, *J. Am. Chem. Soc.* **79**, 531 (1957).
66. P. K. Glasoe and F. R. Long, *J. Phys. Chem.* **64**, 188 (1960).
67. K. Mikkulsen and S. O. Nielson, *J. Phys. Chem.* **64**, 632 (1960).
68. G. A. Rechnitz and S. B. Zamochnick, *Talanta* **11**, 979 (1964).
69. G. Eisenman, *Adv. Anal. Chem. Instrum.* **4**, 215 (1965).
70. R. D. Lanier, *J. Phys. Chem.* **69**, 2697 (1965).
71. R. A. Shiurba and W. L. Jolly, *J. Am. Chem. Soc.* **90**, 5289 (1968).
72. W. M. Baumann and W. Simon, *Z. Anal. Chem.* **216**, 273 (1966).
73. J. E. McClure and T. B. Reddy, *Anal. Chem.* **40**, 2064 (1968).
74. F. Wilcox, Sr., *J. Chem. Educ.* **52**, 123 (1975).
75. K. Notz and A. G. Keenan, *J. Phys. Chem.* **70**, 662 (1966).
76. B. Karlberg, *Anal. Chim. Acta* **66**, 93 (1973).
77. J. T. Alessi, D. G. Bush, and J. A. Van Allan, *Anal. Chem.* **46**, 443 (1974).
78. A. I. Busev, B. E. Zaitsen, V. K. Akisnov, Ya. Chelikhovskii, and F. Kopetski, *Zh. Obshch. Khim.* **38**, 534 (1968).
79. A. I. Busev, V. K. Akimov, and I. A. Emelyanova, *Zh. Anal. Khim.* **23**, 616 (1968).
80. W. Huber, "Titrations in Nonaqueous Solvents." Academic Press, New York, 1967.
81. J. Kucharsky and L. Safarik, "Titrations in Nonaqueous Solvents." Am. Elsevier, New York, 1965.

82. I. Gyenes, "Titration in Nonaqueous Media." Akadémia Press, Budapest, Iliffe Books, Ltd., London and D. Van Nostrand Company Inc., New Jersey, 1967.
83. K. Hozumi and N. Akimoto, *Anal. Chem.* **42**, 1312 (1970).
84. N. Akimoto and K. Hozumi, *Anal. Chem.* **46**, 776 (1974).
85. G. Eisenman, "Glass Electrodes for Hydrogen and Other Cations." Dekker, New York, 1967.
86. G. A. Rechnitz and G. C. Kugler, *Anal. Chem.* **39**, 1682 (1967).
87. G. A. Rechnitz, *Natl. Bur. Stand. (U.S.), Spec. Publ.* **314** (1969).
88. J. Berčik, *Chem. Zvesti* **14**, 372 (1970).
89. S. S. Lord and B. Rogers, *Anal. Chem.* **26**, 284 (1954).
90. V. G. Gayor, A. L. Conrad, and I. H. Landerl, *Anal. Chem.* **29**, 224 (1957).
91. V. G. Gayor, A. L. Conrad, and I. H. Landerl, *Anal. Chem.* **29**, 228 (1957).
92. J. Berčik and Z. Hladky, *Proc. Conf. Appl. Phys.-Chem. Methods Chem. Anal., 1966* p. 99 (1966).
93. J. Berčik and Z. Hladky, *Chem. Zvesti* **22**, 768 (1968).
94. J. Berčik and M. Čakrt, *Chem. Zvesti* **22**, 755 (1968).
95. J. Berčik, M. Čakrt, and K. Derzsiova, *Chem. Zvesti* **22**, 761 (1968).
96. F. J. Miller, *Anal. Chem.* **35**, 929 (1963).
97. F. J. Miller and H. E. Zittel, *Anal. Chem.* **35**, 1866 (1963).
98. J. Doležal and K. Štulik, *J. Electroanal. Chem.* **17**, 87 (1968).
99. E. Pungor and É. Szepesváry, *Anal. Chim. Acta* **43**, 289 (1968).
100. É. Szepesváry and E. Pungor, *Anal. Chim. Acta* **54**, 199 (1971).
101. E. Pungor and É. Szepesváry, *Period. Polytech., Chem. Eng.* **16**, No. 4, 323 (1973).
102. N. A. Kazarjan and E. Pungor, *Acta Chim. Acad. Sci. Hung.* **66**, 183 (1970).
103. N. A. Kazarjan and E. Pungor, *Magy. Kem. Foly.* **77**, 186 (1971).
104. N. A. Kazarjan and E. Pungor, *Acta Chim. Acad. Sci. Hung.* **76**, 339 (1973).
105. N. A. Kazarjan and E. Pungor, *Anal. Chim. Acta* **60**, 193 (1972).
106. N. A. Kazarjan, A. P. Kreshkov, and T. M. Syrykh, *Zh. Fiz. Khim.* **47**, 2590 (1973).
107. A. P. Kreshkov, N. A. Kazarjan, and T. M. Syrykh, *Zh. Anal. Khim.* **29**, 1025 (1974).
108. G. Lemachieu, C. Lemahieu-Hodé, and B. Résibois, *Analusis* **1**, 110 (1972).
109. W. H. Ficklin and W. C. Gotschall, *Anal. Lett.* **6**, 217 (1973).
110. R. Gyenge and J. Lipták, *Prof. Conf. Appl. Phys. Chem., 2nd, 1971*, Vol. 1, p. 639 (1971).
111. P. Szarvas, I. Korondán, and M. Szabó, *Magy. Kem. Foly.* **80**, 207 (1974).
112. J. J. Lingane, *Anal. Chem.* **40**, 935 (1968).
113. J. N. Butler, *Natl. Bur. Stand. (U.S.), Spec. Publ.* **314** (1969).
114. S. S. M. Hassan, *Microchim. Acta* No. 5, p. 889 (1974).
115. G. Muto, Y. K. Lee, K. J. Whang, and K. Nozaki, *Bunseki Kagaku* **20**, 1271 (1971).
116. A. Burdin, J. Mesplede, and M. Porthault, *C.R. Hebd. Seances Acad. Sci., Ser. C* **276**, 173 (1973).
117. S. Ikeda, J. Hirata, and H. Satake, *Nippon Kagaku Kaishi* **8**, 1473 (1973).
118. W. Selig and A. Salamon, *Microchim. Acta* No. 4, p. 663 (1974).
119. J. W. Ross, Jr., and M. S. Frant, *Anal. Chem.* **41**, 967 (1969).
120. G. A. Rechnitz and N. C. Kenny, *Anal. Lett.* **2**, 395 (1969).

⁓ 5 ⁓

Pyridine as a Nonaqueous Solvent

∽

JEAN-MAXIME NIGRETTO AND MARCEL JOZEFOWICZ

Department of Chemistry, University of Paris
Villetaneuse, France

I. Introduction

Of all the organic solvents, pyridine ranks among the most important, judging by the volume of literature devoted to it. The reasons for this interest are linked with its remarkable physical and chemical properties. (1) The polarity of the constituent molecule promotes the solubility of a large number of compounds, in spite of a rather low dielectric constant. (2) Pyridine is generally considered to be the only truly aprotic organic solvent because of its stable aromatic character and the high resonance energy of the heterocycle. From this point of view, pyridine offers an appreciable advantage over protogenic solvents by affording a wider domain of stability and hence by reducing the likelihood of chemical interference. In addition, since the proton availability can be kept at very low levels, organic reactions involving the exchange of electrons can be studied with little or no complication due to protonation. (3) The presence of the heteroatom fused into the conjugated ring system makes pyridine very important as a basic solvent catalyst, especially for nucleophilic sustitution reactions.

In view of the importance of its role in preparative chemistry, many electrochemists, apparently not discouraged by the awful odor of pyridine, have attempted to determine the quantitative features of this solvent. Although it has not yet acquired as much notoriety as other solvents, such as acetonitrile or dimethylformamide, pyridine should now gain popularity in the field of analytical application.

II. Purification of Pyridine

In order to compare and standardize the data available in the literature, we must first define the experimental conditions and, in particular, the purity of the solvent. However, the distinction between a pure solvent and a solvent mixture cannot always be clearly made. Since substantial quantities are often required, a pure solvent is actually an abstract notion that experiments cannot afford. The criterion of purity therefore has to be extended to a solvent in which any remaining impurities do not interfere with measurement.

Industrially, pyridine is isolated by the destructive distillation of coal, from petroleum or from so-called " bone oil." Commercially, it is available in several grades of various qualities. Because of the variety of sources of pyridine and of the solvent brands, there is no single mode of purification. In addition, it should be noted that some experiments can afford the presence of certain contaminants (e.g., nonelectroactive compounds in electrochemical studies) since others may be of greater inconvenience. However it is obtained, a further purification of the solvent is necessary to ensure its quality before use. Most methods associate a general preparation of the solvent (distillation, fractional freezing, etc.) with a more particular technique for removing specific impurities, e.g., water or compounds with similar boiling points.

Coulson et al.[1] reported that the purity of the best grades of "pyridine A.R." samples is above 99%. In attempting to improve on this figure, the authors found that even repeated distillations in a high-efficiency column did not suffice to remove small amounts of impurities, although pyridine has no homologs of similar boiling point. However, pyrrole derivatives, sulfur compounds (thiophenes), hydrocarbons (xylenes), phenols,[1] or weak acids[2] may nevertheless be present.

Repeated fractional freezing techniques lead to good results for electro-chemical purposes since unidentified impurities (characterized by the prod-ucts of reaction with the diphenylpicrylhydrazy radical) were able to be eliminated.[2] Other processes involving the chemical isolation of pyridine have been used: after crystallization of the pyridinium perchlorate,[3,4] oxalate,[5] or zinc chloride–pyridine complex,[6,7] followed by a careful purification of the salt, the base was liberated.

To remove particular contaminants, quite a few procedures of refluxing pyridine over various reagents have been reported. These consisted of treat-ing even crude pyridine with cerium sulfate,[8,9] barium oxide,[10-12] cadmium perchlorate,[13,14] potassium permanganate,[15,16] mostly in the presence of potassium hydroxide or, as regards sulfur-containing impurities, sulfuric acid.[1] In some cases, purity was established by densitometric,[11,16] electrical conductance,[17-19] polarographic,[2] and other measurement techniques.[1,9,20]

Generally, owing to its marked hygroscopic character, pyridine can be specially treated to eliminate water and carbon dioxide, usually in an atmosphere-tight device.[18] Drying reagents such as alumina oxide,[7] silica gel,[21] 3Å or 4Å Linde molecular sieves,[20,22-24] have been conveniently used and their efficiency tested. This operation may be effectively completed if combined with the distillation of water as a component either of the ternary azeotropes pyridine–water–benzene (bp 69.3°C[15]) or pyridine–water–toluene (bp 84.1°C[25]). Hence, the water content, estimated by using the Karl Fischer procedure, may be reduced to less than 0.01%.

III. PHYSICAL PROPERTIES OF PYRIDINE

It is clear that observations or measurements made on impure materials may induce discrepancies or erroneous statements. Ascertainment of the final quality and even quantitative estimation of residual impurities are crucial and must therefore be carried out before use of the solvent. The most important physical properties of pyridine are listed in Table I.[1,17,18,19,26–32]

TABLE I

PHYSICAL PROPERTIES OF PYRIDINE

Property	Value	Reference
Melting point	−41.55°C	1
Boiling point (760 mm Hg)	115.256°C	1
Viscosity		
(0°C)	13.60 mP	26
(20°C)	9.58 mP	26
(25°C)	8.882 mP	14
(30°C)	8.29 mP	26
Density		
(20°C)	0.98310 ± 0.00000	1
(25°C)	0.97806	1
	0.97801 ± 0.00003	18
(30°C)	0.97301 ± 0.00000	1
Vapor density	2.73	27
Refractive index		
n_d^{20}	1.51020	1
n_d^{25}	1.50763 ± 0.00005	18
Specific conductance (25°C)	$3.10^{-10} \ \Omega^{-1} \ cm^{-1}$	19
Heat of fusion	22.4 cal/g	1
Heat of vaporization (760 mm Hg)	109.95 ± 0.04 cal/g	1
Heat of combustion (25°C)	−664.58 kcal/mole	28
Vapor pressures		
115.287°C	760.58 mm Hg	29
107.169°C	597.64 mm Hg	29
100.994°C	493.23 mm Hg	29
82.728°C	267.00 mm Hg	29
Cryoscopic constant	4.75°C	1
Dielectric constant (25°C)	12.3	30
	12.01	31
Dipole moment	2.23 D	32

At room temperature, pyridine is a colorless and slightly oily liquid. It has a characteristic, unpleasant, and penetrating odor, in spite of its relatively low vapor pressure. Its density and viscosity are similar to those of water. Due to the presence of the electronegative atom in the molecule, pyridine behaves quite differently from benzene, a parent species of the same molecular size and weight. Its appreciable intrinsic polarity is responsible for its lower melting point and higher boiling point; thus, pyridine covers a fairly large liquid range. Pyridine is a good solvent for a relatively large number of organic and inorganic substances. Its solvation property is principally related to the electron-donating ability of the heteroatom in such a way that solute–solvent interactions can be described on the basis of its strong Lewis base character. This also supports the hygroscopic nature of pyridine and its complete solubility in water,[33,34] thus enhancing the difficulty of preparing anhydrous solvent samples.

The presence of impurities, e.g., residual water, has a marked influence on the specific electrical conductance of the solvent. In the investigation of Kortüm and Wilski,[17] this was found to be lowered to the order of 10^{-10} ohm^{-1} cm^{-1}, making solvent correction unnecessary even in the most dilute solutions. The heat of combustion and other precise thermodynamic data are available in the work of Cox, Challoner, and Meetham.[28]

The ultraviolet absorption spectra of pyridine in the vapor and liquid phases have frequently been described[35] and can be found in general reviews.[36] The data available on pyridine and the accurate investigations of the closely related benzene spectra make possible a satisfactory analysis of the pyridine bands.[37] The spectra in heptane solutions have been reported by Spiers and Wibaut,[38] and those in aqueous solutions by Hughes, Jellinek, and Ambrose.[39] Authorative data obtained from highly purified samples under different conditions have been discussed by Herington from the point of view of chemical analysis.[40]

Bellamy has reviewed the infrared data of pyridine and its homologs.[41] The C–H stretching frequencies are located in the 3020–3070 cm^{-1} region and the corresponding C=C and C=N bands range from 1590 to 1660 cm^{-1}. Vibrational bonds of C–H occur in the 1000–1100 and 650–900 cm^{-1} region. Out-of-plane C–H deformation bands appear at 750 cm^{-1}. Only rare studies are available concerning the Raman spectra of pyridine.[37,42]

Toxicologically speaking, pyridine is known to exert action on the nervous system, causing neurodigestive troubles.[43] The upper tolerance limit in air was fixed at 5 ppm (15 mg/m^3) by American hygienists in 1969.

IV. PYRIDINE AS A NONAQUEOUS SOLVENT FOR ANALYTICAL STUDIES

In recent years, interest in the behavior of nonaqueous media for analytical purposes has greatly increased as a result of practical and theoretical demand. From this point of view, pyridine has guaranteed fruitful investigation right from the early stages. Owing to its chemical structure and physical properties, pyridine shows evidence of providing a fairly wide domain for analytical applications. The proton availability is very low so that pyridine is usually classified as a non-proton-releasing solvent. It should, however, be noted that pyridine may not be considered as inert, for example, as benzene, since in some cases, reactions with more or less reactive compounds may occur. Thus, its chemical stability, postulated as the ease with which different reagents will react with it, defines the absolute limitations of its range of practical use.

The following sections aim at successively describing the chemical behavior of pyridine with respect to the conditions and to the reagents commonly used or involved in analytical studies and to the working conditions required for these studies.

A. Solvent Stability Range

In relation to chemical oxidation one would expect pyridine to remain less reactive relative to benzene. This fact is not surprising for it carries an electron-withdrawing atom which tends to diminish the electron density across the ring. Accordingly, reducing agents will have a relatively higher readiness to react since they are electron donors. A detailed review dealing with these properties is available.[36]

1. REDUCTION OF PYRIDINE

Depending upon the strength of the reductive attack, ring opening reactions may or may not occur. Usually, the latter is true as long as N-substitution is not concerned.

The stability of pyridine toward the alkali earth metals has often been tested in the course of attempts to plate out these metals from their salts in that solvent. Since 1897, deposition of lithium,[44-47] sodium,[44] and potassium[47,48] were reported. Similar attempts were made with alkali earth metals by von Hevesy.[46] Most of these authors claimed to have succeeded in their attempts. In fact, a chemical reaction following the electron transfer and proceeding relatively slowly at room temperature has been observed by Emmert for metals of the alkali series.[49] On heating, this reaction yields a mixture of tetrahydropyridine, piperidine, and bipyridyls,[36,50,51] the relative amounts of which depend on the proton availability in the medium. The reduction is generally believed to occur via the successive addition of two metal atoms onto the ring.

Partial reduction of the heterocyclic ring also takes place with reactive Grignard reagents, such as lithium compounds. At room temperature, lithium hydride in ethereal solution reacts in a self-sustained, exothermic reaction with dissolution. Unstable dihydropyridine complexes are formed,[52] which have been conveniently used as selective reducing agents for electrophilic carbonyl compounds.[53,54] The dihydropyridine structure (I) can be assigned on the basis of general nucleophilic addition schemes of organometallics to N-heteroaromatics[52,55].

$$\text{(pyridine)} + \text{LiAlH}_4 \longrightarrow \text{Li}^+ \left[\text{(dihydropyridine)} \begin{matrix} H \\ H \end{matrix} \right]^- \tag{1}$$

AlH₃

(I)

Under normal conditions, the electrochemical reduction of pyridine has been described mostly in proton-rich media. It generally does not produce a ring cleavage of the pyridinium species. In aqueous concentrated solutions of sulfuric acid, the macroscale reduction of pyridine is well known. It leads to a mixture of piperidine, tetrahydropyridine,[56] and α,α'- and γ,γ'-dipiperidyls,[57] i.e., to more or less hydrogenated pyridine species. The mechanistic aspects of this reduction have been made clear by several polarographic investigations in both buffered and unbuffered solutions.[58] Two different steps are involved, the reduction of the pyridinium ion itself occurring after the catalytic evolution of hydrogen.[59,60] In much more concentrated solutions of pyridine, the first step disappears.[61,62]

Recently, Elving reviewed this topic and extended the study to mixtures of gradually concentrated water–pyridine solutions ranging up to the anhydrous solvent.[63,64] Macroscale electrolysis and coulometry indicated a one-electron reduction of the pyridinium ion. The transient formation of a colored free radical was observed visually. On the basis of these results, the following reaction scheme was suggested to hold for the formation of tetrahydrobipyridyls(II and III) after dimerization of a free radical. The abbreviation Py designates pyridine.

$$\text{PyH}^+ + e^- \rightarrow \text{PyH}^0 \tag{2}$$

$$2\text{PyH}^0 \rightarrow \text{(II), (III), etc.} \tag{3}$$

(II) (III)

Thus it appears that the N-substitution by a proton is apparently not sufficient to produce ring fission by electroreduction. However, such an eventuality must not be discarded if pyridinium compounds resulting from the addition of stronger electron acceptors are formed: such electron acceptor groups, R, capable of depleting the electron density of the ring, will facilitate further attack by a base, B^-, according to the general process:

$$\text{(structure)} + B^- \longrightarrow \text{(structure)} \qquad (4)$$

This type of reaction is the primary step of the Zincke type synthesis of dyes.[65,66] A wide range of variations exists, depending upon the conditions employed and upon the chemical nature of R and B^-. Elving reported on the chemical reduction of pyridine as a solvent using lithium aluminum hydride.[67] Spectroscopic measurements, analytical data, and subsequent organic analysis supported evidence for the formation of radical anions prior to dimerization or polymerization reactions on exposure to air and water. After alkaline hydrolysis, polyamide and polyurethane unsaturated chains were identified, resulting from ring opening of intermediately formed lactams. Other suitable experiments, in the presence of aluminum trichloride as electron acceptor, showed that analogous reaction products are yielded when the corresponding pyridinium compound is electrochemically reduced.

2. OXIDATION OF PYRIDINE

Electrolysis experiments have been carried out in pyridine as a solvent, at potentials where oxidative decomposition takes place. On pyrolytic graphite electrodes, formation of N-pyridylpyridinium species **(IV)** was suggested.[68] Analogous intermediates were characterized during chemical oxidation under somewhat different conditions when pyridine was treated with persulfate salts,[69,70] according to the reaction:

$$S_2O_8^{2-} + \text{(structure)} \longrightarrow \text{(structure)} + 2\,SO_4^{2-} + H^+ \qquad (5)$$

(IV)

It has also been noticed that pyridine may undergo some electrophilic ring substitution, even at room temperature.[71] Since this type of reaction was induced by halogens, it will be discussed later on in Section VIII,B.

B. Practical Working Conditions

Experimental conditions which involved electrochemical methods using pyridine are discussed in the following sections.

1. SOLVENT RANGE OF ANALYTICAL APPLICATION

Practically speaking, it is of interest to know the width of the domain of electrochemical application in pyridine, in connection with the nature of supporting electrolyte and of electrode material. Voltammetric and polarographic data are collected in Table II.[9,23,72][74] They show that the absolute anodic limit in pyridine can be readily reached, provided that anions of sufficient resistance toward oxidation, e.g., perchlorates, are used. Alternatively, the potential where cathodic decomposition of the solvent occurs has not yet been estimated with electrochemical methods because of the acid–base properties of the cations.

2. SUPPORTING ELECTROLYTES

Since the apparent ionic radii of alkali metal ions are smaller than those of tetraalkylammonium cations, ion pairing is less likely to occur with the latter.[75] In addition, it is known that their acid–base properties are rather poor[9] and that they do not alter the proper functioning of glass membranes if glass electrodes are used.[76] For these reasons, the choice of tetraalkylammonium salts should therefore be recommended as the more suitable for analytical purposes. Results presented in Table II show that the corresponding perchlorates give the most extended span of available potentials in pyridine.

3. REFERENCE ELECTRODES

The hydrogen electrode is universally adopted as the primary standard with which all other electrodes are compared.[77] Yet, in the past, the use of this electrode appeared less convenient for frequent measurements.[78] Recent papers, however, have displayed the possibility of setting up this electrode in pyridine solutions. Since its efficiency decreased rapidly with time, it was found essential to use freshly platinized electrodes. After this treatment, rotating hydrogen electrodes manifested a quite reversible behavior, as shown by the features of voltammograms,[9] although some failures have been mentioned in the course of studies involving mixtures of concentrated organic acids.[79] The potentials of various reference electrodes were determined against the potential of the normal hydrogen electrode (NHE)

TABLE II

POLAROGRAPHIC AND VOLTAMMETRIC AVAILABLE SPAN OF POTENTIALS IN PYRIDINE RELATED TO THE USE OF VARIOUS BACKGROUND ELECTROLYTES AND TO THE NATURE OF ELECTRODE MATERIALS[a]

Electrolyte (0.1 M)	Mercury (dme) Cathodic	Pyrolytic graphite[c] Anodic	Bright platinum[d] Cathodic	Anodic	Bright silver[d] Cathodic	Anodic	Platinized platinum[d] Cathodic	Anodic	Limiting reaction Reduction	Oxidation
NBu_4ClO_4	—	+1.3	−2.15	+1.25	−2.10	+0.20	−2.10	+1.20	NBu_4^+, H_2O[d]	Pyridine[c]
NBu_4Pic	—	+1.3	—	—	—	—	—	—	NBu_4^+, H_2O[d]	Pic[c]
NBu_4Cl	—	~ +0.8	—	—	—	—	—	—	NBu_4^+, H_2O[d]	Cl^-[c]
NBu_4Br	—	+0.5	—	—	—	—	—	—	NBu_4^+, H_2O[d]	Br^-[c]
NBu_4I	−2.3[f]	+0.2	−2.15	+0.20	−2.10	+0.20	−2.10	+0.20	NBu_4^+, H_2O[d]	I^-[c]
NEt_4ClO_4	−2.4[b]	+1.3	−2.05	+1.25	−2.10	+0.20	−2.05	+1.20	NEt_4^+, H_2O[d]	Pyridine[c]
NEt_4I	—	—	−2.05	+0.20	−2.10	+0.20	−2.05	+0.20	NEt_4^+, H_2O[d]	I^-[c]
$LiClO_4$	−2.0[b]	+1.3	−3.55	+1.25	−3.55	+0.20	−1.45	+1.20	Li^+, H_2O[e]	Pyridine[c]
$LiNO_3$	—	—	−3.55	+1.25	−3.55	+0.20	−1.45	+1.20	Li^+, H_2O[e]	Pyridine[c]
$LiCl$	−1.95[f]	—	−3.55	+1.10	−3.55	+0.20	−1.45	+0.90	Li^+, H_2O[e]	Pyridine[c]
$NaClO_4$, H_2O	—	—	−1.65	+1.25	−1.65	+0.20	−1.30	+1.20	Na^+, H_2O[e]	Pyridine[c]
$NaB(C_6H_6)_4$	—	—	−1.75	+0.75	—	—	−1.50	+0.55	Na^+, H_2O[e]	$B(C_6H_6)_4^-$[d]

[a] Underlined figures correspond to cases where a maximum appears. All potentials are referred vs. the normal silver reference electrode in pyridine.[23] From Nigretto and Jozefowicz[9] (with permission).
[b] Ref. 72.
[c] Ref. 73.
[d] Ref. 9.
[e] Ref. 74.
[f] Ref. 23.

$E^0(H^+/H_2)$ using two distinct procedures, either based on methods of extrapolating potentials measured in diluted solutions,[9,15] or on calculations from known dissociation constants of acids.[19] Table III presents the data relative to these electrodes as well as to those discussed later on.

All potentials mentioned in fact ignore liquid junction potentials and ionic strength corrections. Thus, correlations of some of these potentials will therefore be tainted with some uncertainty.[80]

On the assumption that slightly polarizable metal ions of low charge should provide a means to establish an absolute scale of potentials in diverse media, Strehlow suggested that the ferrocene–ferricenium standard potential could be selected as an absolute reference.[81] In pyridine very little work has been done on its estimation. However, the observation of inconsistent data resulting from polarographic[68] and voltammetric[9,82] measurements on the one hand, and potentiometric[83] measurements on the other, seems to indicate that the use of this standard is still premature and that further investigations are required. (See Section VII,B,1 for a more detailed discussion.)

Although they are generally considered to be inadequate, the direct employment of aqueous reference electrodes has often been reported; in fact, they were generally used for routine work, i.e., when accurate or reproducible measurements were not necessary. To minimize contamination with

TABLE III

POTENTIALS (IN V) OF SOME COMMONLY USED REFERENCE ELECTRODES
IN PYRIDINE, REFERRED TO THAT OF THE NORMAL
HYDROGEN ELECTRODE (NHE) IN THE SAME SOLVENT[a]

Reference electrode	Normal hydrogen electrode in pyridine	Reference
Cu(II)/Cu(I), py	0.500 ± 0.030	9
Ag/Ag(I), 1 M, py	0.500 ± 0.030^b	9
Ag/Ag(I), 0.048 M, py	0.534	78
Zn(Hg)/ZnCl$_2$ (sat.), py	-0.788 ± 0.002	19
Ferricenium/ferrocene	0.628	83
	0.970^c	9
	0.930^d	73
SCE (aqueous)	$+0.430 \pm 0.030$	9
	$+0.460$	15
	$+0.476; +0.505$	95

[a] Values relative to the aqueous saturated calomel electrode (SCE) include liquid junction potentials.

[b] Potential of the normal Ag reference electrode: Ag/AgNO$_3$, 1 M, NBu$_4$ClO$_4$.

[c] Oxidation half-wave potential of ferrocene.

[d] Oxidation half-wave potential of ferrocene referred to the NHE, according to Nigretto and Jozefowicz.[9]

water, the use of a modified-methanol saturated calomel electrode (SCE) was proposed, in which the filling solution was replaced by a methanolic solution.[15,84,85] Differences in junction potentials between the solution and the SCE may, however, introduce sources of discrepancies. As it was impossible to saturate pyridine solutions with tetraalkylammonium salts in the construction of a calomel electrode because of the decomposition of mercurous chloride, the alternative was either to develop a salt bridge to avoid this disadvantage[86] or to opt for another reference system for which the solvent chosen was the same in both cell and reference compartment.[79,87]

The silver Ag/Ag(I) redox couple was found to behave reversibly and reproducibly,[88] at least when water contamination was excluded.[89] The electrode was constituted of a silver wire immersed in a 1 M,[23,90] 0.1 M,[89] or 0.048 M[78] solution of silver nitrate in anhydrous pyridine. However, loss of silver through an aging process caused a potential shift over a longer or shorter period, so that other reference electrodes were sought.

The copper(I) and copper(II) chlorides in pyridine form a redox couple whose potential stability and reversibility were checked by micropolarization tests.[78,91] In practical terms, a suitable reference electrode was prepared using a platinum sheet in a pyridine solution of 2.5 mM copper(I) chloride, 25 mM copper(II) chloride, and 0.1 M tetramethylammonium chloride.[92] Mukherjee has reported other types of reference electrodes of the second kind, represented by the two half cells: $Hg/HgCl_2$(sat.), LiCl(sat.) in pyridine, and $Zn(Hg)/ZnCl_2$(sat.) in pyridine. The last was prepared by adding a saturated solution of $ZnCl_2$–pyridine salt to an amalgam containing 3% of its weight in zinc.[19]

4. INDICATING ELECTRODES

A wide variety of indicating electrodes has been used in pyridine. This fact is not surprising due to the large number of electrical titrations carried out in this solvent. Although the use of the hydrogen electrode as a reference system has not been widespread because of its inconvenience, it nevertheless appears useful as a proton availability indicator. Yvernault and More introduced it for emf measurements in pyridine.[93] Since then, several cells have been studied potentiometrically and voltammetrically and relationships between the emf and the proton activity established.[9,94-96]

The earliest work concerned potentiometric titrations performed using polarized bimetallic electrode pairs. These indicating electrode-reference electrode couples included gold, platinum, platinized platinum, and antimony as electrode material. The influence of electrode pretreatment has been discussed by Harlow.[97] In addition, some investigators have found that methods combining the above-listed electrode materials with the modified methanol electrode exhibited steady potentials.[98,99] The description of a new device for microtitrations has recently been published.[100]

In spite of a higher sensitivity toward surface-active ions that may be present in solution, the conventional aqueous glass electrode still remains the most widely used for potentiometric determinations of acidic compounds or acidic mixtures in pyridine.[79,86] In experiments conducted by Cundiff and Markunas, the resulting potentials were found to be reached rapidly, remaining constant and reproducible,[101] except in some cases where anomalous titration curves were obtained; the phenomenon was partially attributed to adsorption effects onto the glass membrane.[102] Nevertheless, quite satisfactory results were noted when the electrode was periodically calibrated in standard solutions and provided that convenient storage conditions were respected.[15,87] A nonaqueous glass electrode, made with a filling solution of a redox couple [Cu(II)/Cu(I)] and an acid-base buffer (acetic acid/tetrabutylammonium acetate) in a pyridine solution, was tested in several buffer solutions. Over the whole span of available potentials in pyridine, the slope of its linear response was close to 58 mV per pH unit.[9]

5. TITRANT SOLUTIONS

Acidic titrants are readily available in the form of solutions of soluble strong acids in pyridine: perchloric and p-toluenesulfonic acids meet the general requirements of titrants in this solvent.[103,104] On the other hand, basic titrants are much less easy to obtain. Although they have been extensively used in numerous investigations,[105-107] alkali metal bases are inadequate in pyridine because of their relative weakness and their poor solubility. Attention was, therefore, directed toward more suitable bases which would overcome a number of the limitations of previously suggested bases. Thus, ammonium hydroxides (methyltributyl-, tetrabutyl-, tetraethyl-, triethylbutyl-, and trimethylbenzylammonium[106]) in isopropanol or benzene–methanol solutions were proposed.[84] The advantages of these bases over alkali bases are straightforward, as shown by comparison of the neutralization curves of weak acids.[108] The considerably larger precision and better resolution obtained for mixtures of weak acids[85,101,108,109] can be related to the stronger acid–base properties of the NR_4^+, H_2O/NR_4OH couple.[9]

Tetrabutylammonium hydroxide and tributylmethylammonium hydroxide solutions in pyridine have been prepared in two ways. One method required conversion of the corresponding iodide into the OH form through a basic anion exchange resin; thus, a 0.2 N titrant with only 0.5% water was obtained.[108,110] By the other method, the corresponding iodide underwent reaction with silver oxide in methanol.[85] Unfortunately, erronous results, attributed to the presence of unidentified impurities, were reported.[85,109,111] Since then, a great deal of effort has gone into the investigation of the sources of these impurities[112] and into improving the preparation of these bases. At present, tetraalkylammonium bases are available commercially in satisfactory grades of purity. Great care must be taken, however, since it is known that pyridine favors a possible Hofmann degradation of these types of compounds. Nonetheless, they can be kept without noticeable decomposition for several months if stored in a cool place. Ideally, all contact with air and moisture should be avoided.

V. BEHAVIOR OF SALTS IN PYRIDINE

Considering only its relatively low dielectric constant ($\varepsilon = 12$), one would predict that pyridine would exhibit rather poor solvent properties and that association phenomena would be pronounced. The former tendency is, however, favorably counterbalanced by its complexing ability, which promotes the solubility of a large number of organic and inorganic substances. For instance, silver halides, well known for their insolubility in water, readily dissolve in pyridine and even noticeable conductances occur. The considerable heat arising from dissolution in pyridine of salts containing strong Lewis cations, e.g., mercury(II), copper(II), or aluminum(III) cations, also reveals its basic character.[113-115]

A. Solubility

Table IV summarizes the qualitative solubility data available for inorganic salts in pyridine. Temperature dependences are given for some salts in Muller[116]; relations of solubilities to osmotic properties are compiled in Nelson.[117]

B. Ionic Equilibria

The solute–solvent interaction between pyridine and a salt, MX, is generally exemplified by

$$m \text{ Py} + \text{MX} \rightleftharpoons m \text{ Py, MX} \rightleftharpoons n \text{ Py, M}^+ + n' \text{ Py, X}^- \tag{6}$$

$$\textbf{(V)}$$

The first equilibrium process produces ionization of MX within the ion pair (V), whereas the second dissociates both ions. As a result of the Lewis-base character of pyridine, the global effect is to enhance the dissociation constant of MX in comparison with that observed in more neutral solvents with the same dielectric constant.[12]

TABLE IV

QUALITATIVE SOLUBILITIES OF INORGANIC SALTS IN PYRIDINE[a]

Very soluble	Soluble	Insoluble
$AgNO_3$[c]	$AgCl$, $AgBr$, $AgCN$, $AgSCN$, $AgNO_2$	Ag citrate,[b] tartrate[b]
	Ag_2SO_4	
$AlCl_3$	$AlBr_3$,[c] AlI_3[c]	AlF_3, $Al(NO_3)_3$, $AlPO_4$
		$Al_2(SO_4)_3$
	$AuCl_3$	
$BaBr_2$	$BaCl_2$, BaI_2[c]	
$BeBr_2$[c]		
	$CaCl_2$,[c] $CaBr_2$, CaI_2, $Ca(NO_3)_2$[c]	
$CdBr_2$, CdI_2[c]		
	$CeCl_3$,[c] $CeBr_3$	
$Co(NO_3)_2$	$CoBr_2$, CoI_2	$Co_3(AsO_4)_2$
$Cu(CN)_2$	CuF_2, $CuCl_2$, $CuBr_2$, CuI_2, $Cu(NO_3)_2$	
	CuI,[c] $CuCl$, $Cu(I)$ acetate	
	$Cu(NO_3)_2$	
	$FeCl_2$, $FeBr_2$,[c] $FeCl_3$	
$Hg(NO_3)_2$	$HgCl_2 < HgBr_2 < HgI_2 < Hg(CN)_2$	
I_2		
KCl, KI, KCN	$KMnO_4$, $KSCN$	K_2CrO_4, KIO_3, KNO_3, KOH
		$KClO_4$, K_2SO_4
	$LaCl_3$	
	$LiCl$, $LiBr$, LiI, $LiNO_3$[c]	Li_2CO_3
	$MgCl_2$, $MgBr_2$,[c] MgI_2	
	$MnCl_2$[c]	
NH_4, NH_4NO_3[c]	NH_4SCN	
	$NaBr$, NaI, $NaNO_2$[c]	
$NiCl_2$, $Ni(NO_3)_2$		
PbI_2	$PbCl_2$, $PbBr_2$, $Pb(NO_3)_2$[c]	$Pb(IO_3)_2$, $PbCrO_4$[b]
	$Pb(SCN)_2$, $Pb(CN)_2$, Pb acetate[b]	Pb tartrate,[b] bromate[b]
		$PbSO_4$[b]
	$PtCl_2$	
	$SbCl_3$	
	$SnCl_2$, $SnBr_2$, $SnCl_3$	
	$SrCl_2$, $Sr(NO_3)_2$[c]	
	UCl_4	
	$ZnCl_2$, $ZnBr_2$,[c] ZnI_2[c]	

[a] Data are from Naumann,[115] with permission. For substances of biological interest, see Nelson.[117] "Insoluble" means that at best only traces go into solution.

[b] See Nelson.[117]

[c] See Muller.[116]

An interesting case is that of silver salts, for which unexpected results have been obtained. Although the conductance of silver ions corresponds to that of moderately large ions, the dissociation constant of silver picrate is surprisingly high, about twice that of the tetrabutylammonium salt (see Tables V[7,14,19,83,118-121] and VI[7,14,79,118,119,122-124]). Calculations of the distance between centers of the ion pairs, according to the Fuoss method, lead to the value 13.2 Å. Mukherjee et al.[121] attempted to elucidate this behavior on the basis of equilibria involving triple ions, dimers, as well as the simple

TABLE V

Limiting Conductances and Dissociation Constants of
Some Silver and Alkali Salts in Pyridine at 25°C

Salt	Limiting conductance		Dissociation constant ($\times 10^4$)	pK	
Silver					
Picrate	66.0[b]	68.0[a]	30.6[a]	2.51[a]	—
Nitrate	85.5[b]	86.9[c]	9.3[c]	3.03[c]	—
Perchlorate	—	81.9[c]	19.1[c]	2.72[c]	—
Chloride	—	—	—	4.08[e]	7.08[g]
Thiocyanate	—	—	0.035[g]	5.46[g]	—
Cyanide	—	—	0.013[g]	5.12[g]	—
Lithium					
Picrate	52[b]	58.6[a]	0.83[a]	4.08[a]	—
Nitrate	—	—	—	4.96[f]	—
Perchlorate	71.8[d]	—	—	4.38[f]	2.77[d]
Acetate	—	—	—	< 7.62[d]	—
Benzoate	—	—	—	< 8.15[d]	—
Sodium					
Picrate	58[b]	60.5[a]	0.43[a]	4.37[a]	—
Iodide	71.8[b]	75.2[a]	3.7[a]	3.43[a]	—
Potassium					
Picrate	60.5[b]	65.7[a]	1.0[a]	4.0[a]	—
Iodide	—	80.4[a]	2.1[a]	3.68[a]	—
Ferricenium					
Picrate	71.48[h]	—	4.0[h]	3.40[h]	—

[a] Ref. 118.
[b] Ref. 14.
[c] Ref. 7.
[d] Ref. 119.
[e] Ref. 120.
[f] Ref. 19.
[g] Ref. 121.
[h] Ref. 83.

TABLE VI   195

LIMITING CONDUCTANCES AND DISSOCIATION CONSTANTS OF SOME ONIUM
SALTS IN PYRIDINE AT 25°C

Salt	Limiting conductance	Dissociation constant ($\times 10^4$)	pK
Ammonium			
Picrate	80.5[a]	2.8[a]	3.55[a]
Iodide	95.2[a]	2.4[a]	3.62[a]
Tetramethylammonium			
Picrate	75.5[b]; 76.7[a]	6.7[a]; 25.00[c]	3.17[a]; 2.60[c]
2,4-Dinitrophenolate	—	23[c]	2.64[c]
2,5-Dinitrophenolate	—	10[c]	3.00[c]
Acetate	—	41[c]	2.39[c]
Benzoate	—	0.31[c]	4.51[c]
Tetraethylammonium			
Picrate	72[b]; 73.31[a]	10.4[a]	2.98[a]
Chloride	85[b]	—	—
Bromide	87[b]	—	—
Iodide	86.1[b]	—	—
Perchlorate	85.7[b]	—	—
Nitrate	91.1[b]	—	—
Benzoate	72[h]	—	—
Tetrapropylammonium			
Perchlorate	77.5[b]	—	—
Picrate	63[b]; 62.11[c]	11.2[c]	2.95[c]
Tetrabutylammonium			
Benzoate	—	1.99[f]	3.70[f]
Picrate	57.7[d]	12.8[d]	2.89[d]
Bromide	75.3[d]	2.5[d]	3.60[d]
Iodide	73.1[d]	4.1[d]	3.39[d]
Nitrate	76.6[a]	3.7[a]	3.43[a]
Acetate	76[a]	1.7[a]	3.77[a]; 3.80[f]
Perchlorate	66.2[g]	—	—
Triphenylfluoroborate	48[a]	13.2[a]	2.88[a]
Tetraisoamylammonium			
Picrate	56.4[b]	—	—
Perchlorate	70.7[b]	—	—
$C_5H_5NHNO_3$	102.2[a]	0.51[a]	4.29[a]
$C_5H_{10}NH_2NO_3$	91.1[a]	0.18[a]	4.74[a]
$(C_2H_5)(CH_3)_3N$ picrate	75.5[a]	8.2[a]	3.09[a]
$(HOC_2H_4)(CH_3)_3N$ picrate	67.0[a]	1.5[a]	3.82[a]
$(BrC_2H_4)(CH_3)_3N$ picrate	67.1[a]	5.8[a]	3.24[a]
$(BrCH_2)(CH_3)_3N$ picrate	71.5[a]	4.8[a]	3.32[a]
$C_5H_5NC_6H_5$ picrate	66.3[a]	11.5[a]	2.94[a]
$(C_6H_5)(CH_3)_2OHN$ picrate	62.3[a]	12.3[a]	2.91[a]

[a] Ref. 118.
[b] Ref. 14.
[c] Ref. 122.
[d] Ref. 7.
[e] Ref. 123 (from DVP measurements at 37°C).
[f] Ref. 79.
[g] Ref. 124.
[h] Ref. 119.

dissociation as monomers. The following reactions of complexation were postulated:

$$Ag^+ + X^- \rightleftharpoons AgX \qquad K_1 = [AgX]/[Ag^+][X^-] \qquad (7)$$

$$AgX + Ag^+ \rightleftharpoons Ag_2X^+ \qquad K_{2a} = [Ag_2X^+]/[AgX][Ag^+] \qquad (8)$$

$$AgX + X^- \rightleftharpoons AgX_2^- \qquad K_{2b} = [AgX_2^-]/[AgX][X^-] \qquad (9)$$

$$Ag_2X^+ + X^- \rightleftharpoons Ag_2X_2 \qquad K_{3a} = [Ag_2X_2]/[Ag_2X^+][X^-] \qquad (10)$$

$$AgX_2^- + Ag^+ \rightleftharpoons Ag_2X_2 \qquad K_{3b} = [Ag_2X_2]/[AgX_2^-][Ag^+] \qquad (11)$$

For silver chloride, cyanide, and thiocyanate, satisfactory correlations with potentiometric measurements resulted. Assuming $K_{2a} = K_{2b} = K_2$ and setting $K_{3a} = K_{3b} = K_3$, the constants were evaluated by comparing experimental plots with theoretical ones. The best fit yielded the respective values given in Table VII. Based on the observed value of K_1, the calculated monomeric dissociation constant of AgCl, $K[AgCl] = 8.3 \times 10^{-8}$, was, however, not corroborated by an earlier study in which Bruckenstein and Osugi[120] made a spectrophotometric analysis of pyridine solutions of silver chloride in the 300–400 nm range. The authors observed that the spectrum was markedly concentration dependent. The dissociation constant was determined from the variation of the apparent molar absorptivity of silver chloride at 310 nm. While in aqueous solutions of pyridine absorption bands of $PyAg^+$ and Py_2Ag^+ occur below 280 nm, involving $\pi \rightarrow \pi^*$ transitions, the appearance of the 310 nm band probably resulted from transitions within the perturbed solvent molecules bound to silver chloride or to the silver ion. The value $K[AgCl] = 8.4 \times 10^{-5}$ was found. For silver bromide, concentration dependence of conductances suggested the existence of aggregates beyond dimers, i.e., even trimers,[121] according to Eq. 12:

$$3 AgBr \rightleftharpoons Ag_3Br_3 \qquad K_4 = [Ag_3Br_3]/[AgBr]^3 \qquad (12)$$

Additional equations for the formation of such species were therefore introduced into the calculations. K_4 was found equal to 10^{6}.[125]

Alkali or tetraalkylammonium salts (MX salts) are quite often involved in acid–base titrations as reactants, when not required in electrochemical studies in the form of background electrolytes. It therefore seems important to know the nature of the ionic equilibria occurring in the presence of these salts. (Available dissociation data were given in Tables V and VI.) Thus, it can be noted that most dissociation constants of quaternary ammonium salts are higher than 10^{-4}. On the other hand, only lithium salts formed with anions of strong Brønsted (HA) acids, e.g., perchlorate or picrate, behave similarly, whereas the others are rather weak electrolytes. Consequently, especially for acid–base titrations, two limiting cases must be envisaged:

TABLE VII

CALCULATED EQUILIBRIUM CONSTANTS OF SOME SILVER SALTS IN PYRIDINE[a]

Equilibrium	Constant	AgCl	AgBr	AgCN	AgSCN
$\rightleftharpoons Ag^+ + X^-$	$K_1 = [AgX]/[Ag^+][X^-]$	1.2×10^7	5×10^6	7.5×10^5	2.9×10^5
$X^+ \rightleftharpoons Ag^+ + AgX$	$K_{2a} = [Ag_2X^+]/[AgX][Ag^+]$	$1.6 \times 10^{4\,b}$	$3 \times 10^{3\,b}$	$2.0 \times 10^{4\,b}$	$10^{2\,b}$
$_2{}^- \rightleftharpoons X^- + AgX$	$K_{2b} = [AgX_2{}^-]/[AgX][X^-]$				
$X_2 \rightleftharpoons X^- + Ag_2X^+$	$K_{3a} = [Ag_2X_2]/[Ag_2X^+][X^-]$	$6.5 \times 10^{5\,c}$	$10^{6\,c}$	$2.0 \times 10^{4\,c}$	0^c
$X_2 \rightleftharpoons Ag^+ + AgX_2{}^-$	$K_{3b} = [Ag_2X_2]/[AgX_2{}^-][Ag^+]$				
$X_3 \rightleftharpoons 3\,AgX$	$K_4 = [Ag_3X_3]/[AgX]^3$	—	10^6	—	—
$+ HBr \rightleftharpoons Ag\,HXBr$	$K_c = [AgHXBr]/[AgX][HBr]$	—	7.5×10^3	—	—
$\rightleftharpoons Ag^+ + X^-$	$K = [Ag^+][X^-]/[AgX]$	8.3×10^{-8}	—	—	—
(overall)	(overall)	$8.4 \times 10^{-5\,d}$	—	—	—

From Mukherjee et al.[121,125]

Assuming $K_{2a} = K_{2b} = K_2$.

Assuming $K_{3a} = K_{3b} = K_3$.

Ref. 120.

solutes containing A^- anions of either strong or weak acids, in the presence of MX salts. No interference will occur if MA is more dissociated than MX or if both MA and MX are of comparable strength. Otherwise, a large excess of MX will entail the displacement of the ionic equilibria normally taking place toward the preferential formation of MA. This effect, designated as the leveling or salt effect, has been thoroughly investigated by Elving et al. for lithium and tetraethylammonium salts.[119] The authors showed that the fundamental bulk solution equilibrium, rationalizing the effect of MX, can be represented by Eq. 13:

$$SHA + MX \rightleftharpoons SHX + MA \qquad (13)$$

The conditional equilibrium constant for this reaction is

$$K = SHX/MX \times MA/SHA \qquad (14)$$

According to whether the acid–base character of MX is strong or weak relative to that of SHA (the solvated acid HA), Eq. 13 is shifted to the right or to the left, respectively. The extent of dissociation of HA in the MX + solvent system will depend upon the sizes and symmetry of the ions forming MX, especially those of the cation. If M^+ is a large cation, e.g., $NEt_4{}^+$, its interaction in the dissociation of HA will be small; on the other hand, if M^+ is a small cation, which can function as a strong Lewis acid comparatively to SH^+, Eq. 13 will be displaced, thus converting a significant portion of SHA into SHX. This displacement is furthermore accentuated by the degree of analytical concentration of MX, usually 100 times greater than that of HA, if MX is the background electrolyte. Species of Eq. 13 may in turn undergo dissociation according to Eqs. 15a–d:

$$SHA \rightleftharpoons SH^+ + A^- \tag{15a}$$

$$SHX \rightleftharpoons SH^+ + X^- \tag{15b}$$

$$MA \rightleftharpoons M^+ + A^- \tag{15c}$$

$$MX \rightleftharpoons M^+ + X^- \tag{15d}$$

for which respective dissociation constants are $k[SHA]$, $k[SHX]$, $k[MA]$, and $k[MX]$. Neglecting the formation of higher aggregates and inserting Eqs. 15a–d into Eq. 14, one can define the constant K as

$$K = k[SHA]/k[SHX] \times k[MX]/k[MA] \tag{16}$$

in which the second ratio depicts the specific action of MX; the larger this ratio, the greater the displacement of the ion exchange reaction with the background cation:

$$MX + A^- \rightleftharpoons MA + X^- \tag{17}$$

Spectrophotometric measurements in pyridine solutions of appropriate Brønsted acids showed evidence for the replacement of the NEt_4^+ content as Li^+ was progressively added to the solution. In the case of p-nitrophenol as the HA form, an isobestic point could be detected; the equilibrium constant for reaction (13) was found to be 6×10^{-9}.[119] When 0.1 M NEt_4ClO_4 is present, the second ratio of Eq. 16, calculated from conductance results, is not far from unity. Alternatively, when 0.1 M $LiClO_4$ is employed, the second ratio is much greater. Elving attempted to evaluate this ratio by assuming the dissociation constant $k[SHA]$ to be about 10^{-10}, which is typical of a weak acid in pyridine. Setting the value 10^{-3} for each $k[SHX]$, $k[MA]$, and $k[MX]$, deduced from earlier published data,[21] the ratio $k[LiClO_4]/k[LiA]$ was thereafter found to be on the order of 10^5.[119] The results obtained by Mukherjee lead to 10^4.[19]

In conclusion, these results show that great care must be exercised when weak electrolytes are involved in quantitative determinations.

C. Conductance Studies

Of the earliest conductance investigations undertaken in pyridine, only those dealing with strong electrolytes can lead to interpretable experimental data, since traces of impurities in the solutes or the solvent have a marked effect on the limiting conductance estimations. The measurements were often carried out under more or less strictly defined conditions, especially as far as water content and solvent purity are concerned. As a result, discrepancies frequently arose when comparisons with different sources of literature on the subject were made. Alkali metal salts[48,126 128] and transition metal salts[48,129,130] have often been studied. Recently, more rigorous working conditions led to the production of pyridine, the specific conductance of which ranged around 10^{-10} ohm^{-1} cm^{-1}. This limit has been considered as low enough to reflect the solvent limit.[12,17] Hence, accurate determinations could be carried out, yielding valuable and new information as to the nature of the ions and their interactions in pyridine solutions. Data illustrating limiting conductances will be found in Tables V, VI, VIII, and IX.[14,83,118,122,124,131]

TABLE VIII

IONIC LIMITING CONDUCTANCES OF SOME
ANIONS IN PYRIDINE AT 25°C

Anion	Λ_0	Reference
Nitrate	52.6	118
	49.4	14
Picrate	33.7	118
	30.3	14
Benzoate	32.4	7
	20.6	79
Fluoride	32.0	124
Chloride	51.4	124
Bromide	51.3	124
Iodide	48.4	118
Perchlorate	47.6	124
Acetate	51.8	124
Chloroacetate	50.1	124

Walden, Audrieth, and Birr[14] evaluated the limiting conductances of various tetraalkylammonium salts and primary and secondary amine salts, as well as of a number of inorganic salts. The validity of the Kohlrausch law was found to hold in pyridine for strong and medium electrolytes, under the critical concentration 1.25×10^{-4} M. The behavior of weak electrolytes, like silver chloride or diethylammonium chloride, diverged noticeably from what might be expected theoretically. While most of the salts studied behaved normally, limiting conductances obtained for tetraalkylammonium halides were somewhat too low. The authors were unable to determine whether these anomalous results were due to the inadequacy of solvent correction or to the existence of a real phenomenon. The individual conductances of the studied ions fell in the following order: for the cations

$$(CH_3)_4N^+ > (C_2H_5)_4N^+ > Ag^+ > (C_3H_7)_4N^+ >$$

$$K^+, C_5H_5NH^+, C_5H_{10}NH_2^+ > Na^+, \text{iso-}(C_5H_{11})_4N^+ > Li^+$$

TABLE IX

IONIC LIMITING CONDUCTANCES OF SOME CATIONS IN PYRIDINE AT 25°C

Cation	Limiting conductance	Reference
$C_5H_5NH^+$	49.6	118
	30.5	14
$C_5H_{10}NH_2^+$	38.5	118
	31.6	14
$C_5H_5(C_6H_5)N^+$	32.6	118
H_4N^+	46.8	118
$(CH_3)_4N^+$	43.0	118
	45.2	14
$(C_2H_5)_4N^+$	39.6	122
	41.7	14
$(C_3H_7)_4N^+$	24.4	122
	33.35	14
$(C_4H_9)_4N^+$	24.0	118
$(C_5H_{11})_4N^+$	21.6	122
$(iso-C_5H_{11})_4N^+$	26.15	118
$(BrCH_2)(CH_3)_3N^+$	37.8	118
$(C_2H_5)(CH_3)_3N^+$	41.8	118
$(BrC_2H_4)(CH_3)_3N^+$	33.4	118
$(OHC_2H_4)(CH_3)_3N^+$	33.3	118
$(C_6H_5)(CH_3)_2OHN^+$	28.6	118
Octadecyltrimethyl N^+	21.9	124
Octadecyltributyl N^+	17.1	122
Dioctadecyldibutyl N^+	13.3	122
1,3-Diphenylguanidinium	18.6	131
Ag^+	34.3	118
	35.7	14
Li^+	24.9	118
	21.7	14
Na^+	26.8	118
	27.45	14
K^+	30.2	14
Ferricenium	37.8	83

and for the anions

nitrate > bromide > iodide, perchlorate > chloride > picrate

In a comprehensive study, Kraus et al.[118,122,124] redetermined the limiting conductances of inorganic and quaternary ammonium salts using the method of successive approximations described by Fuoss and Kraus.[132,133] Individual ion conductances were deduced according to Fowler's method,[134] with the following values of constants in computation: density of pyridine = 0.97797, viscosity = 8.824 mP, and dielectric constant = 12.01 at 25°C. Generally, the Fuoss plots exhibited linear relationships at low concentrations but deviations became apparent above the critical value $5 \times 10^{-4} M$; the strongest electrolytes deviated toward higher conductances, the weakest toward lower conductances. It seems likely that the failure of the conductance data is due to the inapplicability of the Debye–Hückel law at concentrations above this limit. Except for the nitrate ion, ion conductances derived from several combinations of different salts compared very well. As usually encountered, the conductances of small ions were found to diminish markedly with decreasing dimensions of the radii of unsolvated ions. Thus, the conductance of the strongly solvated lithium ion (see Table IX) is of the same order as that of the less solvated tetrabutylammonium ion.

The bromide and chloride ions have practically the same conductance, which is only 6% higher than that of iodide[7] (Table VIII) but stands appreciably higher than that of fluoride. Significant effects result when the cation carries groups able to interact with the solvent molecules. Thus, the conductances of the ethyltrimethylammonium, the bromomethyltrimethylammonium, and the bromoethyltrimethylammonium ions are respectively 41.8, 37.8, and 33.4; the conductance of the phenyldimethylhydroxyammonium ion is much lower than of the phenylpyridinium ion, showing evidence that hydroxy groups interact strongly with the basic solvent molecule. As shown in Tables VIII and IX, the conductances of the nitrate anion and of the pyridinium cation, respectively 52.6 and 49.6, are the highest thus far measured in pyridine.

Kraus' data compare well with those obtained by Walden. It is, however, significant to note that Walden's conductance values are systematically smaller, although the same physical constants of pyridine were used in the computations. This fact can probably be ascribed to the effect of solvent correction, since Walden reported a noticeably higher specific conductance.

VI. ACID–BASE EQUILIBRIA IN PYRIDINE

Former approaches to the nature of acid–base equilibria in nonaqueous media generally tended to separate the role of the solvent from the electrolytic processes, or ignored it altogether. Fundamentally, modern concepts tend more to associate the solvent with any reaction stage. For convenience, the literature often classifies organic solvents with reference to their acid–base properties. Because of the nature of its constituent molecule, pyridine is essentially considered to be a protophilic, dipolar aprotic solvent.[135] Extending the Brønsted classification, Davis[136] distinguished between solvents with a dielectric constant greater or smaller than 20. For each class, numerous treatments of acid–base reactions were published.[136] [140] According to whether the solvent was considered as sufficiently dissociating or totally inert, simplifying assumptions could be postulated as to the extent of dissociation of evolving species. On the other hand, exact treatment of acid–base reactions in media with an intermediate dielectric constant, such as pyridine, is rather complex. Quite a number of equilibrium schemes, including partially dissociated species, have to be taken into account in a thorough description of processes following dissolution of the reactants. For the sake of completeness, interionic or intermolecular interactions between solutes yielding hydrogen-bonded aggregates may not be disregarded either,[136,141] although high-order aggregates are not expected to form extensively in pyridine because of its basic character.

Clearly, then, a lucid and concise presentation of all of these interactions would be rather complicated. They will therefore be described separately, later on, as the experimental data are reviewed. Under these conditions, it seems that another approach based on thermodynamics would provide a more useful and more appropriate means of rationalizing the data available in the literature. Instead of considering sequences of separate steps, the magnitude of which are governed by the chemical properties of the solvent and solutes, the latter method treats the dissolution process as a whole, integrating the intermediate steps. This theoretical procedure is developed in the following section.

A. Thermodynamics

Let us consider a monophasic system consisting of a solution of i species in pyridine, each of them carrying a charge z_i. The increase of chemical free enthalpy dG, consecutive to the introduction of dn_i species and dq_i charges into the system, under constant temperature and pressure, is conventionally written according to the equation:

$$dG = \sum \mu_i dn_i + \sum \phi_i dq_i \qquad (18)$$

Both intensive quantities, μ_i, the partial Gibbs molal energy, and ϕ_i, the internal potential resulting from the individual presence of charged species, can be expressed by the electrochemical potential $\bar{\mu}_i$,[142,143] as:

$$\bar{\mu}_i = \mu_i + z_i \mathscr{F} \phi_i \qquad (19)$$

$\mathscr{F}$ is the Faraday. Thus

$$dG = \sum \bar{\mu}_i dn_i \qquad (20)$$

The electrochemical potential $\bar{\mu}_i$, the activity a_i, the ionic activity coefficient γ_i, and the concentration c_i of species i are linked by Eq. 21:

$$\bar{\mu}_i = \bar{\mu}_i^{\,0} + RT \ln a_i = \bar{\mu}_i^{\,0} + RT \ln \gamma_i c_i \qquad (21)$$

where $\bar{\mu}_i^0$ is the standard electrochemical potential, R is the gas constant, and T is the temperature on the Kelvin scale. Under equilibrium conditions, dG equals zero; resolution of the preceding equation system leads to the well-known mass law constant, which reflects the choice of the reference state. For convenience, the latter can be arbitrarily defined by an ideal solution of species i in pyridine, at a normal concentration. A complete description of this state still requires the ionic activity coefficients to become unity at infinite dilution in pyridine:

$$a_i = \gamma_i c_i \quad \text{with } \gamma_i \to 1 \quad \text{when } c_i \to 0 \tag{22}$$

B. pH Scale in Pyridine

The hydrogen ion (proton) free energy level in two solutions, 1 and 2, in pyridine is well characterized by its electrochemical potential:

$$\bar{\mu}_{H^+}^{\ 1} = \bar{\mu}_{H^+}^{\ 0,1} + RT \ln a_{H^+}^{\ 1} + \mathscr{F}\phi_{H^+}^{\ 1} \tag{23}$$

and

$$\bar{\mu}_{H^+}^{\ 2} = \bar{\mu}_{H^+}^{\ 0,2} + RT \ln a_{H^+}^{\ 2} + \mathscr{F}\phi_{H^+}^{\ 2} \tag{24}$$

Formally, the hydrogen ion activity can be written $-\log a_{H^+} = $ pH. Hence, assuming the potential ϕ_{H^+} to be identical and constant for the dilute solutions 1 and 2 in pyridine, the subtraction of Eq. 24 from Eq. 23 leads to

$$pH_1 - pH_2 = \frac{\bar{\mu}_{H^+}^{\ 2} - \bar{\mu}_{H^+}^{\ 1}}{2.3RT} = \frac{\mu_{H^+}^{\ 2} - \mu_{H^+}^{\ 1}}{2.3RT} \tag{25}$$

If solution 2 is the reference system defined as above, the pH of an unknown solution x will be expressed as

$$pH_x = \frac{\bar{\mu}_{H^+}^{\ 0} - \bar{\mu}_{H^+}^{\ x}}{2.3RT} = \frac{\mu_{H^+}^{\ 0} - \mu_{H^+}^{\ x}}{2.3RT} \tag{26}$$

The redox potential of the electrochemical process

$$H^+ + e^- \rightleftharpoons \tfrac{1}{2}H_2 \tag{27}$$

is given by applying the Nernst relation:

$$E = E^0(H^+/H_2) - 0.058pH \tag{28}$$

thus, provided that the standard state is defined, the establishment of a scale of hydrogen ion activity accordingly becomes possible. Experimentally, it would simply require the hydrogen electrode to function reversibly and according to the Nernst equation.

C. Acid–Base Constants

In the Brønsted–Lowry concept of acid-base behavior,[144] which is used here, any acid–base reaction can be exemplified by the following reactions:

$$n\text{S} + \text{HA} \rightleftharpoons n\text{SHA} \rightleftharpoons n\text{SH}^+\text{A}^- \tag{29a}$$

$$\text{(VI)} \qquad \text{(VII)} \qquad \text{(VIII)}$$

or

$$\text{HA} \rightleftharpoons \text{H}^+\text{A}^- \tag{29b}$$

and

$$n\text{SH}^+\text{A}^- \rightleftharpoons p\text{SH}^+ + q\text{A}^- \tag{30a}$$

$$\text{(VIII)} \qquad \text{(IX)}$$

or

$$\text{H}^+\text{A}^- \rightleftharpoons \text{H}^+ + \text{A}^- \tag{30b}$$

in which the action of the solvent S during dissolution of monobasic un-charged Brønsted acids HA is arbitrarily decomposed in a two-stage process (Eqs. 29a–30b).[140,145] In the first stage, the electric field induced by the nearby solvent molecules modifies the nature of the H–A bonding. This effect, called solvolysis, emphasizes its ionic character. The solvated species HA undergo more or less pronounced ionization (Eq. 29a); this reaction operates within two limits represented by species **VII** and **VIII**. Its scope is regulated by the polarizability of the H–A bond and by the solvent basicity. Species **VII** designate the solvated (hydrogen-bonded) acid and species **VIII** the ionized but still associated ion pairs, which may, in turn, dissociate according to Eq. 30a. The spread of Eqs. 29a–b is principally linked with the solvent basicity, while dissociation is related to the solvent dielectric constant influence: the higher the latter, the greater the magnitude of dissociation. The respective equilibrium constants are given by

$$K_i = [\text{H}^+\text{A}^-]/[\text{HA}] \tag{31}$$

and

$$K_d = [\text{H}^+][\text{A}^-]/[\text{H}^+\text{A}^-] \tag{32}$$

which combine within the overall dissociation constant K:

$$K = \frac{[\text{H}^+][\text{A}^-]}{[\text{HA}] + [\text{H}^+\text{A}^-]} = \frac{K_i K_d}{1 + K_i} \tag{33}$$

As a matter of fact, in pyridine solutions, other processes, namely, ion pairing,[138] promoted by the particular nature of the solvent, will be expected to intervene within the simple representation of Eqs. 29a–30b; each of them can similarly be described with an acid–base type reaction. As a result, one may see that the equilibrium constant K_a which rules the global reaction of dissociation (for the sake of simplicity, solvent solvation molecules and charges of either charged acid A or conjugate base B are omitted).

$$A \rightleftharpoons H^+ + B \tag{34}$$

actually reflects the extent of a multistage process. It therefore designates an apparent constant of acidity.

D. Experimental Acidity Scale in Pyridine

For its definition, the chosen reference state introduces an arbitrary limitation at the origin of the pH scale. It corresponds to the pH of an ideal normal solution of a strong acid. This limitation reflects the basic properties of the solvent. On the other hand, no theoretical limitation bars the way to high basicities, which are, however, excluded by experimental contingencies. So far, the basic limit of this aprotic solvent has not yet been explored, so that the limiting reaction still remains unknown.

1. THE EXPERIMENTAL DETERMINATION OF THE pH SCALE ORIGIN

As seen above, the determination of the standard state in pyridine requires the estimation of the potential of the normal hydrogen electrode (NHE) in this solvent. While real solutions, in which the activity of hydrogen ions is unity, are necessarily far from the ideal, two outstanding procedures have been suggested to obviate this difficulty. The more widely employed method consists of extrapolating potential values obtained in dilute solutions of strong acids up to unit activities. In low dielectric solvents, electrolytes are expected to deviate greatly from the laws of the dilute solutions[146] so that this method of plotting cannot give a truly straight line for more than a limited range. However, if a region exists in which the Nernst equation applies, then it should be possible to determine E^0 (H^+/H_2) values in this way. The second method involves calculation of this potential from the knowledge of the pK_a of acids, postulating a dissociation pattern without complexity. Since both methods are based on analysis of potentiometric results, their feasibility depends heavily on the reliability of the electrochemical measurements.

In this respect, a few experiments were made by Gupta to explore the possibility of setting up a reversible hydrogen electrode in pyridine.[78] Using the method of extrapolation with hydrochloric acid, he estimated the potential of the Ag/AgNO$_3$ 0.048 M reference electrode vs. the NHE potential at 0.534 V. However, his hydrogen electrode was reported not to function reversibly.

An analogous method, previously described by Hitchcock,[147] was used by Hladky and Vrestal,[15] using dilute solutions of the same acid. The authors carried out emf measurements at 25°C with the cell:

$$\text{Pt, } H_2 \text{ in pyridine}|HCl, \text{ Py} \parallel KCl(\text{sat.}), CH_3OH \parallel KCl(\text{sat.}), H_2O|SCE \qquad (35)$$

The results yielded linear plots, from which the NHE potential was deduced: $E^0 (H^+/H_2) = -0.460$ V. The slope of the straight section of the curves was close to 58 mV, as confirmed by determinations using glass electrodes. Direct comparison of this trial $E^0 (H^+/H_2)$ value with data obtained from cells excluding the presence of any extraneous solvent will indeed be hazardous, on account of the existence of nonnegligible junction potentials. It is, however, worth noting that this value, as well as others obtained in a similar manner,[9,95] are not in great discordance since they fall within an 80 mV range (see Table III).

The alternative determination of the NHE potential was effected by evaluation of the pK_a's of Brønsted acids in pyridine. Mukherjee combined spectrophotometric and differential vapor pressure measurements with potentiometric data in acid–salt mixtures. Emf measurements were recorded from the cell:

$$\text{Zn(Hg), ZnCl}_2(\text{sat.}) \text{ in Py}|HA + MA \text{ in Py}|Pt, H_2 (1 \text{ atm}) \qquad (36)$$

Based on the potentiometric results, the pK_a's of perchloric, hydrobromic, and nitric acids were calculated to be 3.26, 4.36, and 4.06, respectively. Thus, the NHE potential was estimated at $+0.788 \pm 0.002$ V vs. the reference electrode used.[19]

Recent studies were conducted with perchloric acid,[9] using the cell:

rotating black platinum disk electrode H_2 (1 atm)	$HClO_4$ at various concentrations in pyridine 0.1 M NBu$_4$ClO$_4$	Ref. 1: Pt/CuCl$_2$ (25 mM), CuCl (2.5 mM), and NMe$_4$Cl (0.1 M) in pyridine or Ref. 2: Ag/AgNO$_3$ (1 M), NBu$_4$ClO$_4$ (0.1 M) in pyridine	(37)

Analyzing the voltametric reduction waves produced by this acid, the authors at the same time ascertained the reversibility of the reduction of hydrogen ions on the hydrogen electrode and the validity of the Nernst equation in this solvent. The linear relationship of the potential vs. $-\log (C_{HClO_4})$ plots led to an estimation by extrapolation of a potential difference of 0.500 ± 0.030 V between the NHE and both reference electrodes used (their potentials are comparable within 10 mV.[74]

The NHE potentials may be compared with the standard potentials of redox couples which used to be employed as reference systems in nonaqueous solvents, namely Ag(I)/Ag and ferricenium/ferrocene (Fc$^+$/Fc), according to the Strehlow theory.[81]

$E^0(Ag^+/Ag)$ was calculated by Mukherjee on the basis of conductance and potentiometric data with various silver salts.[121] He found that $E^0(Ag^+/Ag) = 0.551 \pm 0.002$ V vs. the NHE. It should be noticed that, although the dissociation constant of silver chloride was not corroborated by earlier studies,[120] the author nevertheless used its numerical value to evaluate $E^0(H^+/H_2)$. More recently, he analyzed the potentials of half cells consisting of equimolar mixtures of ferrocene and ferricenium picrate at several concentrations. Assuming that the salt dissociates simply, an average value of 0.628 V vs. the NHE was yielded for $E^0(Fc^+/Fc)$, i.e., 0.077 V vs. $E^0(Ag^+/Ag)$. On the other hand, Elving[73] observed a quite different quantitative behavior; the first polarographic oxidation wave of ferrocene was found to occur at 0.43 V against the normal Ag reference electrode. Despite the fact that the working procedures followed were not very similar, this value was nonetheless confirmed by voltammetric studies using rotating hydrogen electrodes.[9] Furthermore, extrapolation of voltammetric data obtained in water-pyridine mixtures gradually enriched to 100% of amine led to the same result.[82,148]

Whichever standard potential these E^0 are referred to, a significant divergence is to be observed between these two sets of data, as was illustrated in Table III. Without debating, at this time, the reliability of the Fc^+/Fc couple as an absolute reference electrode in pyridine (this point is discussed in Section VII,B,1), it is nonetheless interesting that the inconsistency of the numerical values occurs, whether the calculations are deduced from oxidation half-wave potentials of the ferrocene alone or from equilibrium potentials of Fc^+/Fc mixtures. Consequently, any determination directly deduced from either $E^0(H^+/H_2)$ value, or referred to it, will systematically be controversial, as is seen in Section VI,D,2.

2. CORRELATIONS BETWEEN THE AQUEOUS AND PYRIDINE pH SCALES

Correlation of the pH scale in pyridine with the aqueous standard state would indeed provide a useful means of estimating the medium effect of pyridine relative to water. This problem is one of the traditional unsolved questions of analytical chemistry. Recently, its current status has been comprehensively reviewed by Popovych.[149] Previous attempts required some extrathermodynamic assumptions, expressed schematically, either in terms of junction potentials[150,151] or in terms of transfer activity coefficients of a single ion from a reference solvent "O" to another solvent "S." The latter procedure was suggested mainly by Pleskov, postulating that the chemical potential of rubidium Rb^+ takes the same value in any solvent[152]; Popovych, who considered the variation of the chemical potentials of the tetraphenylboride and tetraphenylarsonium ions to be the same between solvents " O " and " S "[149]; and Strehlow, for whom the chemical free energy of transfer of the ferricenium ion Fc^+ should be very nearly that of the uncharged ferrocene Fc species in any solvent.[81]

Calculation of the activity coefficient of the transfer of protons from water (solvent " O ") to pyridine (solvent " S "), noted $\log_{||}{}^w\gamma_{H^+}{}^{py}$, has been derived from the evaluations of the $E^0(H^+/H_2)$ potentials in both solvents, according to the relation:

$$\log {}^w\gamma_{H^+}{}^{py} = [E^0(H^+/H_2)(py) - E^0(H^+/H_2)(H_2O)]\frac{\mathscr{F}}{2.3RT} \quad (38)$$

Therefore, two distinct methods have been used, either involving emf measurements of combined galvanic cells including the silver reference electrode or presuming the standard potential of the Fc^+/Fc system to remain constant and independent of the solvent. Experimental results, compiled in Table X, show evidence for the obtaining of two substantially different sets of data, depending upon the procedure used. Parker[151] recently tested the assumption of the occurrence of negligible liquid junction potentials over a variety of solvents with cells of the type:

$$
\begin{array}{l}
\text{Ag/AgClO}_4 \text{ in } \| \text{ NEt}_4\text{Pic (0.1 } M\text{) with bridge } \| \text{ AgClO}_4 \text{ in S } \| \text{ Ag} \\
\text{acetonitrile} \quad\quad \text{solvents S or acetonitrile}
\end{array} \quad (39)
$$

TABLE X

TRANSFER ACTIVITY COEFFICIENT VALUES ($\log {}^w\gamma_{H^+}{}^{py}$) FROM WATER TO PYRIDINE

Reference system	$\Delta E[\text{NHE(py)} - \text{NHE(H}_2\text{O)}]$	$\log {}^w\gamma_{H^+}{}^{py}$	Reference
Ag(I)/Ag	$-0.500 + 0.320 = -0.170$	-3.0	9
Fc^+/Fc	$-0.628 + 0.400 = -0.228$	-3.85	83
Extrapolated f(H) acidity function		-3.18	153

Results showed that the emf of cell (39) was independent, within 20 mV, of the bridge solvent, except with formamide, if the picrate bridge has a higher conductance than the end compartments and if the bridge electrolyte is equitransferent for both cations and anions. Postulating these requirements to be roughly the same for the cells described in Nigretto and Jozefowicz,[9] from which a $\log {}^w\gamma_{H^+}{}^{py}$ value has been derived, the transfer activity coefficient between water and pyridine should accordingly be of the order of -3.0. The value of -3.18 supplied by extrapolation of emf measurements from cells of the type:

$$\text{Pt, H}_2 \mid \text{HClO}_4 \text{ at various concentrations in S}' \mid \text{KCl(sat.), in S}' \mid \text{SCE} \quad (40)$$

to 100% pure amine and to a normal acid concentration (S' represents aqueous mixtures of pyridine) compares satisfactorily with the preceding estimate.[153] In this connection, however, it is interesting to note moderately good agreement of this value with Mukherjee's potentiometric determinations from the Fc^+/Fc system, log $^w\gamma_{H^+}{}^{py} = -3.85$, taking into account the surprising behavior of this system in pyridine, as opposed to other values reported in Table III. In any event, these values indicate that the basic character of pyridine is not as marked as is usually believed.

3. DETERMINATIONS OF DISSOCIATION CONSTANTS IN PYRIDINE

For over 40 years, quite a number of investigations have shown that pyridine is an excellent differentiative solvent for such analytical applications as acid titrations. Generally, quantitative attempts were simply related to the characteristics of the titration of a reference substance in the same solvent. In pyridine, as well as in other nonaqueous solvents, recent trends have been to rationalize existing data through the establishment of absolute scales of acidities. To this end, calculations of absolute dissociation constants were carried out. In the following sections, unless otherwise stated, only global (apparent) pK_a values will be reported. Equilibrium constants regarding specific interactions, e.g., associations, will be presented and discussed in Section VI,D,4.

a. *Strong Acids.* It is often considered that the class of strong acids includes acids whose pK_a's are less than *ca.* 4, since in dilute solutions they dissociate appreciably. Necessary additional evidence may be found by observing the invariance of the degree of dissociation in relation to the concentration of the conjugated base, at least within a defined range of acid concentration. Acids appreciably dissociated in pyridine are expected to yield a common species, the pyridinium ion. This proceeds from an acid–base reaction on the solvent.

Polarographically, nitric, trifluoroacetic, dichloracetic, oxalic(I), and phthalic(I) acids exhibited an identical reduction wave, leveling at about -1.33 to -1.40 V vs. the normal silver reference electrode.[154] Titrimetrically, with glass electrodes, three acids (p-toluenesulfonic, naphthalenesulfonic acids, and 2,4,6-trinitrophenol) gave similar, leveled-off, half-neutralization potential values.[155,156] With hydrogen and glass electrodes, the extrapolated potential vs. ($-\log C_{HA}$) relationships for several Brønsted acids (HI, H_2SO_4(I)), p-toluenesulfonic and picric acids, and 2,4- and 2,6-dinitrophenols) coincided with that of perchloric acid.[9] Other strong criteria substantiated this agreement and confirmed the conclusion concerning the strength of these acids in a defined concentration range. These criteria include the analysis of the corresponding voltammetric reduction waves and the lack of influence of conjugated base addition on dissociation.

Mukherjee et al. deduced the pK_a's of several Brønsted acids using different techniques. In preliminary determinations, the overall dissociation constants of 2,4- and 2,5-dinitrophenols were measured, followed by calculations based on a theoretical treatment previously developed for ethylenediamine solutions.[157] Investigations of mixtures of these indicators with two nonabsorbing acids, HBr and HNO_3, yielded the respective pK_a's, 4.36 and 4.06, assuming a simple dissociation to occur.[19] The dissociation constant of HI was then determined from differential vapor pressure (DVP) measurements at 37°C. Based on these results, potentiometric studies conducted with the hydrogen electrode, in mixtures of $HClO_4$ and HCl with their salts,[96] led in turn to a calculation of the corresponding pK_a values. These numerical values are presented in Table XI with other available data drawn from other sources and obtained with different techniques. Because of the general lack of knowledge about the temperature coefficient of dissociation constants in nonaqueous solvents, the pK_a value of HI evaluated at two temperatures is reported.

TABLE XI

DISSOCIATION CONSTANTS (pK_a) OF HALIDE ACIDS IN PYRIDINE

Acid	DVP, spectrophotometry, and potentiometry	Conductometry	DVP at 37°C	Spectrophotometry
$HClO_4$	3.26[a]	3.12[b]	3.0[d]	3.0[d]
HI	3.39[a]	3.23[b]	3.11[a]	—
HNO_3	4.06[a]	4.30[b], 4.29[c]	—	—
HBr	4.36[a]	—	—	—
HCl	5.06[a]	6.15[d]	—	—

[a] Ref. 16.
[b] Ref. 21.
[c] Ref. 118.
[d] Ref. 87.

b. *Weak Acids.* In the intermediate range of acidity, the acidity scale can be built up taking into account the activity of hydrogen ions with acid–base buffers. Dissociation constants of acids have been calculated on the basis of the overall dissociation equation:

$$pH = pK_a + \log (C_A/C_B) \tag{41}$$

by determining half-neutralization potentials (hnp), regardless of homo- or heteroconjugation effects. Despite the fact that the shape of the neutralization curves of acids in pyridine can be quite different from that found, for instance, in water, it can be easily shown that the numerical value of the pH at 50% neutralization nevertheless equals the pK_a.[138] Thus, insertion of the estimated standard potential $E^0(H^+/H_2)$ value in Eqs. 28 and 41 yields the following equation, which offers a direct means of converting potentiometric data in terms of pK_a's:

$$pK_a = \frac{E^0(H^+/H_2) - E}{0.058} \tag{42}$$

Various HA/NBu_4A acid–base couples were tested in this way by Nigretto and Jozefowicz.[9] They measured equilibrium potentials of rotating hydrogen electrodes and modified glass electrodes, using tetrabutylammonium hydroxide as a titrant. (Results are given separately in Tables XVI to XIX.) The overall dissociation constants of benzoic and acetic acids were determined by Mukherjee and Schultz[79] by conventional methods. Since preliminary attempts involving the use of the Pt/H_2 electrode were unsatisfactory for treating the data according to the potentiometric procedure previously used[96] the authors developed other experiments. Their work consisted of DVP as well as hydrogen ion activity measurements with glass electrodes, in mixtures of the acids with their salts. To provide independent estimates of dissociation constants of the salts, they considered it more helpful to use conductance values instead of DVP values, because of their relatively low precision. For DVP studies, the Marshall–Grunwald equation was used to calculate the ionic activity coefficients.[158] At 25°C, average pK_a values for acetic and benzoic acid were found to be 10.1 and 9.8, respectively. The former value is close to that determined by Nigretto and Jozefowicz.[9] Though they used the same technique, Bos and Dahmen[123] reported a substantially higher pK_a value, 12.0. For benzoic acid, appreciable discrepancies between the results of these three sources should thus be noted (Table XII). A better agreement is reached by postulating that association effects intervene. This topic is discussed in the next section. However, the relative order of acidities of both acids under consideration remains the same in pyridine as in water.

Acid–base strengths of charged acids of the type HB^+/B, determined by the conventional dissociation constant K_{HB^+}, were evaluated by either hnp measurements[9,159] or by successive adjustments between experimental and calculated titration curves.[87] In the latter procedure, equilibrium constants of dissociation were calculated from the data by repeatedly entering various values for the constants to be determined into the calculation until the best fit was reached. The pK_a values were able to be estimated when bases were titrated with acids of a known dissociation constant. In the calculations, Mukherjee's values were used.[19] Results yielded by both procedures are compiled in Table XIII.

TABLE XII

DISSOCIATION CONSTANTS (pK_a) AND HOMOCONJUGATION CONSTANTS OF SOME ACIDS IN PYRIDINE AT 25°C

Method	Acetic acid	Benzoic acid	2,4-Dinitrophenol	2,5-Dinitrophenol	2,6-Dinitrophenol	2,4,6-Trinitrophenol
Spectrophotometry	—	—	4.00[h]	5.0[a], 5.15[g]	4.2[a]	3.0[a]
Spectrophotometry, DVP, and potentiometry	—	—	4.38[b]	5.76[b]	—	—
Potentiometry and DVP	10.1[c]	9.8[c]	—	—	—	—
Differential vapor pressure	—	—	4.0[a]	4.6[a]	3.7[a]	2.9[a]
Potentiometry	—	—	4.4[a]	5.3[a]	4.7[a]	3.5[a]
Conductometry	—	—	—	—	5.35[d]	—
Titration	10.6[e]	8.5[e]	4.3[a]	6.5[a]	4.7[a]	—
	12.0[f]	11.0[f]	—	—	—	—
$pK_a(HA_2^-)$	−2.0[f]	—	−1.0[f]	−2.0[f]	1.0[f]	—

[a] Ref. 87.
[b] Ref. 19.
[c] Ref. 79.
[d] Ref. 12.
[e] Ref. 9.
[f] Ref. 123.
[g] Ref. 161.
[h] Ref. 161, in 50% ethanol.

TABLE XIII

DISSOCIATION CONSTANTS (pK_a) OF SOME HB^+/B COUPLES IN PYRIDINE

Acids of HB^+/B Acid–base couples	Titration			
	Hydrogen electrode[a]	Glass electrode[b]	hnp(mV) vs. hnp of benzoic acid[c]	pK_a[d]
Triethylaminium	4.15	3.8	—	—
n-Butylaminium	—	5.5	+115	6.5
Dibutylaminium	5.0	—	—	—
Tributylaminium	3.5	—	—	—
Tetramethylguanidinium	—	9.6	—	—
1,3-Diphenylguanidinium	—	5.34[e]	—	—
Morpholinium	—	3.5	—	—
Piperidinium	—	—	+120	6.45
Ammonium	—	—	+155	5.85
Ethylenediaminium(I)	—	—	+230	4.6
Ethylenediaminium(II)	—	—	+100	6.8

[a] Ref. 9.
[b] Ref. 87.
[c] Ref. 159.
[d] Ref. pK_a calculated applying Eq. 42 with $E^0(H^+/H_2) = -0.500$ V and according to the conversion method described in section VI,D,5,b.
[e] Ref. 131.

Finally, the highest pK_a values measured in pyridine so far correspond to the M^+, H_2O/MOH acid–base couples. Indeed, in the presence of a background electrolyte, water necessarily imposes a limitation of the available basic range of the solvent because of its basic character. The voltametric reduction waves of H_2O associated with M^+ cations were observed at rotating hydrogen electrodes.[9] Potentiometric results at zero current yielded the pK_a values of several M^+, H_2O/MOH couples. They may be conveniently compared to that of tetramethylammonium hydroxide obtained by calculation from the known constants of the corresponding salts[123] (Table XIV).

4. ASSOCIATION EFFECTS

Interesting investigations have demonstrated the importance of association effects as far as acid–base behavior is concerned. Some authors looked to the homoconjugation effect to explain the typical form of titration curves as well as the discrepancies in the pK_a values of acids (see Tables XII and XIII), for they were generally analyzed on the basis of simple dissociation

TABLE XIV

Dissociation Constants (pK_a) of Some
M^+, H_2O/MOH Acid–Base Couples
in Pyridine at 25 C

Acid–base couples	pK_a	Reference
Na^+, $H_2O/NaOH$	13.5 ± 2.0	9
Li^+, $H_2O/LiOH$	17 ± 2.0	9
NMe_4^+, H_2O/NMe_4OH	21	123
NEt_4^+, H_2O/NEt_4OH	20.5 ± 0.5	9
NBu_4^+, H_2O/NBu_4OH	22.0 ± 0.5	9

only. If this phenomenon occurs, the additional equilibrium, determined by the homoconjugation constant $K_{HA_2^-}$, must be taken into account:

$$HA + A^- \rightleftharpoons HA_2^- \tag{43}$$

In addition, by postulating homoconjugation rather than a dimerization pattern, a greater agreement between calculated and experimental data was obtained. This fact conforms with what was observed elsewhere in solvents with a comparable proton-accepting tendency.[141]

Bos and Dahmen[87] provided evidence for the occurrence of homoconjugation effects, especially with both the indicator acids that served as references for the pK_a calculations in Mukherjee's work.[19] The authors employed a trial and error treatment of data to reach the best fit between experimental and calculated titration curves. The results, including $pK_{HA_2^-}$ values, were summarized in Table XII.

Kolthoff and Bruckenstein[160] reported that the dissociation and ionization constants, K_d and K_i, of indicator acids were able to be determined spectrophotometrically by means of the equation:

$$\frac{(HIn)}{[\Sigma (In) - K_i(HIn)]^2} = (K_i \cdot K_d)^{-1} \tag{44}$$

where HIn and In are the respective concentrations of the indicator in the acid and base color, determined from treatment of absorbance data. In the case of polynitrophenols and sulfophthaleins, application of Eq. 44 by Bos and Dahmen[87] did not result in a single value for K_i; the results did, however, fit the equation for simple dissociation:

$$\frac{[\Sigma (In)]^2}{(HIn)} = K_a \tag{45}$$

The earliest experiments were performed by Bruss and Harlow,[110] using low-frequency conductance measurements. Striking differences in the shape of conductometric curves appeared, whether unhindered (monosubstituted) or hindered (di- and trisubstituted) phenols were titrated. In the first case

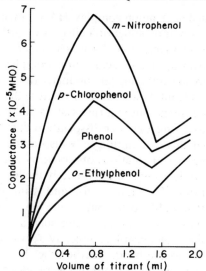

FIG. 1. Conductance curves of unhindered phenols in pyridine. From Bruss and Harlow,[110] reprinted by permission of *Analytical Chemistry*.

(Fig. 1), a conductance maximum was observed at about the midpoint, while this maximum was not apparent when both ortho positions of the phenol were occupied (Fig. 2). From these results, it was postulated that ortho groups prevent the formation of highly conductive species by steric hindrance of the phenolic group. The nature of this effect was attributed to the

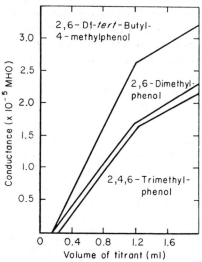

FIG. 2. Conductance curves of hindered phenols in pyridine. From Bruss and Harlow,[110] reprinted by permission of *Analytical Chemistry*.

formation of a bonded acid–anion complex, HAA^-, including the untitrated phenol molecule. This homoconjugation effect was further reinforced if the titrant possessed a relatively low basic character (sodium, potassium hydroxides, or ethoxides). These acid–anion complexes, associated with the titrant cation, should exhibit higher conductivities than either the phenol or the phenolate ion-titrant cation ion pair because the negative charges of the complex are more spread out in a ring system, holding the titrant cation more loosely, and thus giving a greater dissociation than the other species. It should be noted that this assumption may also account for the "anomalous" behavior of picric acid solutions observed by Davies.[21]

Corey attempted to determine approximately the fate of certain phenolic compounds in pyridine.[161] Mono-, di-, and trinitrophenols exhibited absorption bands attributable to the phenoxide ions, at 430 nm, and to the phenol band, at 360 nm. Addition of a base such as N-ethylpiperidine to solutions of 2,6-dinitro-3,4-xylenol completely removed the phenol band and promoted the phenoxide band. On the other hand, when successively larger amounts of acid (pyridinium nitrate) were added, the opposite effect was observed. The changes in intensity of the absorption bands were discussed in terms of relative values of the respective ionization and dissociation constants. His results were in agreement with those anticipated by Kraus from conductance data.[118]

The complexity of the dissociation process may also be illustrated by the conductance studies of 2,6-dinitrophenol solutions carried out by Wilski and Kortüm.[12] The authors observed that the value of its dissociation constant in water–pyridine mixtures of variable composition reached a maximum for solutions having 10% proportion of amine. Though this variation was attributed to specific solvation effects, in connection with a change in basicity and dielectric constant, it seems that the numerical value yielded for high amine proportions cannot satisfactorily account for simple dissociation schemes only, as was shown by the comparative results compiled in Table XII.

5. RELATIVE ACIDITIES

Unlike the small number of systematic investigations performed to determine the dissociation constants of weak acids in pyridine, considerable data dealing with their relative acidities are available.[101,162] Except for data collected by the use of indicators, the techniques were based on the treatment of either titration curves or polarographic reduction waves of acids in pyridine.

a. *Indicators.* Reviews of the literature available up to 1961[163-165] reveal the rather reduced number of indicators which have been proposed for visual titrations in pyridine. As pointed out by Fritz and Gainer,[163] the transition ranges of indicators should be measured in terms of pK_a's in this solvent, so as to allow a selection of an indicator for any particular requirement. Furthermore, great care must be taken regarding the salt effect on the indicator equilibrium: in solvents with a low dielectric constant, this effect may be so pronounced that the correct end point at one concentration is incorrect at another.[146] However, most available papers simply aimed at providing indications for routine experiments. Since these attempts were generally made before absolute dissociation constants were available, only relative transition ranges of indicators have been reported. This fact may be connected with the almost complete absence of investigations aiming to set up Hammett acidity scales in this solvent. In any case it was found necessary first to titrate potentiometrically to ascertain whether a compound has an end point and, if so, the color of this end point. A comparative study of azo violet and thymol blue, the most widely used indicators, was conducted by Maurmeyer, Margosis, and Ma.[166]

In a recent critical investigation, Fritz and Gainer[163] determined the transition ranges of 13 indicators under conditions simulating actual acid–base titrations. The proposed indicators proved to cover almost the whole span of acids in pyridine; they are listed in Table XV[87,101,106,155,163,165] [170] with the corresponding wavelengths of the maximum absorbance of basic and acidic forms. Other data from different sources are also presented in Table XV. Figure 3 shows both the visual and photometric transition ranges of the indicators studied together by Fritz and Gainer, with the colors at each end of the range.

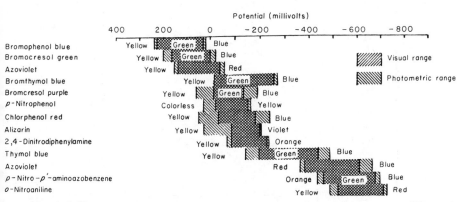

FIG. 3. Indicator transition ranges in pyridine as solvent. From Fritz and Gainer,[163] reprinted by permission of *Talanta*.

TABLE XV

INDICATORS IN PYRIDINE

Indicators	Wavelength (nm) Basic[a]	Wavelength (nm) Acidic[a]	pK_a(py)	Titrable acidic compounds
Thymol blue	622	401	—	Acetic, benzoic, nicotinic, salicylic, ammonium nitrate and acetate, p-nitrophenol, o-nitrophenol, 2,4-dinitrophenol,[d] aminobarbital, barbital, sulfacetamide, sulfadiazine, β-naphthol,[e] o-, m-, p-nitrophenol, 2,4-dichlorophenol, 4-nitro-2,6-dichlorophenol, p-hydroxybenzaldehyde and p-hydroxyacetophenone,[f] trichloroacetic,[g] cyclic nitramines[h]
Azovioliet	555	430	—	N-Arylhydroxamic,[i] resorcinol, phthalimide, sulfacetamide, sulfanilide[e]
Bromocresol purple	608	418	7.5[b]	m-Chlorobenzoic, benzoic;[a] trichloroacetic[g]
Chlorophenol red	605	418	10.1[b]	
p-Nitrophenol	425	—	7.6[c]	
Alizarin	585	435	—	Benzoic, o-chlorophenol[a]
Bromophenol blue	608	430	5.3[b]	Perchloric, salicylic, m-chlorobenzoic, benzoic;[a] trichloroacetic[g]
Bromocresol green	634	428	—	
Bromothymol blue	638	406	—	Salicylic, m-chlorobenzoic, benzoic, o-chlorobenzoic, o-chlorophenol, phenol[a]
2,4-Dinitrodiphenylamine	500	363	—	
p-Nitro-p′-amino-azobenzene	645	460	—	Phenol, o-chlorophenol, 2,4,6-trimethylphenol[a]
o-Nitraniline	510	421	—	Active esters of amino acids[j]
Azovioliet	631	555	5.5[b]	
Bromocresol green	—	—	—	
Phenolphthalein	—	—	—	Substituted ureas (peroral antidiabetics)[k]

[a] Ref. 163 (with permission).
[b] Ref. 87.
[c] Ref. 155.
[d] Ref. 101.
[e] Ref. 166.
[f] Ref. 167.
[g] Ref. 168.
[h] Ref. 169.
[i] Ref. 170.
[j] Ref. 106.
[k] Ref. 165.

b. *Polarographic and Potentiometric Titrations.* Polarographic reduction
half wave potentials $(E_{1/2})$ and titrimetric half-neutralization potentials
$[E(\text{hnp})]$ of numerous acids are compiled in Tables XVI[9,79,123,154,155,171,172]
to XIX. The values are derived principally from the studies carried out by
Elving and Streuli, respectively.

TABLE XVI

HALF-NEUTRALIZATION POTENTIALS, POLAROGRAPHIC HALF-WAVE POTENTIALS, AND DISSOCIA-
TION CONSTANTS (pK_a) OF SOME BENZOIC ACIDS IN PYRIDINE AT 25°C

| Benzoic acids | $pK_a{}^a$ (H_2O) | $\Delta E(\text{hnp})^b$ (mV) | $E_{1/2}{}^c$ (mV) | Calculated | | | |
				$pK_a{}^d$	$pK_a{}^e$	$pK_a{}^f$	$pK_a{}^g$
Benzoic	4.20	0	−1590	8.5	8.5	9.8	11.0
Methylbenzoic							
o-	3.91	−44	—	9.2	—	—	—
m-	4.27	−16	—	8.7	—	—	—
p-	4.37	−3	—	8.55	—	—	—
Methoxybenzoic							
o-	4.09	−33	—	9.0	—	—	—
m-	4.09	5	—	8.4	—	—	—
p-	4.47	−45	—	9.2	—	—	—
Chlorobenzoic							
o-	2.94	65	—	7.4	—	—	—
m-	3.83	49	—	7.65	—	—	—
p-	3.98	25	—	8.0	—	—	—
Bromobenzoic							
o-	2.86	77	—	7.2	—	—	—
m-	3.81	63	—	7.4	—	—	—
p-	3.97	30	—	8.0	—	—	—
Nitrobenzoic							
o-	2.17	142	—	6.1	—	—	—
m-	3.49	97	—	6.85	—	—	—
p-	3.42	110	—	6.6	6.45	—	—
Hydroxybenzoic							
o-	2.98	121	—	6.4	—	—	—
m-	4.08	−23	—	8.8	—	—	—
p-	4.54	−80	—	9.8	—	—	—
Aminobenzoic							
o-	5.00	−10	—	8.6	—	—	—
m-	4.82	−38	—	9.15	—	—	—
p-	4.92	−105	—	10.2	—	—	—

[a] Refs. 171, 172.
[b] Ref. 155 [taking zero for $E(\text{hnp})$ of benzoic acid].
[c] Ref. 154 (vs. the normal Ag reference electrode—see Table III).
[d] Ref. 155, taking $E^0(H^+/H_2) = -0.5$ V vs. the NHE and according to the conversion
method described in Section VI,D,5,b.
[e] Ref. 9.
[f] Ref. 79.
[g] Ref. 123.

TABLE XVII

HALF-NEUTRALIZATION POTENTIALS, POLAROGRAPHIC HALF-WAVE POTENTIALS, AND DISSOCIATION CONSTANTS (pK$_a$) OF VARIOUS ACIDS IN PYRIDINE AT 25°C

Acids	pK$_a$(H$_2$O)[a]	ΔE(hnp)[b] (mV)	ΔE(hnp)[c] (mV)	$E_{1/2}$[d] (mV)	pK$_a$ Measured			pK$_a$ Calculated		
					e	f	g	h	i	j
Acetic	4.70	−97	−90	−1710	10.6	10.1	12.0	10.1	10.5	10.0
Chloroacetic	—	—	—	−1490	6.5	—	—	—	6.8	—
Dichloroacetic	—	—	—	−1380*	—	—	—	—	—	—
Trifluoroacetic	—	—	—	−1350*	—	—	—	—	—	—
Trimethylacetic	—	−102	—	−1740	—	—	—	10.2	11.0	—
Picolinic	1.01	—	—	—	—	—	—	—	7.5	—
Formic	3.77	57	25	−1530	—	—	—	7.5	8.2	8.1
Lactic	3.87	39	—	−1570	—	—	—	7.8	—	—
Crotonic	4.26	−28	—	—	—	—	—	9.0	—	—
Acrylic	4.69	−83	—	—	—	—	—	9.9	—	—
Barbituric	4.0	185	—	—	10.2	—	—	5.3	—	—
Cyanuric	—	−106	—	—	—	—	—	10.3	—	—
Nicotinic	2.07	—	—	−1530	—	—	—	—	7.5	—
p-Toluenesulfonic	—	421*	310	—	—	—	—	—	—	—
Naphthalenesulfonic	—	420*	—	—	—	—	—	—	—	—
Methanesulfonic	—	386	—	—	—	—	—	1.9	—	—
Cyanamide	9.46	−310	—	—	—	—	—	13.7	—	—
Dicyanamide	14.22	−441	—	—	—	—	—	15.9	—	—
m-Benzenedisulfonamide	—	−102	—	—	—	—	—	10.2	—	G
Phthalimide	9.9	−115	—	—	—	—	—	10.4	—	—
Succinimide	9.62	−270	—	—	—	—	—	13.1	—	—

Compound	pKa	E (mV)	E (mV)	E (mV)	pK	pK
α-Cyanacetamide	—	−319	—	—	—	13.9
α-Chloroacetamide	—	−535	—	—	—	17.6
α-α-Dichloro-N-methylacetamide	—	−498	—	—	16.9	—
Benzotriazole	—	−47	—	—	9.3	—
Trinitrotoluene	—	2	—	—	8.5	—
Nitromethane	—	−401	−405	—	15.3	15.4
Purine	8.93	—	—	−1660	9.7	—
6-Hydroxypurine						
(I)	8.94	—	—	−1720	10.7	—
(II)	12.1	—	—	−2150	18.0	—
6-Aminopurine	9.83	—	—	−1810	12.2	—
6-Methoxypurine	9.16	—	—	−1740	11.0	—
6-Mercaptopurine						
(I)	7.4	—	—	−1610	8.85	—
(II)	10.84	—	—	−2020	15.8	—
2,6-Dihydroxypurine						
(I)	7.74	—	—	−1700	10.5	—
(II)	11.86	—	—	−2110	17.5	—

[a] Refs. 154, 156, 171, and 172.
[b] Ref. 154 [taking zero for E(hnp) of benzoic acid].
[c] Ref. 159 [taking zero for E(hnp) of benzoic acid].
[d] Ref. 154 (vs. the normal Ag reference electrode—see Table III).
[e] Ref. 9.
[f] Ref. 79.
[g] Ref. 123.
[h] Ref. 156; calculated pK_a's, taking $E°(H^+/H_2) = -0.5$ V vs. the NHE and according to the conversion method described in Section VI,D,5,b.
[i] Ref. 154, on the same basis as h.
[j] Ref. 159, on the same basis as h.
* Leveled potential.

TABLE XVIII

HALF-NEUTRALIZATION POTENTIALS, POLAROGRAPHIC HALF-WAVE POTENTIALS, AND DISSOCIATION CONSTANTS (pK_a) OF SOME DICARBOXYLIC ACIDS IN PYRIDINE AT 25°C

Dicarboxylic acids		$pK_a(H_2O)^a$	$\Delta E(hnp)^b$ (mV)	$\Delta E(hnp)^c$	$E_{1/2}^d$	pK_a^e	Calculated pK_a		
							f	g	h
Oxalic	I	1.46	196	—	−1370*	4.25	—	5.1	—
	II	4.40	−227	—	−1780	11.8	—	12.3	11.7
Malonic	I	2.83	238	—	—	—	—	4.45	—
	II	—	−280	—	—	—	—	13.2	—
Succinic	I	4.21	115	—	−1500	—	—	6.5	7.0
	II	5.64	−250	—	−1930	—	—	12.7	14.25
Glutaric	I	4.34	30	—	—	—	—	8.0	—
	II	—	−192	—	—	—	—	11.7	—
Adipic	I	4.41	−12	—	−1650	—	—	8.65	9.5
	II	5.28	−111	—	−1760	—	—	10.4	11.3
Pimelic	I	4.48	−44	—	—	—	—	9.2	—
	II	—	−121	—	—	—	—	10.5	—
Azelaic	I	4.55	−40	—	—	—	—	9.1	—
	II	—	−102	—	—	—	—	10.2	—
Malic	I	3.65	225	—	—	—	—	2.5	—
	II	—	−153	—	—	—	—	11.1	—
Tartaric	I	3.03	196	—	−1420	4.25	—	5.1	5.5
	II	4.37	−78	—	−1650	—	—	10.0	9.5
Maleic	I	2.00	352	—	—	—	—	2.6	—
	II	6.26	−303	—	−2010	—	—	13.6	15.5

Compound		pKa							
Fumaric	I	3.02	65	—	−1460	—	—	7.45	6.3
	II	4.45	−92	—	−1780	—	—	10.0	11.7
Phthalic	I	2.96	231	—	−1350*	—	—	4.6	—
	II	5.51	−363	—	—	—	—	14.6	—
Isophthalic	I	3.62	63	—	—	—	—	7.4	—
	II	—	−37	—	—	—	—	9.1	—
Terephtalic	I	3.54	28	—	−1530	—	—	8.0	7.5
	II	4.46	−57	—	−1680	—	—	9.5	10.0
Salicylic	I	2.97	—	—	−1400	—	—	—	5.2
	II	13.0	—	—	—	—	—	—	—
cis-1,4-Dicarboxycyclohexane	I	—	−13	—	—	—	—	8.7	—
	II	—	−37	—	—	—	—	9.2	—
Sulfhydric	I	—	—	−120	—	—	10.5	—	—
	II	—	—	—	—	—	—	—	—
Sulfuric	I	—	—	—	—	—	—	—	—
	II	—	—	−265	—	11.5	13	—	—
Citric	I	—	—	—	—	—	—	—	—
	II	—	—	—	—	10.7	—	—	—

[a] Refs. 171 and 172.
[b] Ref. 155 [taking zero for E(hnp) of benzoic acid].
[c] Ref. 159 [taking zero for E(hnp) of benzoic acid].
[d] Ref. 154 (vs. the normal AG reference electrode—see Table III).
[e] Ref. 9.
[f] Ref. 159 [calculated pK_a, taking $E^0(H^+/H_2) = -0.5$ V vs. the NHE and according to the conversion method described in Section VI,D,5,b].
[g] Ref. 155, as for f.
[h] Ref. 154 as for f.
* Leveled potentials.

TABLE XIX

HALF-NEUTRALIZATION POTENTIALS, POLAROGRAPHIC HALF-WAVE POTENTIALS, AND DISSOCIATION CONSTANTS (pK_a) OF PHENOLS IN PYRIDINE AT 25°C

Acids	$pK_a(H_2O)$	$\Delta E(hnp)^a$ (mV)	$E_{1/2}{}^b$ (mV)	pK_a Measured c	d	e	pK_a Calculated f	g
Phenol	10.0	−340	−2010	—	—	—	14.3	15.6
Catechol	9.4	−194	—	—	—	—	11.8	—
Resorcinol	9.4	−367	—	—	—	—	14.6	—
Hydroquinone	10.0	−345	—	15.6	—	—	14.3	—
tert-Butylcatechol	—	−224	—	—	—	—	12.3	—
Pyrogallol	9.35	−156	—	—	—	—	11.1	—
Phoroglucinol	8.7	−287	—	—	—	—	13.3	—
Thiophenol	—	—	—	—	—	—	8.8	—
o-Cresol	10.2	−386	—	—	—	—	15.0	—
m-Cresol	10.0	−386	—	—	—	—	15.0	—
p-Cresol	10.3	−387	—	14.9	—	—	15.0	—
o-Chlorophenol	8.5	−190	—	12.1	—	—	11.7	—
m-Chlorophenol	9.0	−220	—	—	—	—	12.2	—
p-Chlorophenol	9.4	−248	−1940	—	—	—	12.7	14.5
o-Nitrophenol	7.2	+25	—	—	—	—	8.1	—
m-Nitrophenol	8.4	−139	—	10.3	12.5	—	10.8	—
p-Nitrophenol	7.1	+50	—	—	—	—	7.6	—
o-Aminophenol	9.95	−389	—	—	—	—	14.2	—
m-Aminophenol	—	−438	—	—	—	—	15.8	—
p-Aminophenol	10.7	−463	—	—	—	—	16.0	—
m-Methoxyphenol	9.3	−310	—	—	—	—	13.8	—
p-Methoxyphenol	10.2	−403	—	—	—	—	15.3	—
2,4-Dinitrophenol	3.5	359	—	—	4.4	4.38	2.4	—
2,5-Dinitrophenol	4.7	—	—	6.25	5.3	5.76	—	—
2,6-Dinitrophenol	3.0	—	—	—	4.7	—	—	—
2,4,6-Trinitrophenol	0.4	428	—	—	3.0	—	—	—
2,4-Dichlorophenol	7.85	−78	−1760	—	—	—	9.8	11.5
2,4,6-Trichlorophenol	7.6	49	—	—	—	—	7.7	—
3,4-Dimethylphenol	10.3	—	−2060	—	—	—	—	16.5
2,4-Dimethylphenol	—	—	—	15.6	—	—	—	—
2-Hydroxypyridine	—	—	—	12.9	—	—	—	—
1-Naphthol	9.85	−210	—	—	—	—	12.0	—
2-Naphthol	9.95	−205	—	—	—	—	12.0	—

[a] Refs. 155 and 156 [taking zero for E(hnp) of benzoic acid].

[b] Ref. 154 (vs. the normal Ag reference electrode—see Table III).

[c] Ref. 9.

[d] Ref. 87.

[e] Ref. 19.

[f] Refs. 155 and 156 [calculated pK_a, taking $E^0(H^+/H_2) = -0.5$ V vs. the NHE and according to the conversion method described in Section VI,D,5,b].

[g] Ref. 154, as for f.

It would be of great interest to correlate these potentials with dissociation constants determined by other measurements. One means would consist of using Eq. 42 to express potentials in terms of pK_a's with the compulsory assumption that Eq. 42 holds for each type of acid and for the procedures employed. For convenience, benzoic acid has been chosen as both the potential and pK_a standard. On this basis, pK_a values have been calculated from $E_{1/2}$ and $E(hnp)$ values, by assigning the value 8.5 to the pK_a (py) of benzoic acid. Tables XVI to XIX present pK_a's determined in this way.

The titration behavior of various substances was reported, generally based on potentiometric methods using glass or polarized bimetallic electrodes. Attempts concerned mono- and dialkylphosphoric acids,[173] differently substituted ammonium arenesulfonates,[105] derivatives of benzothiazole,[101,174] neutral and acidic tetraethylammonium salts of dicarboxylic acids,[175] complex thiocyanates,[176] thiobis- and methylenebisphenols,[177] primary and secondary p-bromobenzylamines,[178] several compounds of pharmaceutical interest,[179,180] weak acids deriving from dibenzoylmethane,[100] imides, thiols, enols,[101,167] dihalophenylsilanediols and hexaorganocyclotrisiloxanes,[181] naphthyl esters of carboxylic acids,[99] diethyl- and monoethylhydrogen sulfates in mixtures with sulfuric acid,[182] sodium aminoethylthiosulfates and aminoethylthiosulfuric acid derivatives,[183] acidic groups of lignin,[98] and quite a number of organic compounds including the quantitative determination of hydroxyl groups on the submicroscale.[103,167] The relative acidities of nitrophenols, nitrosamines, oximes, organic substances carrying acidic hydroxyl groups, and Brønsted acids were estimated by conductance measurements.[184]

6. CORRELATIONS BETWEEN ACIDITY SCALES IN WATER AND PYRIDINE

a. *Titrimetric Analysis.* Streuli and Miron[155] undertook extensive studies to determine whether any basis can be established for predicting the titration behavior of organic acids. They therefore intended to classify certain aliphatic acids, phenols, and mono- and dicarboxylic acids in pyridine. Rules for this classification were set up on the basis of their dissociation constants in water, designated by $pK_a(H_2O)$, as a function of the hnp measured in pyridine. Benzoic acid with an assigned hnp of zero was the standard in this solvent. Measurements were carried out with glass electrodes using NBu_4OH in methanol–benzene solutions as a titrant.

Data dealing with the benzoic acids tested are listed in Table XVI. The so-called Δhnp scales thus obtained readily showed striking features, provided by the linear relationships observed for acids of the same types (Fig. 4). This was the case for meta- and para-substituted benzoic acids, on the one hand, and aliphatic monocarboxylic acids on the other hand. In

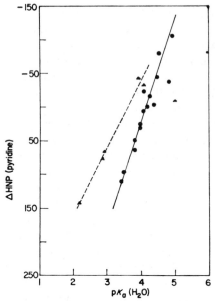

FIG. 4. Acidic strength of substituted benzoic acids in pyridine and in water. ▲, Ortho-substituted acids; ●, meta- and para-substituted acids. (ΔHNP designates the half-neutralization potentials of the acids relative to that of benzoic acid, taken as reference.) From Streuli and Miron,[155] reprinted by permission of *Analytical Chemistry*.

addition, all these acids maintained the same relative relationship in both water and pyridine. The dependence between $pK_a(H_2O)$ and $E(hnp)$ in pyridine was found to be expressed by the relation

$$E(hnp) = 0.647 - 0.156 \, pK_a(H_2O) \qquad (46)$$

To be consistent with the general convention, the sign of $E(hnp)$ potentials used by Streuli has been reversed here and in the following equations. Most ortho acids also showed a linear dependence between their acidity in water and in pyridine, but the slope of the straight line was less steep. The corresponding relationship was described by the equation

$$E(hnp) = 0.361 - 0.1 \, pK_a(H_2O) \qquad (47)$$

From these data, it was concluded that most benzoic acids are weaker in pyridine. If changes in acidity can be ascribed primarily to inductive and electronic effects, the results implied that the so-called ortho effect is less marked in pyridine than in water. Hence, because of the alteration of this

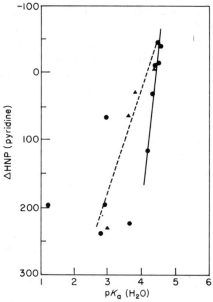

FIG. 5. Acidic strength of dibasic acids in pyridine and water. ●, Aliphatic acids; ▲, aromatic acids. (ΔHNP designates the half-neutralization potentials of the acids relative to that of benzoic acid, taken as reference.) From Streuli and Miron,[155] reprinted by permission of *Analytical Chemistry*.

ortho-para connection, the ability to resolve these pairs by titration appears inferior in pyridine.

A list of the aliphatic acids tested is reported in Table XVII. Relevant data were obtained from the study of dibasic acids (Table XVIII). They indicated that these compounds can readily be subdivided into three classes: those containing a saturated chain between the carboxyls, those with unsaturated linkage between carboxyls, and those with the carboxyl functions attached to an aromatic ring (Fig. 5). All these acids gave two neutralization breaks. Among the former group, the heights of the breaks decreased gradually with increasing chain length. In the third group of acids, enhancement of the acidity of the ortho isomers was noticed. Thus, mixtures of phthalic and isophthalic acid or phthalic and terephthalic acids may be simultaneously titrated in pyridine. However, corresponding relationships could not be found for short-chain dibasic acids. As pointed out by Elving,[119,154] this is due to the lack of stabilization of the open-ring monoanion by the solvent through hydrogen bonding; pyridine, unlike water, favors the formation of the cyclic form, since it lacks a hydrogen-donor character. Consequently, the effect of intramolecular bonding on the first dissociation of such dibasic acids will be more marked in pyridine, thereby increasing the strength of the first carboxyl function. On the other hand, the second neutralization step is accompanied by disruption of the internal hydrogen bonding and thus produces a decrease in the dissociation constant beyond what would occur, for instance, in water. This effect can best be observed in very short-chain acids (e.g., oxalic and malonic) and in rigidly structured (e.g., maleic, fumaric, and phthalic) acids. If the geometric structure of the anion hinders this type of bonding, e.g., terephthalic acid, the E(hnp)-pK_a(H_2O) relationship coincides with that of monocarboxylic acids.

Among the phenols studied,[156] the ortho-substituted compounds manifested a similar enhancement of relative acidity over the meta- and para isomers (Table XIX). Titration curves of varying shapes were noted and related to both acid strength and steric effects. However, no regular gradation was observed. It appeared that any phenol with a pK_a(H_2O) value greater than 7.8 showed a steep titration curve in pyridine, as compared with that of benzoic acid. Plots of Δhnp values vs. pK_a(H_2O) for the monosubstituted phenols studied (Fig. 6) gave a linear relationship according to the equation:

$$E(\text{hnp}) = 1.14 - 0.152 \ pK_a(H_2O) \tag{48}$$

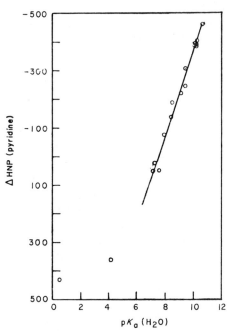

FIG. 6. Acidities of substituted phenols in pyridine and water. (ΔHNP designates the half-neutralization potentials of the phenols relative to that of benzoic acid, taken as reference.) From Streuli,[156] reprinted by permission of *Analytical Chemistry*.

Observation of similar slopes in Eqs. 46 and 48 indicated that the difference in the constants represented the relative enhancement in acidity of phenolic compounds in pyridine over water, i.e., 3.4 pK_a units.

In the same paper, several hydroxy aromatic compounds and acidic organic substances containing neither hydroxyl nor carboxyl groups were tested titrimetrically. (Appropriate data were presented in Table XVII.) There was only occasional agreement with predicted values derived from the preceding equations. As mentioned, this discrepancy is not surprising in view of the importance of structural effects.

b. *Polarographic Analysis.* An alternate approach for exploring the behavior of acids in pyridine solutions was developed by Elving and collaborators. They observed that 29 Brønsted acids, including mono- and dicarboxylic acids, phenols, and purines produced a one-electron diffusion-controlled polarographic wave. Taking into account the additive property of the waves, the electrochemical process was attributed to the reduction of undissociated pyridinium–acid anion complexes.[154] Following the interesting lead suggested by Streuli,[155,156] plots of half-wave potential ($E_{1/2}$) values vs. $pK_a(H_2O)$ were drawn. Experimentally, linear and separate dependences were built up for each type of acid, thus paralleling Streuli's results for $E(hnp)$ vs. $pK_a(H_2O)$ plots (Fig. 7). The straight portions of the relationships were found to correspond to the equations:

$$E_{1/2} = -1.10 - 0.124\ pK_a(H_2O) \qquad (49)$$

$$E_{1/2} = -0.79 - 0.124\ pK_a(H_2O) \qquad (50)$$

$$E_{1/2} = -0.62 - 0.124\ pK_a(H_2O) \qquad (51)$$

which hold for carboxylic acids, phenols, and purines, respectively. A comparison of Eqs. 46 to 48 with Eqs. 49 to 51 makes it clear that $E(hnp)$ vs. $E_{1/2}$ plots will yield an almost 1 : 1 correlation. This suggests that both $E(hnp)$ and $E_{1/2}$ potentials are linked with the strengths of the acids in pyridine. Elving explained the agreement between both procedures in the following manner: the ease of electron addition onto the py/HA complex would be proportional to the electron density at the site attacked. This density would, in turn, depend upon the strength of the $H^{\pm}$–A^- binding. Considering the overall dissociation equilibrium of the py/HA species

$$py/HA \overset{K_a}{\rightleftharpoons} py/H^+ + A^- \qquad (52)$$

it was assumed that reaction 52 was rapid compared with the rate-determining step constituted by the electron transfer reaction. On this basis, $E_{1/2}$ was expected to become more positive with an increasing K_a, in agreement with experimental results.

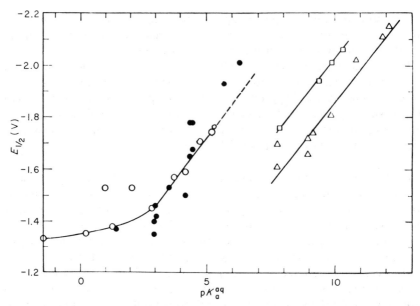

FIG. 7. Relations between half-wave potentials of waves given by solutions of Brønsted acids in pyridine containing 0.1 M tetraethylammonium perchlorate at 25°C and the aqueous pK_a values of the acids. $\bigcirc$, Nitric and monocarboxylic acids; $\bullet$, salicylic and dicarboxylic acids; $\square$, phenols; $\triangle$, purines. ($E_{1/2}$ potentials are referred to the normal silver reference electrode in pyridine.) From Tsuji and Elving,[154] reprinted by permission of *Analytical Chemistry*.

Thus, if $pK_a(H_2O)$ reflects the mutual interactions in the $H_2O-H^{\pm}-A^-$ system, either $E(\text{hnp})$ or $E_{1/2}$ would show the extent of the ionic character in the coordinated $py/H^{\pm}-A^-$ system, which is directly connected with the actual pK_a in pyridine.

Deviations of the acid strengths from the $E_{1/2}$ vs. $pK_a(H_2O)$ relations were clearly explained in terms of solvent change.[154,162] In particular, as seen above for dicarboxylic acids, the relative enhancement in the first acidity function appears qualitatively consistent with considerations based on nonelectrical effects. On the other hand, the observation of smaller second dissociation constants best accounts for the effect of the dielectric constant on the more negatively charged HA^-/A^{2-} couple than the preceding H_2A/HA^-.

It should be noted that there is no obstacle to building up a convergent system of straight lines, including the preceding $E(\text{hnp})$ or $E_{1/2}$ values, on the basis of the conversions described in Section VI,D,5,b. Such a $pK_a(Py)$ vs. $pK_a(H_2O)$ diagram was drawn up by Nigretto and Jozefowicz[9] using their dissociation constant estimates (Fig. 8).

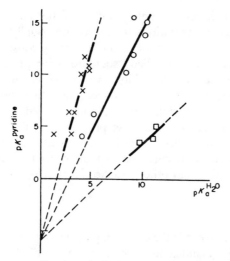

FIG. 8. Relations between pK_a values of acids in pyridine and water. ×, Carboxylic acids; ○, phenols; □, HB^+/B acid–base couples. From Nigretto and Jozefowicz,[9] reprinted by permission of *Electrochimica Acta*.

c. *Interpretation of the Correlations.* Elving[154] attempted to explain the occurrence of parallel and distinctly separate relationships between $E_{1/2}$ and $pK_a(H_2O)$, which had not been discussed previously for $E(hnp)$ vs. $pK_a(H_2O)$ plots. He finally emphasized two compatible interpretations, proposed initially by Grunwald and Price[185] and by Glover.[186] These were based either on dispersion effects due to London interactions or on changes in solvation numbers. Accordingly, the relative strengths of interactions through hydrogen bonding could be predicted to decrease in the following order: carboxylate > phenolate > purine anion, and to be much lower still for neutral bases such as substitute amines. However, lack of thorough knowledge of the solvation numbers of electrolytes, of the nature of the dissociation processes, and of a sufficiently wide range of data makes any quantitative verifications of the proposed theories in any but the most typical cases questionable.

VII. REDOX REACTIONS IN PYRIDINE

Redox systems in nonaqueous media are generally expected to follow a Nernstian behavior. Such organic or inorganic systems depend, however, on the proton availability of the medium, i.e., the pH. Thus, redox data inherently include two variables, the potential and the acidity. These parameters can conveniently be correlated by setting forth potential–pH diagrams of the thermodynamic stability of the corresponding species. Potentials affected by variations in acidity are usually designated by the term "apparent potentials" while those remaining constant over a more or less restricted pH range are called "standard redox potentials."

In fact, this procedure has seldom been followed over a wide range of acidity in pyridine, except for data concerning iodine and its redox derivatives. Consequently, the published data must be used with caution because they generally ignore the acid–base properties of the species. However, especially for inorganic species, results have mostly been collected in media sufficiently buffered by strong acids for the potentials given to express standard values, provided that other specific effects are controlled.

A. Earlier Studies

Earlier works concerned the electrodeposition of metal ions. Laszcynski[48] and Kahlenberg[44] claimed to have succeeded in discharging alkali metal ions from their salts on mercury cathodes. Von Hevesy[40] reported on similar attempts with alkali earth metal ions. However, Muller et al.[187] disputed some of the preceding results on the basis of polarization curves recorded in a cleaner solvent. They determined the decomposition potentials of various metallic salts and the single potentials of the corresponding metals, including alkalis, alkali earths, aluminum, and cerium.[45,187–192] They demonstrated that most of these ions could not be discharged in pyridine since they decomposed the solvent through the creation of adducts. With beryllium, the formation of gas was observed. On the other hand, silver, zinc, copper, and iron were isolated.[192,193] A survey of the literature up to 1932 has been published by Audrieth and Nelson.[193]

B. Recent Studies

More precise data obtained by recent techniques and under more rigorous working conditions are now available. Most of the interest has focused on redox couples involved as potential standards in pyridine[194]; these will therefore be reviewed separately.

1. REDOX COUPLES USED AS POTENTIAL REFERENCES

Hladky and Vrestal[15] conducted a series of potentiometric investigations to estimate the standard redox potentials E^0 of the Cu(II)/Cu(I), Fe(III)/Fe(II), Ti(IV)/Ti(III), and V(IV)/V(III) couples in hydrochloric pyridine solutions. Emf measurements were carried out using the cell:

$$Hg\|Hg_2Cl_2, KCl(sat.), H_2O\|KCl(sat.), MeOH\|HCl, Red, Ox, Py\|Pt \qquad (53)$$

that is, with potentials being referred to the aqueous SCE through a methanolic bridging solution, probably to avoid inconvenient precipitation of the salts. The numerical values were readily determined by considering the half-neutralization potentials of the titration curves of conjugate elements of the different couples listed above. These values are given in Table XX. All the potentials were expressed in relation to the NHE potential in pyridine. The potential of the latter electrode was indicated to be 460 mV lower than that of the aqueous SCE in water. There is no doubt about the occurrence of junction potentials which, in this case, are undetermined. To make useful comparisons with data published elsewhere, relative E^0 values can be calculated (see Table XX). Surprisingly enough, the corresponding values do not differ greatly, even though various dispersion effects would be expected to operate extensively (associations, complexations, and influences due to

TABLE XX

STANDARD REDOX POTENTIALS IN PYRIDINE (RELATIVE TO THE NHE)

Redox couples	E^0	Medium, HCl (mM)	E^0 [Fe(III)/Fe(II)] $- E^0$ [Cu(II)/Cu(I)]
Fe(III)/Fe(II)	0.695^a	50	—
			$0.270–0.285^a$
			0.290^b
			0.3^c
Cu(II)/Cu(I)	0.509^d (18°C)	—	—
	0.560^d (95°C)	—	—
	0.425^a	50	—
	0.405^a	500	—
Cu(I)/Cu(0)	-0.004^d (18°C)	—	—
	0.057^d (95°C)	—	—
Cu(II)/Cu(0)	0.253^d (18°C)	—	—
	0.309^d (95°C)	—	—
Ti(IV)/Ti(III)	-0.190^a	200	—
V(IV)/V(III)	0.200^a	200	—
Ag(I)/Ag	0.551^e	—	—
	0.500^f	—	—
Fc$^+$/Fc	0.628^g	—	—
	0.970^h	—	—

[a] Ref. 15.
[b] Ref. 90.
[c] Refs. 197 and 198.
[d] Ref. 78.
[e] Ref. 121.
[f] Ref. 9 (potential of the Ag/Ag NO$_3$ 1 M, NBu$_4$ClO$_4$ 0.1 M reference electrode).
[g] Ref. 83.
[h] Ref. 9 (oxidation half-wave potential of ferrocene).

the differences in the nature of the salts used). Unfortunately, Hladky and Vrestal failed to apply the Nernst equation for these systems, despite the fact that the Nernstian plots exhibited linear relationships.

Unlike Hladky and Vrestal, Gupta[78] succeeded in his experiments and obtained straight lines with a slope close to the theoretical value 58 mV over the 2–50 mM range of concentration for the Cu(II)/Cu(I) couple. Emf measurements were performed with platinum electrodes in pyridine containing various amounts of cupric and cuprous chloride. The quantity $E^0[Cu(II)/Cu(I)]$ was determined by extrapolation, using the Nernst equation. However, $E^0[Cu(I)/Cu(O)]$, calculated from the corresponding Nernst equation, showed a definite drift in relation to the CuCl concentration, due to unreachable steady potentials. Nevertheless, extrapolation of the straight sections of the Nernst plots to zero concentrations in CuCl led to the value of -0.004 V vs. the NHE in pyridine. From the measured values of the Cu(II)/Cu(I) and Cu(I)/Cu(0) systems, the equilibrium constant and the enthalpy of the reaction

$$2\,Cu^+ \rightleftharpoons Cu^{2+} + Cu^0 \qquad (54)$$

were deduced; these were, respectively, $K = 1.4.10^{-9}$ at 18°C ($K = 1.3.10^{-7}$ at 95°C) and $\Delta H = 12.6$ kcal/mole. These results show that the relative stability of the cuprous and cupric systems is completely reversed in pyridine relative to water. This effect is due to the difference in the free solvation energies of the two ions. The comparative dependencies of potentials on chloride ion concentration in either cuprous or cupric solutions indicated that the former had a smaller tendency to promote complexes than cupric ion. However, insufficient conductance or thermodynamic data are available for further comparisons with the behavior of other transition metal salts in pyridine to be made.

In the absence of further cross checking with other results, some controversy surrounds the determination of the $E^0[Ag(I)/Ag(0)]$ potential. It has, however, been widely used in the past as a reference system, in particular in concentration cells,[195,196] or simply in comparative emf measurements with extraneous solvents.[90,197,198] However, only a few noteworthy attempts have been made to estimate the position of this system on a potential scale in pyridine,[9,121] and the resulting data do not entirely correspond, as shown in Table XX. This divergence may be ascribed to unconsidered complexing effects.[195]

In pyridine solutions, the solubility of transition metal salts and the oxidation potential of mercury are lowered, relative to water. For these reasons, the electrochemical behavior of relatively few metallic cations has been examined polarographically. Essential work has been done by Elving and Cizak[23] (Table XXI). The polarographic reduction waves of thallous(I) salts

TABLE XXI

POLAROGRAPHIC HALF-WAVE POTENTIALS OF SOME METAL IONS IN PYRIDINE[a]

Metals	$E_{1/2}$ (V)	Medium
Tl(I)	-0.48	Nitrate
Pb(II)	-0.61	Nitrate
	-0.96	Chloride
Zn(II)	-1.61 (1st wave)	Chloride
	-1.70 (2nd wave)	Chloride
Al(III)	-1.30 (1st wave)	Chloride
	-1.45 (2nd wave)	Chloride

[a] From Cisak and Elving.[23] All potentials are relative to the normal Ag reference electrode in pyridine.

were found to develop reversibly, at least for concentrations of less than 0.5 mM. Beyond that limit, first-kind maxima appeared. Likewise, typical two-electron reduction waves were observed for lead(II) solutions. Two waves were obtained for zinc(II), separated by an inflection point instead of a diffusion plateau. After prolonged electrolysis, the presence of metallic zinc in the mercury cathode was established. It was suggested that the electrochemical process consisted of a two-stage reaction, involving undissociated pyridine–zinc salt ion pairs, as follows:

$$ZnCl_2(Py)_4 + e^- \rightleftharpoons ZnCl_2(Py)_4^- \tag{55}$$

$$ZnCl_2(Py)_4^- + e^- + Hg \rightleftharpoons Zn(Hg) + 2\,Cl^- + Py \tag{56}$$

However, no cyclic voltammetric studies were ever reported to confirm this reaction scheme.

The ferricenium/ferrocene system (Fc^+/Fc) was first suggested by Strehlow and later used by many electrochemists as an absolute reference for potential scales in solvents.[81] In pyridine solutions, this system was examined either voltammetrically, by oxidizing the Fc form, or potentiometrically, with half cells consisting of equimolar mixtures of both the Fc^+ and Fc forms. A consistent voltammetric half-wave potential value for ferrocene oxidation was found by several authors to be 0.43 V[73] or 0.47 V[9] vs. the same reference electrode (the normal silver reference electrode) in pyridine, (respectively 0.93 and 0.97 V vs. the NHE in pyridine, according to Nigretto and Jozefowicz[9]). These values were confirmed by Brisset[82,148] with measurements carried out in water–pyridine mixtures of graded composition. Extrapolation of the half-wave potentials to a pure amine composition yielded the value 0.778 V vs. the aqueous NHE, or 0.458 V if converted to the normal Ag reference electrode, according to Nigretto and Jozefowicz[9] and Turner and Elving.[73] However, the most striking feature observed by Brisset was the formation of the Fc_2^+ form in mixtures containing more than 70% pyridine. This result has been paralleled by studies conducted in DMSO mixtures.[199] It should also be noted that solutions of ferricenium ions in methanol (and probably in other solvents) exhibit the same behavior as in pyridine.[200] This seems to imply that a different redox system is involved in pyridine solutions.

On the other hand, Mukherjee,[83] on the basis of DVP, conductance, and potentiometric measurements, claimed that ferrocene existed as a stable monomer and that ferricenium picrate dissociated as a simple weak electrolyte. His experiments yielded the average value of 0.628 V vs. the NHE in pyridine.

There will obviously be a marked discrepancy between the two sets of results obtained (see Table XX). It seems unlikely that the magnitude of liquid junction potentials is responsible for this difference, because the cells involved no other solvent but pyridine. Nor can the discrepancy be attributed to the insufficient solubility of the Fc^+ form. Rather, it should be sought in the complexity of the Fc^+/Fc system which is less simple than commonly expected. In this respect, it may be noted that Parker[201] recommended slow cyclic voltammetry or polarography rather than potentiometry for the study of ferrocene and ferricenium picrate solutions in DMSO or DMF.

2. ALKALI METALS

Lithium salts have attracted special attention, probably because of their use in analytical experiments. Mandell, McNabb, and Hazel[47] redetermined the decomposition potential of lithium chloride (4.00 V), which coincided with earlier results.[202] Other original values for lithium bromide and lithium iodide dihydrate were similar. The decomposition potential of potassium thiocyanate (3.65–3.70 V) was found to be different than that previously recorded. However, no attempt was made to explain this discrepancy.

Except for the brief report by Willeboordse[203] about the polarography of cadmium, zinc, and cobalt ions, Elving et al. were the first to consider pyridine as a potentially interesting solvent for polarographic studies. Their earliest work concerned the definition of optimum working conditions.[2,23] Systematic investigations, including cyclic voltametry, coulometry, and controlled potential electrolysis were conducted in this solvent. At mercury electrodes, Na(I), K(I), Rb(I), and Cs(I) were polarographically reduced in the 10^{-5}–4×10^{-3} M range of concentrations. Though reversible, the reduction waves exhibited anomalous diffusion current constants. This fact, along with the observation of the low ionic conductances of the corresponding salts, was attributed to strong solvation effects. Alkali ions also manifested a reversible cyclic voltammetric behavior, except for the Li(I) reduction which showed some degree of irreversibility. In addition, the reduction occurred at considerably more negative potentials than that of the other members of the series. This fact demonstrated the complicated nature of the Li(I) reduction. It should be noted that such behavior is similar in many other nonaqueous solvents.

The lithium ion actually displayed a double reduction peak; the first peak was interpreted by the occurrence of the solvent reduction in the coordinated $(Li-Py)^+$ species in preference to a direct reduction of the metal ion.[74] Owing to its smaller ionic radius relative to those of other alkali ions, lithium is more strongly solvated and is therefore highly likely to behave differently. The Lewis character of the lithium ion is strong enough to weaken the electronic density in the aromatic ring, so as to facilitate electron addition to some sites. A similar process was described for aluminum compounds in pyridine solutions, with formation of radical anion–lithium(I) complexes.[67] The second reduction peak was suggested to be due to the reduction of a pyridine molecule attached to Li(I) via a water molecule

(X)

although the reduction of Li(I) in the Li^+,H_2O complex may not be discarded either. Table XXII summarizes the polarographic features of alkali metal ions determined by Elving.

3. ORGANIC SUBSTANCES

Pyridine provides a useful medium in which hydrogen activity can be controlled by means of buffered solutions over a wide domain.[9] In particular, its truly aprotic character[204] enables very low acidity levels to be reached. In addition, its reactivity toward positively charged transient species, which may be formed during electrochemical or chemical reactions, favors their subsequent identification, thus making pyridine a suitable medium for the study of organic processes.

a. *Kinetic Studies.* In view of the importance of hydrogen ion activity, the electrochemical behavior of various organic compounds has been investigated: organochlorosilanes,[205] maleic anhydride, diethyl esters of maleic and fumaric acids,[206] indigo,[207] as well as the redox pattern of stable organic free radicals.[208]

TABLE XXII

POLAROGRAPHIC AND CYCLIC VOLTAMMETRIC BEHAVIOR OF ALKALI METAL IONS IN 0.1 M NEt$_4$ClO$_4$ PYRIDINE[a]

	$E_{1/2}{}^{b}$ (V)	I^c	Wave[d] slope (mV)	Ep^e (V)	D^f	D^g	D^h
Li(I)	-2.17 ± 0.01	1.3 ± 0.1	62	-2.17	0.45	0.41	0.61
Na(I)	-1.89 ± 0.01	1.6 ± 0.1	56	-1.89	0.67	0.60	0.65
K(I)	-1.91 ± 0.01	1.5 ± 0.15	55	-1.91	0.62	0.58	0.78
Rb(I)	-1.91 ± 0.01	1.6 ± 0.1	50	-1.91	0.66	0.59	—
Cs(I)	-1.89 ± 0.01	1.4 ± 0.2	63	-1.89	0.56	0.51	—

[a] Refs. 72 and 74 (with permission).
[b] Reduction half-wave potentials, vs. the NAgE reference electrode in pyridine.
[c] Diffusion current constants (μA mM^{-1}mg$^{-2/3}$ sec$^{1/2}$).
[d] Wave slope calculated from the relation: slope $= E_{1/4} - E_{3/4}$.
[e] Cyclic voltammetric reduction peak potentials, vs. the NAgE reference electrode in pyridine.
[f] Diffusion coefficient.
[g] Diffusion coefficient calculated from the modified Ilkovic equation

$$I = 607nD^{1/2}\left(1 + \frac{39D^{1/2}t^{1/6}}{m^{1/3}}\right)$$

[h] Diffusion coefficient calculated from conductance measurements of Burgess and Krauss.[118]

The polarographic and voltammetric behavior of benzophenone in 0.1 M EtClO$_4$ pyridine solutions was compared with that in protogenic solvents. Benzophenone produced two polarographic one-electron reduction waves, coalescing in a single two-electron step in the presence of 3,4-dimethylphenol added as a proton source.[209] On the first scan, cyclic voltammograms showed two principal cathodic peaks and three anodic waves; on succeeding scans, a third cathodic peak appeared at more positive potentials. With increasing proton availability, changes in the voltammogram pattern were interpreted by postulating the following reaction scheme. In the absence of protons, benzophenone was successively reduced in the free radical–anion (XI) and in the dicarbanion (XII):

$$(C_6H_5)_2CO + e^- \rightleftharpoons (C_6H_5)_2CO^{\overline{\cdot}} \tag{57}$$

(XI)

$$(C_6H_5)_2CO^{\overline{\cdot}} + e^- \rightleftharpoons (C_6H_5)_2CO^{2-} \tag{58}$$

(XI)   (XII)

The final products yielded were either salts of the benzhydrol anion or pyridyl alcohol species resulting from the attack on the solvent by the strong dianion base. Yet, additional pathways involving secondary dimerization, disproportionation, or proton addition on species may be superimposed. A third cathodic wave was attributed to the reduction of the carbonyl bond.

The resulting compounds in turn attacked the pyridine ring to form pyridyl alcohols. Anodic peaks represented the reoxidations of species **(XI)** and **(XII)** along with those of protonated intermediates, $(C_6H_5)_2CHOH$. In pyridine, the most striking difference from studies of this compound performed in easily proton-donating media was noninvolvement of any pinacolic species.

On pyrolytic graphite electrodes,[73,210] the cyclic voltammetric data of the quinone/quinhydrone/hydroquinone system, symbolized by $Q/QH/QH_2$, were interpreted on the basis of pyridinium ion involvement in proton-rich media. In pure pyridine, the reduction of quinone proceeded reversibly and yielded the radical anion $Q^{\bar{\cdot}}$. However, on reverse anodic scans, in the presence of pyridinium ions, the formation of N-dihydroxyphenylpyridinium ions **(XIII)** was considered to constitute the irreversible rate-determining step:

No proton source $Q + e^- \rightleftharpoons Q^{\bar{\cdot}}$ (1 e^- cathodic peak)   (59)

With PyH$^+$ $Q^{\bar{\cdot}} + H^+ \rightleftharpoons QH\cdot$ (fast)   (60)

$QH\cdot + Py \rightarrow e^- + $ **(XIII)** (1 e^- anodic peak)   (61)

(XIII)

These results correspond to recent investigations on the reduction of benzoquinone by tertiary heterocyclic amines.[211] Also, species **(XIII)** was formed in this way by the two-electron voltammetric oxidation of hydroquinone.[73]

b. *Potentiometric Studies.* Since it is of particular interest in testing proposed relationships between structure and redox potentials, a number of works have dealt with the redox behavior of various substituted quinones in acidic pyridine solutions. Gupta[212] developed a technique for potentiometric measurements, which proved generally more difficult in pyridine than in water and could therefore not be carried out with the same accuracy. Nevertheless, sufficient agreement was found between measured and calculated values by applying the Nernst equation to potential-composition curves. In basic conditions, the potential-composition curves of anthraquinone (AQ) showed two distinct one-electron steps, indicating the split of the AQ/AQ^{2-} system into two separate systems, AQ/AQ^- and AQ^-/AQ^{2-}. The disproportionation equilibrium constant K,

$$K = [AQ^-]^2/[AQ][AQ^{2-}]$$ (62)

TABLE XXIII

REDOX POTENTIALS OF SOME QUINONES IN ACIDIC PYRIDINE[a]

Quinones	$E^0(QH_2/Q)$ Temperature (°C)	ΔS (cal)
p-Benzoquinone	0.199	—
1,4-Naphthaquinone	−0.023	—
9,10-Anthraquinone	−0.332	51
9,10-Phenanthraquinone	−0.021	25
3,10-Perylenequinone	0.000	41
Anthanthrone	−0.238	—
3,4,8,9-Dibenzopyrenequinone	−0.242	41.5
Pyranthrone	−0.330	—
Dibenzanthrone	−0.222	—
1,4-Dibenzamidoanthranquinone	−0.410	—
16,17-Dimethoxydibenzanthrone	−0.277	25
Indanthrone	−0.223	—
Indigo	−0.220	—

[a] Relative to the NHE in pyridine (from Gupta,[212] with permission).

was found to be ca. 5×10^4. Table XXIII gives experimental data relative to the various other systems investigated. Unfortunately, no data are available concerning the precise estimation of acidity levels.

Solutions of inorganic salts have been used to titrate the thiolfunction content of organic substances in pyridine. With the Fe(III) and Cu(II) chlorides, the titration breaks of thiophenols and thioorganic acids RSH have been attributed to the reaction[213,214]

$$2 \text{ R-SH} \rightarrow \text{R-S-S-R} + 2 \text{ H}^+ + 2 e^- \tag{63}$$

VIII. HALOGENS AND HALOGEN COMPLEXES IN PYRIDINE

A. Spectrophotometric and Physical Properties

The behavior of halogen solutions in nonaqueous media has intrigued many chemists in the past and has been subjected to numerous and intensive investigations. It is, however, significant to note that the most essential work in pyridine has been devoted to iodine solutions. Several reviews of the physical,[215] spectrophotometric,[216,217] and electrochemical[13,218] properties of these solutions are available.

It is well known that iodine solutions in any solvent fall into two classes according to their color; they may be either violet or brown, depending upon the nature of the solvent molecule. Solvents capable of acting as donors in the formation of coordinate bonds, i.e., compounds containing nitrogen, sulfur, or oxygen atoms, readily give brown solutions. Iodine solutions in pyridine fall into this category. The physical properties of these brown solutions are "irregular," since they differ markedly from those of the violet. It is thus readily apparent that in its violet solutions iodine remains essentially free, while brown solutions show evidence of the occurrence of strong solute–solvent interactions. Isolation of especially stable 1 : 1 molecular complexes of the PyI_2 form was reported by Waentig[219] and Chatelet,[220] in accordance with molecular weight determinations. Further features of the interactions peculiar to iodine and pyridine supply evidence not only for complex formation but also for complexes in which the valence structure of iodine has changed appreciably. The characteristics of these complexes have been specified mainly by thermodynamic and spectrophotometric studies. Hartly and Skinner[221] estimated the formation heat of PyI_2 to be 7.95 kcal/mole, a far higher value than that of the other iodine–solvent complexes. Additional evidence may also be found in the enhancement of the dipole moment of pyridine ($\mu = 2.23$ D) in the formation of PyI_2, which becomes exceptionally large as shown by measurements carried out in hexane solutions ($\mu = 4.8$ $D^{16,218,222}$).

The spectrophotometric behavior of iodine solutions in pyridine reveals a characteristic and very intense peak, interpreted as a charge-transfer band.[216,217] Mulliken suggested that, when iodine is dissolved in pyridine, the following set of reactions should be considered to operate:

$$Py + I_2 \quad \underset{\rightleftharpoons}{\text{fast}} \quad PyI_2 \text{ "outer complex"} \tag{64}$$

$$PyI_2 \quad \rightleftharpoons \quad (PyI)^+I^- \text{ "inner complex"} \tag{65}$$

$$(PyI)^+I^- \quad \underset{\rightleftharpoons}{\text{fast}} \quad PyI^+ + I^- \tag{66}$$

Here, pyridine plays a double role, acting as an electron donor in Eq. 64 and as a polar medium in assisting Eqs. 65 and 66. In heptane solutions, the outer complex PyI_2 was identified by its band, located at 422 nm (free iodine lies at 520 nm), and the expected charge-transfer band at 235 nm. Assuming the geometrical structure (XIV) the outer complex was estimated to have a

(XIV)

dative character of about 25%.[223] In the asymmetrical structure (XIV), pyridine would act chiefly as an n donor and only to a slight extent as a π

donor, like benzene in its iodine complexes. It is of some interest to note that measurements of the equilibrium constant

$$K = [PyI_2]/[Py][I_2] \qquad (67)$$

conducted at several temperatures lead to the calculation of an activation energy value of 7.8 ± 0.2 kcal/mole, which closely agreed with that estimated from the dissolution heat of I_2 in pure pyridine.[221] At 25°C, K was found to be nearly 240 l/mole.

The displacement of equilibrium in Eq. 64 through Eqs. 65 and 66 was carried out in heptane solutions of iodine with gradually increasing concentrations of pyridine. Reversible changes occurred in position (the 422 nm PyI_2 iodine band began to shift toward a limiting value of 389 nm) and in intensity (the extinction coefficient increased to 2120 in pure pyridine), in agreement with earlier results.[224] This gradual shift was attributed to a progressive clustering of the polar solvent around the PyI_2 complex. Unfortunately, the correlative fate of the charge-transfer band could not be ascertained above 1.5 M pyridine. No appreciable change in position or intensity of this band could be detected by varying the pyridine concentration. This suggested that PyI_2 remains predominant for the iodine concentrations used (0.005 M), even in pure pyridine, and that the inner complex $(PyI)^+I^-$, if present, is so only in small amounts. Moreover, Mulliken[217] pointed out that further comparison with other iodine–solvent systems indicated that the inner complex of PyI_2 is not so low in energy as to prevent the interposition of a pyridine molecule between the PyI^+ and I^- ions, thus forming iodo-dipyridinium ions.[225]

B. Stability in Time

In unbuffered pyridine solutions, concentration-time dependencies were observed, in particular with low concentrations of iodine. Audrieth and Birr[13] reported that the electrical conductance of these solutions in pyridine increased with time. At very high dilutions, an asymptotic limiting value of about 130–132 was attained; the more dilute the solutions, the more rapidly the equilibrium values were approached. This extremely high value was about twice as great as might be expected on the basis of a simple dissociation of iodine into positive and negative ions. Accordingly, the authors proposed the formation of a ternary electrolyte, postulating the occurrence of a chemical interaction between solute and solvent:

$$PyI_2 \rightleftharpoons PyI^+ + I^- \rightleftharpoons Py^{2+} + 2\,I^- \qquad (68)$$

$$2\,PyI_2 \rightleftharpoons PyI^+ + I_3^- \qquad (69)$$

While recent observations have left no doubt about the intermediate involvement of triodide ions, no reliable knowledge exists as to the exact nature of the chemical stages responsible for this time effect. At high dilutions, almost all investigators detected a decay of the titratable iodine content, concomitant with a rise in I^- and hence in I_3^-. Under rigorous moisture exclusion, Kortüm and Wilski[17] found this effect to be independent of the presence of water, but strongly catalyzed by a platinum sponge. In addition, their results partially invalidated the determinations of Audrieth and Birr, since freshly prepared solutions initially gave relatively low conductances. Moreover, no definite time limit was attained, as was also shown by another study.[226] From the results of Kleinberg,[71,224,227] who reported the isolation of a small percentage of mono- and diiodo-substituted pyridines in iodine solutions, Kortüm and Wilski concluded that the conductance and spectrophotometric behavior of iodine could best be explained by an electrophilic attack on pyridine of positively charged iodine–pyridine complexes. Without disputing these conclusions, Mulliken made allowance for the possible involvement of impurities which could be responsible for the degradation of iodine solutions with time.[217]

Side-reaction effects were, however, not observed by Nigretto and Jozefowicz,[194] in the course of their electrochemical studies. Using several acid–base buffers ranging up to pH = 13, the authors showed that iodine solutions, in the usual analytical concentrations, could be kept unaltered over several weeks. These observations stress the importance of the acidity level for the kinetics of these side reactions, if they exist.

In the absence of any macroscale evidence, it appears questionable whether pyridine affords any method of purification capable of lowering the level of impurities below the critical concentration, so as to prove the existence of any side reaction. Additional investigations are therefore needed into reactions consuming positive iodine under the normal conditions for analytical studies, as would occur if the iodine were acting upon pyridine under pressure[228] or after heating.[229]

C. Diagram of the Thermodynamic Stability of Iodine

As shown in the preceding section, iodine has been extensively studied in unbuffered pyridine solutions. Surprisingly, there have been very few reports on the fate of these solutions in buffered pH regions in this solvent. This scarcity is conspicuous in view of the role played by iodine species in pyridine during the titration of water, commonly effected according to the Karl Fischer method, as well as because of the importance of pH effects on the nature of dissolved iodine species.[230]

Nigretto and Jozefowicz[194] established the potential–pH (or Pourbaix) diagram of the thermodynamic stability of iodine and related species (Fig. 9). This diagram was set up using the voltammetric analysis of curves recorded with rotating platinized platinum disk electrodes, in 0.1 M

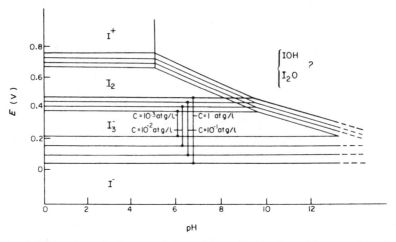

FIG. 9. Thermodynamic diagram of the stability of iodine in pyridine containing 0.1 M tetrabutylammonium perchlorate at 25°C. (E potentials are referred to the normal silver reference electrode or to the copper(II)-copper(I) reference electrode.) From Nigretto and Jozefowicz,[194] reprinted by permission of *Electrochimica Acta*.

NBu_4ClO_4. In acidic media, three distinct redox couples, I_3^-/I^-, I_2/I_3^-, and I^+/I_2 were characterized by their standard potentials, respectively, 0.536, 0.970, and 1.255 V vs. the NHE in pyridine. From these values, the apparent disproportionation constants K_1', K_2', and K_3' of the following equilibria

$$I_2 \rightleftharpoons I^+ + I^- \qquad K_1' = [I^+][I^-]/[I_2] \tag{70}$$

$$2\,I_2 \rightleftharpoons I^+ + I_3^- \qquad K_2' = [I^+][I_3^-]/[I_2]^2 \tag{71}$$

$$I_3^- \rightleftharpoons I_2 + I^- \qquad K_3' = [I_2][I^-]/[I_3^-] \tag{72}$$

were deduced, 1.8×10^{-10}, 1.4×10^{-5}, and 1.25×10^{-5} moles/liter, respectively. The first value almost coincides with that calculated by Pezzatini and Guidelli[231] from the voltammetric analysis of iodine solutions in acetonitrile in the presence of pyridine, but differs noticeably from that obtained from conductance measurements carried out by Kortüm and Wilski, $K_1' = 4.6.10^{-8}$ m/liter.[17] Above ca. pH 5, I_2 began to disproportionate into triiodide and a unipositive iodine species characterized by titration. It is probable that the latter species was the hypothetical base IOH or its anhydride I_2O, as substantiated by the investigations of Carlsohn.[232-234] At higher pH values, the triiodide ions disproportionated in turn into iodide and unipositive iodine. Due to the low likely disproportionation rates and the insolubility of iodine species of a higher oxidation degree, no upper redox state was to be detected with this electrochemical method. However,

titrations with sodium thiosulfate supplied evidence for the further disproportionation of $I(+I)$ iodine into $I(+V)$ and $I(-I)$ beyond ca. pH 13, in
agreement with Carlsohn's observations.[232]

REFERENCES

1. D. P. Biddiscombe, E. A. Coulson, R. Handley, and E. F. G. Herington, *J. Chem. Soc.* p. 1957 (1954).
2. D. A. Hall and P. J. Elving, *Anal. Chim. Acta* **39**, 141 (1967).
3. F. Arndt and P. Nachtwey, *Ber. Dtsch. Chem. Ges.* **59**, 448 (1926).
4. F. Arndt and T. Severge, *Chem. Ztg.* **74**, 140 (1950).
5. A. S. Rozhdestvenskii, B. V. Pukirev, and R. S. Maslova, *Trans. Inst. Pure Chem. Reagents, Moscow* **14**, 58 (1935).
6. J. G. Heap, W. J. Jones, and J. B. Speakman, *J. Am. Chem. Soc.* **43**, 1936 (1921).
7. W. F. Luder and C. A. Kraus, *J. Am. Chem. Soc.* **69**, 2481 (1947).
8. H. N. Wilson and W. C. Hughes, *J. Soc. Chem. Ind. London* **58**, 74 (1939).
9. J. M. Nigretto and M. Jozefowicz, *Electrochim. Acta* **18**, 145 (1973).
10. E. K. Ralph and W. R. Gilkerson, *J. Am. Chem. Soc.* **86**, 4783 (1964).
11. D. G. Leis and B. C. Currans, *J. Am. Chem. Soc.* **67**, 79 (1945).
12. H. Wilski and G. Kortüm, *Z. Phys. Chem.* **5**, 333 (1955).
13. L. F. Audrieth and E. J. Birr, *J. Am. Chem. Soc.* **55**, 668 (1933).
14. P. Walden, L. F. Audrieth, and E. J. Birr, *Z. Phys. Chem., Abt. A* **160**, 337 (1932).
15. Z. Hladky and J. Vrestal, *Collect. Czech. Chem. Commun.* **34**, 984 (1969).
16. G. Kortüm and H. Walz, *Z. Elektrochem.* **57**, 73 (1953).
17. G. Kortüm and H. Wilski, *Z. Phys. Chem.* **202**, 35 (1953).
18. U. Bertocci, *Z. Elektrochem.* **61**, 431 (1957).
19. L. M. Mukherjee, J. J. Kelly, W. Baranetzky, and J. Sicca, *J. Phys. Chem.* **72**, 3410 (1968).
20. M. Redford-Ellis and J. E. Kench, *Anal. Chem.* **32**, 1803 (1960).
21. M. M. Davies, *Trans. Faraday Soc.* **31**, 1561 (1935).
22. K. Varga and H. Beyer, *Acta Chim. Acad. Sci. Hung.* **52**, 69 (1967).
23. A. Cisak and P. J. Elving, *J. Electrochem. Soc.* **110**, 160 (1963).
24. J. J. Kipling and E. H. M. Wright, *Trans. Faraday Soc.* **55**, 1185 (1959).
25. L. Berg, J. M. Harrison, and C. W. Montgomery, *Ind. Eng. Chem.* **37**, (1945).
26. P. Dutoit and H. Duperthuis, *J. Chim. Phys.* **6**, 729 (1908).
27. C. Marden, "Solvents Guide," 2nd ed. Cleaver Hume, London, 1963.
28. J. D. Cox, A. R. Challoner, and A. R. Meetham, *J. Chem. Soc.* p. 265 (1954).
29. E. F. G. Herington and J. F. Martin, *Trans. Faraday Soc.* **49**, 154 (1953).
30. A. A. Maryott and E. R. Smith, *Natl. Bur. Stand. (U.S.), Circ.* **514** (1951).
31. Le Fevre, C. G. *J. Chem. Soc.* **1**, 776 (1935).
32. L. E. Orgel, T. L. Cottrel, W. Dick, and L. E. Sutton, *Trans. Faraday Soc.* **47**, 113 (1951).
33. R. J. L. Andon and J. D. Cox, *J. Chem. Soc.* p. 4601 (1952).
34. J. D. Cox, *J. Chem. Soc.* p. 4606 (1952).
35. V. Henri and P. Angenot, *J. Chim. Phys.* **33**, 641 (1936).
36. R. A. Barnes, *in* "The Chemistry of Heterocyclic Compounds—Pyridine and its deriva-tives" (E. Klinsberg, ed.), Part I, pp. 48–58. Wiley (Interscience), New York, 1960.
37. C. H. Kline, Jr. and J. Turkevich, *J. Chem. Phys.* **12**, 300 (1944).
38. C. W. F. Spiers and J. P. Wibaut, *Recl. Trav. Chim. Pays-Bas* **56**, 573 (1937).
39. E. B. Hughes, H. H. G. Jellinek, and B. A. Ambrose, *J. Phys. Chem.* **53**, 410 (1949).
40. E. F. G. Herington, *Discuss. Faraday Soc.* **9**, 26 (1950).

41. L. J. Bellamy, *in* "The Infra Red Spectra of Complex Molecules," p. 233. Methuen, London, 1954.
42. J. H. Hibben, *Chem. Rev.* **18**, 110 and 204 (1936).
43. F. A. Patty, ed., "Industrial Hygiene and Toxicology," 2nd ed., Vol. 2, p. 2189. Wiley (Interscience), New York, 1963.
44. L. Kahlenberg, *J. Phys. Chem.* **3**, 602 (1899).
45. R. Muller, F. Holzl, A. Pontoni, and O. Wintersteiner, *Monatsh. Chem.* **43**, 419 (1922).
46. G. von Hevesy, *Z. Elektrochem.* **16**, 672 (1910).
47. H. C. Mandell, Jr., W. M. McNabb, and J. F. Hazel, *J. Electrochem. Soc.* **102**, 263 (1955).
48. V. Laszczynski and S. Gorki, *Z. Elektrochem.* **4**, 290 (1897).
49. B. Emmert, *Ber. Dtsch. Chem. Ges.* **49**, 1060 (1916).
50. W. Lipp, *Ber. Dtsch. Chem. Ges.* **25**, 2196 (1892).
51. E. Koenigs, *Ber. Dtsch. Chem. Ges.* **40**, 3199 (1907).
52. F. Bohlmann, *Chem. Ber.* **85**, 390 (1952).
53. P. T. Lansbury and J. O. Peterson, *J. Am. Chem. Soc.* **83**, 3537 (1961).
54. P. T. Lansbury, *J. Am. Chem. Soc.* **83**, 429 (1961).
55. K. Ziegler and R. Zeiser, *Ber. Dtsch. Chem. Ges. B* **63**, 1847 (1930).
56. C. Marie and G. Lejeune, *J. Chim. Phys.* **22**, 59 (1925).
57. B. Emmert, *Ber. Dtsch. Chem. Ges.* **42**, 5150 (1909).
58. M. Shikata and I. Tachi, *Mem. Coll. Agric., Kyoto Imp. Univ.* **4**, 19 (1927).
59. J. J. Lingane and O. L. Davis, *J. Biol. Chem.* **137**, 567 (1941).
60. P. C. Tompkins and C. L. A. Schmidt, *J. Biol. Chem.* **137**, 643 (1941).
61. G. A. Tedoradze, S. G. Mairanovskii, and L. D. Klyukina, *Izv. Akad. Nauk SSSR* 7, 1352 (1961).
62. J. Kuta and J. Drabek, *Collect. Czech. Chem. Commun.* **20**, 902 (1955).
63. M. S. Spritzer, J. M. Costa, and P. J. Elving, *Anal. Chem.* **37**, 211 (1965).
64. L. Floch, M. S. Spritzer, and P. J. Elving, *Anal. Chem.* **38**, 1074 (1966).
65. T. Zincke and G. Mülhausen, *Ber. Dtsch. Chem. Ges.* **38**, 3824 (1905).
66. P. Baumgarten, *Ber. Dtsch. Chem. Ges.* **57**, 1622 (1933).
67. A. Cisak and P. J. Elving, *Electrochim. Acta* **10**, 935 (1965).
68. W. R. Turner and P. J. Elving, *Anal. Chem.* **37**, 467 (1965).
69. P. Baumgarten, *Ber. Dtsch. Chem. Ges. B* **69**, 1938 (1936).
70. P. Baumgarten and E. Dammann, *Ber. Dtsch. Chem. Ges.*   **66**, 1633 (1933).
71. R. A. Zingaro, C. A. Vanderwerf, and J. Kleinberg, *J. Am. Chem. Soc.* **72**, 5341 (1950).
72. J. Broadhead and P. J. Elving, *J. Electrochem. Soc.* **118**, 63 (1971).
73. W. R. Turner and P. J. Elving, *Anal. Chem.* **37**, 467 (1965).
74. J. Broadhead and P. J. Elving, *Anal. Chem.* **41**, 1814 (1969).
75. J. E. Hickey, M. S. Spritzer, and P. J. Elving, *Anal. Chim. Acta* **35**, 277 (1966).
76. V. Z. Deal and G. Wyld, *Anal. Chem.* **27**, 47 (1955).
77. D. J. G. Ives and G. J. Janz, "Reference Electrodes: Theory and Practice." Academic Press, New York, 1961.
78. A. K. Gupta, *J. Chem. Soc.* p. 3473 (1952).
79. L. M. Mukherjee and R. S. Schultz, *Talanta* **19**, 707 (1972).
80. K. Tsuji and P. J. Elving, *Anal. Chem.* **41**, 216 (1969).
81. H. Strehlow, *in* "The Chemistry of Non-Aqueous Solvents" (J. J. Lagowski, ed.), Vol. 1, p. 129. Academic Press, New York, 1966.
82. J. L. Brisset, *J. Solution. Chem.* **5**, 587 (1976).
83. L. M. Mukherjee, *J. Phys. Chem.* **76**, 243 (1972).
84. J. S. Fritz, A. J. Moye, and M. J. Richard, *Anal. Chem.* **29**, 1685 (1957).
85. R. H. Cundiff and P. C. Markunas, *Anal. Chem.* **30**, 1450 (1958).

86. L. W. Marple and J. S. Fritz, *Anal. Chem.* **34**, 796 (1962).
87. M. Bos and A. M. F. Dahmen, *Anal. Chim. Acta* **16**, 392 (1957).
88. U. Bertocci, *Z. Elektrochem.* **61**, 434 (1957).
89. R. Muller, *Z. Anorg. Allg. Chem.* **142**, 130 (1925).
90. R. Abegg and J. Neustadt, *Z. Elektrochem.* **15**, 264 (1909).
91. Y. G. Bochkov and S. V. Gorbachev, *Russ. J. Phys. Chem.* **40**, 393 (1966).
92. J. Broadhead and P. J. Elving, *Anal. Chem.* **48**, 433 (1969).
93. T. Yvernault, J. More, and M. Durand, *Bull. Soc. Chim., Beograd* **16**, 542 (1949).
94. H. Angerstein, *Rocz. Chem.* **30**, 855 (1956).
95. O. Tomicek and S. Krepelka, *Chem. Listy* **47**, 526 (1953).
96. L. M. Mukherjee and J. J. Kelly, *J. Phys. Chem.* **71**, 2348 (1967).
97. G. A. Harlow, C. M. Noble, and G. E. A. Wyld, *Anal. Chem.* **28**, 784 (1956).
98. S. O. Thompson and G. Chesters, *Anal. Chem.* **36**, 655 (1964).
99. A. Groagova and V. Chromy, *Analyst* **95**, 548 (1970).
100. D. A. Lee, *Anal. Chem.* **38**, 1168 (1966).
101. R. H. Cundiff and P. C. Markunas, *Anal. Chem.* **28**, 792 (1956).
102. H. B. van der Heijde, *Anal. Chim. Acta* **16**, 392 (1957).
103. J. S. Fritz and G. H. Schenk, *Anal. Chem.* **31**, 1808 (1959).
104. H. B. van der Heijde, *Anal. Chim. Acta* **17**, 512 (1957).
105. T. Jasinski and R. Korewa, *Chem. Anal. (Warsaw)* **13**, 1319 (1968); *C.A.* **71**, 9489g (1969).
106. M. Wilchek, M. Fridkin, and A. Patchornik, *Anal. Chem.* **42**, 275 (1970).
107. H. Smagowski and T. Jasinski, *Zesz. Nauk. Wydz. Mat., Fiz. Chem., Univv. Gdanski, Chem.* **1**, 37 (1971); *C.A.* **78**, 131833g (1973).
108. G. A. Harlow, C. M. Noble, and G. E. A. Wyld, *Anal. Chem.* **28**, 787 (1956).
109. R. H. Cundiff and P. C. Markunas, *Anal. Chem.* **30**, 1447 (1958).
110. D. B. Bruss and G. A. Harlow, *Anal. Chem.* **30**, 1836 (1958).
111. T. Jasinski and H. Smagowski, *Chem. Anal. (Warsaw)* **14**, 917 (1969); *C.A.* **72**, 9055r (1970).
112. R. H. Cundiff and P. C. Markunas, *Anal. Chem.* **34**, 584 (1962).
113. M. S. Walton and R. F. Judd, *J. Am. Chem. Soc.* **33**, 1026 (1911).
114. J. Schroeder, *Z. Anorg. Chem.* **44**, 1 (1905).
115. A. Naumann, *Ber. Dtsch. Chem. Ges.* **37**, 4609 (1904).
116. R. Muller, *Z. Anorg. Allg. Chem.* **142**, 130 (1925).
117. C. F. Nelson, *J. Am. Chem. Soc.* **35**, 658 (1913).
118. D. S. Burgess and C. A. Kraus, *J. Am. Chem. Soc.* **70**, 706 (1948).
119. K. Tsuji and P. J. Elving, *Anal. Chem.* **41**, 1571 (1969).
120. S. Bruckenstein and J. Osugi, *J. Phys. Chem.* **65**, 1868 (1961).
121. L. M. Mukherjee, J. J. Kelly, McD. Richards, and J. M. Lukacs, Jr., *J. Phys. Chem.* **73**, 580 (1969).
122. H L. Pickering and C. A. Kraus, *J. Am. Chem. Soc.* **71**, 3288 (1949).
123. M. Bos and A. M. F. Dahmen, *Anal. Chem.* **55**, 285 (1971).
124. E. J. Bair and C. A. Kraus, *J. Am. Chem. Soc.* **73**, 2459 (1951).
125. L. M. Mukherjee and J. M. Lukacs, Jr., *J. Phys. Chem.* **73**, 3115 (1969).
126. P. Dutoit and H. Duperthuis, *J. Chim. Phys.* **6**, 699 (1909).
127. E. X. Anderson, *J. Phys. Chem.* **19**, 753 (1915).
128. E. Carrara and P. Levi, *Gazz. Chim. Ital.* **32**, 44 (1902).
129. S. Lincoln, *J. Phys. Chem.* **3**, 469 (1899).
130. J. H. Mathews and A. J. Johnson, *J. Phys. Chem.* **21**, 294 (1917).
131. L. M. Mukherjee, D. W. Suwala, and R. S. Schultz, *Anal. Chim. Acta* **60**, 247 (1972).
132. R. M. Fuoss and C. A. Kraus, *J. Am. Chem. Soc.* **55**, 476 (1933).
133. R. M. Fuoss, *J. Am. Chem. Soc.* **57**, 488 (1935).

134. D. L. Fowler and C. A. Kraus, *J. Am. Chem. Soc.* **62**, 2237 (1940).
135. M. M. Davis, *in* "The Chemistry of Non-Aqueous Solvents" (J. J. Lagowski, ed.), Vol. 3, Academic Press, New York, 1970.
136. M. M. Davis, "Acid-Base Behavior in Agnotic Organic Solvents," *Natl. Bur. Stand. (U.S.).* Monograph 105, (U.S.) 1968.
137. "Titrations in Non Aqueous Media," Anal. Chem., Fundam. Rev., 1964, 1966, 1968, 1970, 1972, and 1974.
138. I. M. Kolthoff, *Anal. Chem.* **46**, 1992 (1974).
139. J. S. Fritz, "Acid Base Titrations in Non Aqueous Solvents." Allyn & Bacon, Boston, Massachusetts, 1973.
140. G. Charlot and B. Tremillon, *in* "Les réactions chimiques dans les solvants et les sels fondus." Gauthiers-Villars, Paris, 1963.
141. G. Allen and E. F. Caldin, *Q. Rev., Chem. Soc.* **7**, 255 (1953).
142. R. Gaboriaud, *J. Chim. Phys.* **66**, 10 (1969).
143. V. A. Rabinovitch, A. E. Nikerov, and V. P. Rotstein, *Electrochim. Acta* **12**, 155 (1962).
144. J. N. Brønsted, *Recl. Trav. Chim. Pays-Bas* **42**, 718 (1923).
145. E. Price, *in* "The Chemistry of Non-Aqueous Solvents" (J. J. Lagowski, ed), Vol. 1, p. 67. Academic Press, New York, 1966.
146. E. Grunwald, *Anal. Chem.* **26**, 1696 (1954).
147. D. I. Hitchcock, *J. Am. Chem. Soc.* **50**, 2076 (1928).
148. J. L. Brisset, Ph.D. Thesis, University of Paris, 1973.
149. O. Popovych, *Anal. Chem.* **38**, 558 (1966); *Crit. Rev. Anal. Chem.* **1**, 73 (1970).
150. R. Gaboriaud, *C. R. Hebd. Seances Acad. Sci., Ser. C* **270**, 1925 (1970).
151. B. G. Cox, A. J. Parker, and W. E. Waghorne, *J. Am. Chem. Soc.* **95**, 1010 (1973).
152. V. A. Pleskov, *Zh. Fiz. Khim.* **22**, 351 (1948).
153. J. L. Brisset, R. Gaboriaud, and R. Schaal, *J. Chim. Phys.* **10**, 1506 (1971).
154. K. Tsuji and P. J. Elving, *Anal. Chem.* **41**, 286 (1969).
155. C. A. Streuli and R. R. Miron, *Anal. Chem.* **30**, 1978 (1958).
156. C. A. Streuli, *Anal. Chem.* **32**, 407 (1960).
157. L. M. Mukherjee, S. Bruckenstein, and F. A. K. Badawi, *J. Phys. Chem.* **69**, 2537 (1965).
158. H. P. Marshall and E. Grunwald, *J. Phys. Chem.* **21**, 2143 (1953).
159. H. B. van der Heijde and E. A. M. F. Dahmen, *Anal. Chim. Acta* **16**, 378 (1957).
160. I. M. Kolthoff and S. Bruckenstein, *J. Am. Chem. Soc.* **78**, 1 (1956).
161. E. J. Corey, *J. Am. Chem. Soc.* **75**, 1172 (1953).
162. P. J. Elving, *J. Electroanal. Chem.* **29**, 55 (1971).
163. J. S. Fritz and F. E. Gainer, *Talanta* **13**, 939 (1966).
164. J. T. Stock and W. C. Purdy, *Chem. Anal.* **48**, 22, 50, and 55 (1959).
165. W. C. Purdy and J. T. Stock, *Chem. Anal.* **50**, 88 (1961).
166. R. K. Maurmeyer, M. Margosis, and T. S. Ma, *Mikrochim. Acta.* Vol. 2 p. 177 (1959).
167. R. Belcher, G. Dryhurst, and A. M. G. MacDonald, *Anal. Chim. Acta* **38**, 435 (1967).
168. L. E. Krohn and V. K. La Mer, *J. Am. Chem. Soc.* **53**, 1163 (1931).
169. M. I. Fauth, M. Frandsen, and B. R. Havlik, *Anal. Chem.* **36**, 380 (1964).
170. R. P. Jaiswal and S. G. Tandon, *J. Indian Chem. Soc.* **47**, 755 (1970).
171. L. F. Fieser and M. Fieser, *in* "Organic Chemistry," 3rd ed. Heath, Boston, Massachusetts, 1956.
172. J. F. J. Dippy, *Chem. Rev.* **25**, 151 (1939).
173. V. V. Rublev and Y. A. Bulatova, *Zh. Anal. Khim.* **24**, 1106 (1969); *C.A.* **71**, 108931j (1969).
174. V. Kapisinska and A. Plachova, *Chem. Prum.* **18**, 607 (1968); *C. A.* **70**, 84145g (1969).
175. H. Smagowski and T. Jasinski, *Zesz. Nauk. Wydz. Mat., Fiz. Chem., Univv. Gdanski, Chem.* **1**, 29 (1971); *C. A.* **78**, 118977n (1973).

176. R. Korewa and T. Jasinski, *Chem. Anal. (Warsaw)* **15**, 127 (1970); *C. A.* **73**, 10448y (1970).
177. G. M. Gal'pern, Y. A. Gurvich, and V. A. Il'ina, *Tr. Konf. Anal. Khim. Nevodnkh Rastvorov Ikh. Fiz.-Khim. Svoistvam, 1st, 1965* Vol. 1, p. 267 (1968); *C. A.* **73**, 137156g (1970).
178. J. Kalamar and D. Mravec, *Zb. Pr. Chemickotechnol. Fak. SVST* p. 183 (1969–1970); *C. A.* **76**, 67969k (1972).
179. N. P. Dzyuba, *Khim.-Farm. Zh.* **5**, 39 (1971); *C. A.* **76**, 117534n (1972).
180. S. Ueoka, S. Okada, S. Iga, H. Isaka, and K. Yoshimura, *Bunseki Kagaku* **20**, 1196 (1971); *C. A.* **75**, 144042m (1971).
181. G. Schott and E. Popowski, *Z. Chem.* **10**, 468 (1970).
182. D. K. Banerjee, M. J. Fuller, and H. Y. Chen, *Anal. Chem.* **36**, 2016 (1964).
183. J. C. MacDonald, *Anal. Chem.* **37**, 1170 (1965).
184. A. Hantzsch and K. S. Caldwell, *Z. Phys. Chem.* **61**, 227 (1908).
185. E. Grunwald and E. Price, *J. Am. Chem. Soc.* **86**, 4517 (1964).
186. D. J. Glover, *J. Am. Chem. Soc.* **87**, 5275 and 5279 (1965).
187. R. Muller, F. Holzl, W. Knaus, F. Planiszig, and K. Prett, *Monatsh. Chem.* **44**, 219 (1924).
188. R. Muller and H. J. Schmidt, *Monatsh. Chem.* **54**, 224 (1929).
189. R. Muller, *Monatsh. Chem.* **53**, 215 (1929).
190. R. Muller, *Monatsh. Chem.* **43**, 67 (1922).
191. R. Muller, *Monatsh. Chem.* **43**, 75 (1922).
192. R. Muller, *Z. Anorg. Allg. Chem.* **156**, 65 (1926).
193. L. F. Audrieth and H. W. Nelson, *Chem. Rev.* **8**, 335 (1931).
194. J. M. Nigretto and M. Jozefowicz, *Electrochim. Acta.* **19**, 809 (1974).
195. J. Neustadt and R. Abegg, *Z. Phys. Chem.* **69**, 486 (1909).
196. L. Kahlenberg, *J. Phys. Chem.* **4**, 709 (1900).
197. J. R. Partington and J. W. Skeen, *Trans. Faraday Soc.* **30**, 1062 (1934).
198. J. R. Partington and J. W. Skeen, *Trans. Faraday Soc.* **32**, 975 (1936).
199. J. Lelièvre, C. Le Feuvre, and R. Gaboriaud, *C. R. Hebd. Seances Acad. Sci., Ser. C* **275**, 1455 (1972).
200. R. Alexander, A. J. Parker, J. H. Sharp, and W. E. Waghorne, *J. Am. Chem. Soc.* **94**, 1148 (1972).
201. J. W. Diggle and A. J. Parker, *Electrochim. Acta* **18**, 975 (1973).
202. H. E. Patten and W. R. Mott, *J. Phys. Chem.* **12**, 49 (1908).
203. F. Willeboordse, Ph.D. Thesis, University of Amsterdam, 1959.
204. I. M. Kolthoff, *J. Polarogr. Soc.* **10**, 22 (1964).
205. E. A. A. Abrahamson, Jr. and C. A. Reynolds, *Anal. Chem.* **24**, 1827 (1952).
206. R. Takahashi and P. J. Elving, *Electrochim. Acta* **12**, 213 (1967).
207. P. J. Elving and D. A. Hall, *Isr. J. Chem.* **8**, 839 (1970).
208. D. A. Hall and P. J. Elving, *Electrochim. Acta* **12**, 1363 (1967).
209. R. F. Michielli and P. J. Elving, *J. Am. Chem. Soc.* **90**, 1989 (1968).
210. W. R. Turner and P. J. Elving, *J. Electrochem. Soc.* **112**, 1215 (1965).
211. T. Hase, *Acta Chem. Scand.* **17**, 2250 (1963); **18**, 1806 (1964).
212. A. K. Gupta, *J. Chem. Soc.* p. 3479 (1952).
213. Z. Hladky, *Z. Chem.* **5**, 424 (1965).
214. Z. Hladky and J. Vrestal, *Collect. Czech. Chem. Commun.* **34**, 1098 (1964).
215. J. Kleinberg and A. W. Davidson, *Chem. Rev.* **42**, 601 (1948).
216. D. L. Glusker, H. W. Thompson, and R. S. Mulliken, *J. Chem. Phys.* **21**, 1407 (1953).
217. C. Reid and R. S. Mulliken, *J. Am. Chem. Soc.* **76**, 3869 (1954).
218. G. Kortüm, *J. Chim. Phys.* **49**, C127 (1952).
219. P. Waentig, *Z. Phys. Chem.* **68**, 513 (1909).
220. M. Chatelet, *Ann. Chim. (Paris)* **2**, 5 (1934).

221. K. Hartley and H. A. Skinner, *Trans. Faraday Soc.* **26**, 621 (1950).
222. Y. K. Syrkin and K. M. Anisimova, *Dokl. Akad. Nauk. SSSR* **59**, 1457 (1948).
223. R. S. Mulliken, *J. Am. Chem. Soc.* **72**, 600 (1950); **74**, 811 (1952); *J. Phys. Chem.* **56**, 801 (1952).
224. J. Kleinberg, E. Colton, J. Sattizahn, and C. A. Van DerWerf, *J. Am. Chem. Soc.* **75**, 442 (1953).
225. J. A. Creighton, I. Haque, and J. L. Wood, *Chem. Commun.* **8**, 229 (1966).
226. R. Gopal and T. N. Srivastava, *J. Indian Chem. Soc.* **29**, 898 (1952).
227. R. A. Zingaro, C. A. Van DerWerf, and J. Kleinberg, *J. Am. Chem. Soc.* **73**, 88 (1951).
228. H. D. Tjeenk Willink, Jr. and J. P. Wibaut, *Recl. Trav. Chim., Pays-Bas* **54**, 275 (1935).
229. H. E. Mertel, *in* "Pyridine and its Derivatives" (E. Klinsberg, ed.), Part II, p. 318. Wiley (Interscience), New York, 1961.
230. M. G. Brown, "Reports of the 7th CITCE Meeting, Lindau, 1955." Butterworth, London, 1957.
231. G. Pezzatini and R. Guidelli, *Electrochim. Acta* **16**, 1415 (1971).
232. H. Carlsohn, "Uber eine neue Klasse von Verbindungen des positiv einwertiges Jods." Hirzel, Leipzig, 1952.
233. H. Carlsohn, *Angew. Chem.* **45**, 580 (1932).
234. H. Carlsohn, *Angew. Chem.* **47**, 747 (1933).

~ 6 ~

Anhydrous Hydrazine and Water–
Hydrazine Mixtures

ૡૅ

DENISE BAUER AND PHILIPPE GAILLOCHET

Laboratoire de Chimie Analytique
École Supérieure de Physique et de Chimie de Paris
10 Rue Vauquelin 75231 Paris-Cedex 05,
France

I. Introduction

Hydrazine is an important chemical product and its total production in the United States, Japan, and Europe is about 30,000 tons a year. Because of its properties it has many chemical uses and it is used as an intermediary in synthesis. The substituted hydrazines or hydrazides exhibit pharmaceutical properties with antibiotic action.

Hydrazine has outstanding reducing properties and its decomposition leads only to gases which are easily eliminated from solution. Some of its main uses taken from the recent literature include: as a propellant for rockets; in alimentation of fuel cells and for rechargeable fuel cells; for removal of oxygen from water, of nitrogen oxides (NO and NO_2) from waste gases, and of chlorine from water or gases; as an inhibitor of corrosion; for production of small-size particles of metal such as silver; as an additive to electrolytic baths for the deposit of metals, such as iron, tin, zinc, or for the electrolytic coloring of anodized aluminum; as a collector in cationic flotation; as a reagent in the preparation of pesticides; as an additive, with H_2O_2, to obtain light-weight cements; as a foaming agent in rubber; for the bleaching of natural fibers (cotton or wool) and the treatment of synthetic ones (acrylic fibers becoming fire-resistant); and as an aid in the raising of sunken objects up to 200 long tons in the ocean.

New plants are to be built in Europe by Bayer and Ugine Kuhlmann in order to increase the production of hydrazine; the demand of this product is increasing for use in foam polymer synthesis.

Because of its various applications and the great number of papers dealing with hydrazine properties, we have selected those which might be helpful in using hydrazine as a solvent. Audrieth and Ogg's[1] book still remains the basic publication concerning hydrazine and is recommended for those who are interested in a general survey of the chemistry of this compound.

II. Preparation

Because of its great instability, the preparation of hydrazine is the object of a large number of patents, one of their purposes being to improve the safety during preparation. Several preparations are proposed every year but the most common one remains the Rashig process (1907). In the first step, ammonia is oxidized by sodium hypochlorite according to Eq. 1.

$$NH_3 + NaOCl \rightarrow NH_2Cl + NaOH \tag{1}$$

Then chloramine (NH_2Cl) reacts with the ammonia in excess to give hydrazine

$$NH_2Cl + NH_3 + NaOH \rightarrow N_2H_4 + NaCl + H_2O \tag{2}$$

These diluted aqueous solutions may be distilled to an azeotrope, with the formula $N_2H_4 \cdot H_2O$.[2] Ammonia can be replaced by urea, and the equation then becomes

$$CO(NH_2)_2 + NaClO + 2\,NaOH \rightarrow NH_2-NH_2 + NaCl + Na_2CO_3 + H_2O \qquad (3)$$

New methods have recently been developed. N_2H_4 can be separated from ammonia and chlorine, charged in a reaction chamber kept at $-30°C$, giving a yield of 99.95%.[3] Hydrazine has been manufactured from NH_3, Me_2CO, MeCN, and H_2O_2.[4] The reaction requires a stabilizer (Na_2EDTA) and a catalyst (NH_4OAc); it leads to acetone azine which is converted to $N_2H_4-H_2O$ in 73–75% yield. Other manufacturing processes are described in Powell.[5]

The electrolytic production of hydrazine is described in McGriff and McRae.[6] The cell has an anode and a cathode chamber separated by an ion-selective membrane. The solvent is a mixture of HCO_2NMe_2, $AcONMe_2$, and Me_2SO; hydrazine forms at the anode and is recovered by distillation. Features of the design and operation of electrochemical reactors are given in Speeding.[7]

Various attempts have been made to prepare hydrazine by ionizing radiation,[8] γ-radiolysis,[9,10] and irradiation by pulsed electrons.[11] The heterogeneous catalytic decomposition of ammonia in a glow discharge has been investigated in water[12] in the presence of solid NaOH at 5 torr[13] and in the presence of solid NaOH or solid KOH at 5 mm Hg.[14]

In a practical way (laboratory uses), 95% N_2H_4 may be obtained by heating $N_2H_4-H_2O$ with NaOH and with subsequent distillation.[15] According to Linke and Taubert[15] anhydrous hydrazine can be obtained after treating 95% N_2H_4 with Ba_3N_4 at 5°C; after distillation the reaction yield is 81% and the purity is 100%. Ba_3N_4 is prepared by heating $Ba(N_3)_2$ in decalin. Other barium products such as barium oxide were formerly used.[16]

Hydrazine is formed in many chemical and electrochemical reactions. Gedye found it as a product in ammonia synthesis.[17] Products from the electrolysis of liquid ammonia at $-40°C$ are H_2, N_2, and liquids including hydrazine (10–30%).[18]

A. Purification

The purity of hydrazine can be investigated with gas chromatography. Various stationary phases are used; a comparison of these was made by Bighi et al.[19–21] and Spigarelli.[22] The decomposition products of hydrazine can also be identified by gas chromatography.[23] Purification is carried by acidifying hydrazine with HCl and then employing cation-exchange chromatography. The resin is then eluted with aqueous soda.[24] Liquid-phase chromatography is a method often used to prepare high-purity hydrazine.[25,26] One can also purify hydrazine by distillation on NaOH in an atmosphere of N_2,[27] or by several fractional crystallizations.[28]

The preparation of anhydrous hydrazine necessitates the removal of all traces of water. The water content is determined by titration with metallic barium[29] or by a potentiometric determination of the equivalent point in its titration by $N_2H_3{}^-$.[30]

A convenient method for obtaining anhydrous hydrazine involves the use of an activated alumina drying agent; this simple method yields a product containing 0.05% H_2O.[31]

B. Toxicology

One must be aware of the high toxicity of hydrazine. A review has recently been published on that topic.[32] The results of the pharmacological studies show that hydrazine induces an increase in protein synthesis in the liver, and is phototoxic.[33] Hydrazine also exhibits a carcinogenic action on the liver,[34] lungs,[35] and intestines.[36]

It has been shown that hydrazine induces the inhibition of amino acid incorporation into rat liver protein.[37] Hydrazine causes depletion of the B6 vitamins; as the latter are frequent enzymatic cofactors, the causes of hydrazine toxicity appear to be clear.

1,1-Dimethylhydrazine (1,1-DMH) is used as a rocket propellant. Those working with this product exhibit irritation. Absorption of 1,1-DMH induces an excitation of the central nervous system[38]; 23% of the workers affected showed a fatty degeneration of the liver with an increase of serum glutamic pyruvic transaminase (SGPT). The ambient concentration of this product must not be greater than 0.5 ppm (an odor detection gives 6–14 ppm). A case of absorption of hydrazine producing coma and collapse has been described.[39] Forced ventilation and B6 vitamins improved the health of the patient.

Silverman[40] has described sensors for detecting reducing vapors. This method may be useful in the laboratory. Toxic vapors of hydrazine are detected by the variations in the resistivity of a metallic film incorporated in a Wheatstone bridge circuit.[40]

It may be useful to have one of the various extinguishing agents proposed by Back and Thomas[41] available for hydrazine fires.

III. PROPERTIES

A. Physical Properties

Hydrazine is a molecular liquid at room temperature. Its main physical characteristics are summarized in Table I.[42–47] The high values for viscosity

TABLE I

PROPERTIES OF HYDRAZINE

Boiling point	113.5°C
Melting point	1.3°C
Density d_{15}^{15} [16,42]	1.014
Surface tension at 35°C[43]	63.32 dyn/cm
Viscosity (30°C)[44]	4.920 cP
Refraction index at 22.3°C[45,46]	$n_{H_z} = 1.46675$
	$n_D = 1.46979$
	$n_{Hg} = 1.48327$
Dielectric constant at 25°C[47]	51.7

and surface tension indicate a strong association of the N_2H_4 molecules in the liquid. The critical point is at 380°C and 145 atm.[48] At low temperature, the vapor pressure is small (9.34×10^{-2} atm at 56°C). The thermodynamic properties of the compressed gas N_2H_4 have recently been published.[49]

Nowadays, a large quantity of spectroscopic data is available, either from calculations or measurements. The IR absorption spectrum exhibits bands at 11.3, 9.1, 7.55, 6.2, 3.13, 2.28, 2.19, 2.07, and 1.57 μm which coincide with Raman lines.[50] The same spectrum in the gas phase has been recorded.[51] The proton magnetic relaxation of solid N_2H_4 has been studied at 240–273 K.[52] From electrodiffraction, the internuclear distances have been calculated.[53]

$$d(N - N) = 1.47 \pm 0.02 \text{ Å}$$

$$d(N - H) = 1.04 \pm 0.06 \text{ Å}$$

$$\angle H - N - H = 108° \pm 10°$$

From IR and Raman spectra, it is shown that a rotation barrier exists between the two NH_2 groups.[53,56] From LCAO–MO–SCF calculations in its equilibrium conformation, the N_2H_4 molecule has a dihedral angle of 95° and the barrier to pyramidal inversion at one N is 6.1 kcal/mole.[57,58] From photoelectron spectra, the first ionization potential has been calculated[59] to be 8.93 eV.

The thermodynamic properties of N_2H_4–H_2O solutions have been published.[60] The great heat which develops on mixing H_2O and N_2H_4 indicates strong interactions between the two molecules.

The ternary system N_2H_4–NH_3–H_2O has been studied[61] from -40°C to -90°C. Seven solid phases exist. Anhydrous solid hydrazine may be prepared from N_2H_4–H_2O and NH_3 mixtures at -80°C.

The vapor pressure and activity coefficient of the NaI–N_2H_4–H_2O system has been published.[62] NaI is more solvated by N_2H_4 than by water.

B. General Chemical Properties

From a chemical point of view, hydrazine is characterized by its great instability and its basic reducing and complexing properties. Hydrazine, which formally contains N(-II), is a metastable compound, the only stable oxidation states of nitrogen being N(-III), N(O), and N(V).

Hydrazine decomposes spontaneously. Two reactions are possible.[48]

$$N_2H_4 \rightarrow N_2 + 2H_2 \qquad (22.4 \text{ kcal}) \qquad (4)$$

$$N_2H_4 \rightarrow NH_3 + \tfrac{1}{2}N_2 + \tfrac{1}{2}H_2 \qquad (34.2 \text{ kcal}) \qquad (5)$$

Thermal decomposition occurs mainly in the second reaction.[63]

A great number of the uses of hydrazine are derived from the fact that only gases result from its decomposition. An abundant literature, which is not reported here, deals with the preparation of catalysts used for the control and/or the acceleration of these reactions. These catalysts are either metallic powders or organometallic compounds.

Because of its high instability, hydrazine has to be handled with caution; the presence of oxidants must be prohibited.

IV. Hydrazine as a Reagent

A. Acid–Base Properties

Hydrazine is a weak base, the pK_A (Eq. 6) is equal to 8.0 at 25°C (ionic strength equal to 0).[64]

$$HN_2H_4^+ \rightleftharpoons N_2H_4 + H^+ \qquad (6)$$

Thus, hydrazine is a base weaker than NH_3, which is predictable if one considers that hydrazine is formally derived from ammonia by the replacement of one hydrogen atom by an NH_2 group.

One can also find salts of $N_2H_6^+$ that are extensively hydrolyzed in water, except when the solution contains an excess of acid. The pK_A of the couple $N_2H_6^{2+}/N_2H_5^+$ is equal to -0.9.[64] When hydrazine is used as a solvent, the only solvent cation to consider is $N_2H_5^+$.

B. Redox Properties

Hydrazine is well known as a reducing agent but most reactions are slow, especially in acidic media. Its oxidation product is mainly N_2 and sometimes nitrogen oxides,[64] or NH_3 and HN_3.[65] According to Eq. 7

$$N_2H_4 \rightleftharpoons N_2 + 4H^+ + 4e^- \qquad (7)$$

The potential–pH diagram is given by Fig. 1.

As a matter of fact, in acidic media hydrazine can reduce Fe^{3+} [$E_0(Fe^{2+}/Fe^{3+}) = +0.77$ V][64] and halogens[65] [for Cl_2, $E_0(Cl^-/Cl_2) = +1.39$ V].

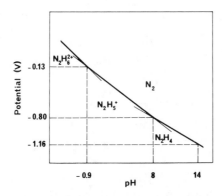

FIG. 1. Potential–pH diagram for hydrazine.

Hydrazine is used as a reducing agent in the potentiometric titration of Cr(VI) in 5 M HCl.[66] Hydrazine is also titrated by Ce(IV) in 6 N sulfuric acid.[67] Gupta and Mishra[68] have studied the stoichiometry and the kinetics of the reaction between Ce(IV) and hydrazine.

One can also use I_2 or IO_3^- to titrate hydrazine.[64] R. C. Paul[69] has reported the potentiometric titration of N_2H_4 in acetic acid. The oxidant is BrCN and the reaction may be written according to Eq. 8.

$$2N_2H_4 + 2Br^+ \rightarrow NH_3 + N_3H + 4H^+ + 2Br^- \tag{8}$$

Hydrazine can be used to reduce Ni^{2+}, Co^{3+}, Pu(IV), V(V), and Ag^{2+}.[70,71]

Hydrazine also reveals oxidizing abilities. For instance, platinum is oxidized by hydrazine in boiling 2 N HCl.[72]

A consequence of its reducing properties is its use in gas removal. In an aqueous basic medium, a few milligrams of N_2H_4 allows the reduction of the dissolved oxygen.[73,74] Chlorine, NO, and NO_2 can be removed from gases and waste gases.[75,76]

Hydrazine is well known as an inhibitor of corrosion in various metals.[77,78] It depresses the quantity of copper which goes into solution in vapor condensors.[79]

C. Electrochemical Properties

The utilization of hydrazine in fuel cells justifies the large number of papers dealing with the electrochemical oxidation of hydrazine.

The most usual medium for this reaction is an alkaline one, generally aqueous KOH. For recent work in that field, see Heitbaum and Vielstich,[80] Nakata et al.,[81] and Korovin and Kicheev.[82] The anodic oxidation of hydrazine occurs via dehydrogenation[80] according to the following equilibria

$$N_2H_{4ad} \rightarrow N_2H_{3ad} + H_{ad}$$

$$H_{ad} + OH^- \rightleftharpoons H_2O + e^-$$

$$N_2H_{3ad} \rightarrow N_{2ad} + 3H_{ad}$$

$$3[H_{ad} + OH^- \rightleftharpoons H_2O + e^-]$$

$$N_{2ad} \rightleftharpoons N_2\uparrow$$

$$N_2H_4 + 4OH^- \rightarrow N_2\uparrow + 4H_2O + 4e^-$$

Under nonstationary conditions, at high current, the removal of the first hydrogen atom is the rate-determining step which results in a limiting current for high overvoltages. At stationary currents, all electrochemical reaction steps (the oxidation of the four hydrogen atoms) occur simultaneously at a platinized platinum electrode.

Kodera had previously studied the concentration dependence of the mixed potential taken by a platinum electrode immersed in an aqueous hydrazine solution.[83] Various types of electrodes may be used in order to electrooxidize hydrazine. Vitvitskaya[84] proposed platinum-coated electrodes. Shmachkova studied the electrocatalytic properties of palladium black deposited on various supports.[85]

Kohmueller examined the influence of addition agents on the anodic oxidation of N_2H_4. The presence of KSCN gives a shift toward more positive potentials (oxidation on Pt and Ni anodes).[86] It has been shown by Milles and Kellet[87] that hydrazine can be electrochemically oxidized in liquid ammonia electrolytes on Pt, Ir, Rh, Re, Pd, V, Mo, and Au electrodes. Nitrocarbides and nitrides of transition metals also show activity for hydrazine oxidation.

Current–potential curves obtained for the electrochemical oxidation of hydrazine on Pt in liquid ammonia depend on the nature of the electrolyte anion; the adsorption effect may be related to the hard and soft acid–base principle. The harder the electrolyte anion, the better will be the performance of the hydrazine anode. The electrochemical oxidation of N_2H_4 on Pt in liquid ammonia appears to be a diffusion-controlled, highly irreversible reaction. Hydrazine, in the same way as its derivatives and hydroxylamines, can be titrated polarographically in alkaline sulfite media.[88] The electrochemical fluorination of hydrazine leads to N_2F_4.[89]

V. ANHYDROUS HYDRAZINE AS A SOLVENT

As indicated by its general chemical properties, hydrazine is a molecular solvent with strong basic properties and solvating abilities. From these two points of view it may be compared to liquid ammonia as a solvent but it radically differs from ammonia in reducing properties.

The relatively high value of its dielectric constant (51.7 at $25°C$[47]) classes it among the dissociating solvents. That is to say, solutions of salts are dissociated and exhibit good electrical conductivity. Ion pairs need not be taken into account in such a solvent.[90]

A. Solubility

A number of inorganic salts are fairly soluble in anhydrous hydrazine because of the high solvation energies of cations and also to a lesser degree of anions; Table II[91-94] summarizes the published values. The relative order of the solubilities is near that observed in liquid ammonia. The particular case of ammonium salts is complicated by the acidic properties of this cation in hydrazine. A part of the ammonia is released.

The inert gases are soluble (Table III) in anhydrous hydrazine. Henry's law is obeyed, thermodynamic calculations are verified for N_2H_4, and its mixtures with substituted hydrazine.[95-97]

The solubilities of NH_3 at various temperatures and pressures have been published.[98] It is, respectively, 0.14 and 0.05 in mole fraction at 278K and 313K at 1.0 atm. The thermodynamic properties of the NH_3–N_2H_4 system are summarized in Table IV. Hydrazine dissolves alkaline metals but the solutions are unstable; hydrogen is liberated. The solutions of sulfur are more stable; they are yellow or red according to the sulfur concentration. Nitrogen evolves slowly and sulfur is reduced to the hydrazinium salt of HS^-. The reaction is complete in 24 hr.[48]

B. Self-Ionization, Ion Product, and Acid–Base Properties

The N_2H_4 molecule is slightly dissociated according to the following equilibrium

$$2N_2H_4 \rightleftharpoons N_2H_5^+ + N_2H_3^-$$

$N_2H_5^+$, the hydrazinium cation, is the solvated form of the proton in hydrazine. Therefore one can write the above equilibrium as

$$N_2H_4 \rightleftharpoons H_{solv}^+ + N_2H_3^-$$

TABLE II

SOLUBILITIES OF SOME INORGANIC PRODUCTS IN ANHYDROUS HYDRAZINE[a] FROM WALDEN AND HILBERT,[91] WELSH AND BRODERSON,[92] AND SAKK AND ROSOLOVSKII[93]

	Cl^-	Br^-	I^-	NO_3^-	ClO_4^-	ClO_3^-	CO_3^{2-}	SO_4^{2-}	BO_3^{3-}
H^+	—	—	—	—	—	—	—	—	55
Li^+	16	—	—	61.2	54.4 47.1[b]	—	—	—	—
Na^+	12.2	—	—	26.6 117.8[30]	85.1 74.3[b]	66	insol.	insol.	—
K^+	8.5	56.4	135.7	28	30.9 28.5[b]	—	—	—	—
Rb^+	5	—	—	35.8	20.5	—	—	—	—
Cs^+	—	—	—	23.1	32.4	—	—	—	—
NH_4^+	75	—	—	—	—	—	—	—	—
Mg^{2+}	—	—	—	39.8	69	—	—	insol.	—
Ca^{2+}	16	—	—	40.0	86.8	—	—	—	—
Sr^{2+}	—	—	—	61.2	88.0	—	—	—	—
Ba^{2+}	—	—	—	88.5	109.3	—	—	—	—
Ni^{2+}	8	—	—	—	—	—	—	—	—
Zn^{2+}	8	—	—	—	—	—	—	—	—
Cd^{2+}	—	40	—	—	—	—	—	—	—
Cr^{3+}	13	—	—	—	—	—	—	—	—

[a] In g/100 g N_2H_4 at 25°C.
[b] At 0°C.[94]

The conductivity of the pure solvent is $1.6\ 10^{-6}\ \Omega^{-1}\ cm^{-1}$ at 25°C[91,99] versus $7.10^{-8}\ \Omega^{-1}\ cm^{-1}$ for water under the same conditions. The mass law applied to these equilibria gives the ion product K_i, the activity of the solvent being taken as equal to unity.

$$K_i = [N_2H_5^+][N_2H_3^-] \quad \text{or} \quad K_i = [H_{solv}^+][N_2H_3^-]$$

TABLE III

SOLUBILITIES OF INERT GASES IN HYDRAZINE[a]

Solvent	He	N_2	Ar
Ideal	16	100	100
N_2H_4	0.52	0.72	1.24
Water	0.7	1.2	2.5
Methylhydrazine	2.51	9.17	17.9
Dimethylhydrazine (unsymmetrical)	9.88	37.5	68.9

[a] From Chang et al.[97] expressed as mole fraction $\times 10^5$ at 25°C and 1 atm.

TABLE IV

THERMODYNAMIC PROPERTIES OF NH_3 AND N_2H_4 IN A LIQUID NH_3–N_2H_4 SYSTEM[a]

| X_{NH_3} | NH_3[a] | | N_2H_4 | | 298.15°K | | 373.15°K | |
	$\Delta \bar{H}$	$\Delta \bar{S}$	$\Delta \bar{H}$	$\Delta \bar{S}$	a_{NH_3}	$a_{N_2H_4}$	a_{NH_3}	$a_{N_2H_4}$
0.0	—	—	0.0	0.0	0.000	1.000	0.000	1.000
0.1	−1	4.01	−178	−0.39	0.133	0.902	0.163	0.958
0.2	−1	2.63	−178	−0.16	0.266	0.802	0.326	0.852
0.3	−1	1.83	−183	+0.09	0.398	0.701	0.398	0.746
0.4	258	2.10	−298	+0.01	0.537	0.601	0.493	0.665
0.5	400	2.19	−441	−0.13	0.653	0.508	0.570	0.590
0.6	264	1.55	−267	−0.68	0.718	0.453	0.656	0.496
0.7	150	1.02	−59	+1.65	0.773	0.395	0.734	0.407
0.9	127	0.56	+375	+4.62	0.931	0.184	0.905	0.162
1.0	0.0	0.0	—	—	1.000	0.000	1.000	0.000

[a] At 298.15K and 373.15K.

in such an equality the square brackets indicate molar concentrations; the value of K_i is an apparent one which depends on the ionic strength.

The K_i value calculated by Pleskov[100] is 2×10^{-25}; another estimation was found to be 10^{-13}. This latter value seems more realistic, considering the value of the conductivity of the pure solvent, unless the conductivity is also too high, owing to the presence of ionic impurities.

Because of the self-ionization of hydrazine, the strongest acid in such a medium is $N_2H_5^+$, the hydrazinium cation. Every acid stronger than $N_2H_5^+$ reacts on the solvent N_2H_4 to form $N_2H_5^+$, which is the case for benzoic, fumaric, maleic, and carbonic (couple H_2CO_3/HCO_3^-) acids and probably the ammonium cation.[91,99] The addition of strong acid to hydrazine is largely exothermic; for example, the heat of formation of $N_2H_5ClO_4$ is −40.7 kcal/mole.[101]

The strongest base is $N_2H_3^-$. However, this base is not used as often as OH^-. From the titration of hydrazinium salt by tetramethylammonium hydroxide,[102] we have estimated the acidity constant of the H_2O/OH^- couple; the pK is $\simeq 11$. Water is a very weak acid in hydrazine in accordance with conductometric studies[99] which showed that water is slightly dissociated in hydrazine.

The acidic properties of methanol are very similar to those of water, as shown by either potentiometric titration[102] or conductometric measurements.[99] Some acidity constants are given in Table V. pH indicators may be used in hydrazine, but some—diphenylamines and carbazoles containing 2,4-dinitro groups—have unstable colors.[103]

TABLE V

ACIDITY CONSTANT (pK) OF SOME WEAK
ACIDS IN ANHYDROUS HYDRAZINE[99]

Acid	pK_{H_2O}	$pK_{N_2H_4}$
p-Chlorophenol	9.4	2.18
Phenol	9.9	3.18
m-Cresol	10.0	3.39
o-Cresol	10.2	3.44
p-Cresol	10.2	3.75

C. Formation of Complexes

No study about the formation of complexes in anhydrous hydrazine has been published to date to the best of our knowledge.

Owing to the high value of the solvation energies (see Table XII), one can predict that the complexes of cations such as Zn^{2+}, Cd^{2+}, and even Ca^{2+}, which are strongly solvated by hydrazine, will be much less stable in hydrazine than in water.

D. Redox Reactions

Hydrazine exhibits reducing properties. Thus, most elements can only exist in this solvent at their lower oxidation degree, for example, Cu(I). In

TABLE VI

NORMAL POTENTIALS OF SOME REDOX COUPLES[a]

	E^0 in N_2H_4		E^0 in H_2O
Metal	Ref. 104	Refs. 105, 106	Ref. 107
Li^+	-0.19	—	-0.10
K^+	-0.01	—	0.00
Ca^{2+}	$+0.10$	—	$+0.06$
Na^+	$+0.18$	—	$+0.17$
Co^{2+}	—	$+1.48$	$+2.65$
Zn^{2+}	$+1.60$	$+1.55$	$+2.17$
Ni^{2+}	—	$+1.66$	$+2.70$
Cd^{2+}	$+1.91$	$+1.99$	$+2.53$
H^+	$+2.01$	—	$+2.93$
Cu^+	$+2.23$	—	$+3.45$
Pb^{2+}	$+2.35$	$+2.31$	$+2.80$
Ag^+	$+2.78$	—	$+3.73$
Hg_2^{2+}	$+2.88$	—	$+3.72$

[a] The reference potential is the potential for Rb^+/Rb.

the same way, Hg_2^{2+} and Ag^+ are partially reduced to their metallic forms.[1]

Some normal potentials have been determined in this solvent. The values are summarized in Table VI and compared to the analogous ones in water. The difference between the normal potential of alkaline metals and other metals is smaller in hydrazine because of the higher solvation of the corresponding cation which renders their reduction more difficult.

Some inversions in the normal potential order are observed; CO^{2+} and Ni^{2+}, which are strongly solvated, become more difficult to reduce than Zn^{2+} and Cd^{2+}. The proton is also highly solvated and become less of an oxidant. For example, lead is attacked by acids in water but not in hydrazine. The same remarks are true for liquid ammonia as a solvent.[104]

VI. WATER–HYDRAZINE MIXTURES AS SOLVENTS

The water–hydrazine mixtures have been studied more extensively than anhydrous hydrazine, perhaps because they are not so unstable. Therefore they are more convenient for practical utilization.

A. Acid–Base Reactions

Hydrazine is more basic than water, and water is more acidic than hydrazine. So it is the basic properties of hydrazine which impose the limit on the acidic medium, by the formation of the hydrazinium cation $N_2H_5^+$, and it is the acidic properties of water which impose the limit on the basic media by the formation of the hydroxyl ion, OH^-. The self ionization of the H_2O–N_2H_4 mixtures can be written as

$$N_2H_4 + H_2O \rightleftharpoons N_2H_5^+ + OH^-$$

The mass law applied to the above equilibrium gives the ion product expression, the activities of hydrazine and water being constant for a given mixture

$$K_i = [N_2H_5^+][OH^-]$$

The experimental determination of K_i has been determined between 5 and 93.5% N_2H_4 (see Table VII).[104–107] The maximum value of K_i (minimum for pK_i) occurs for about 25% N_2H_4. In such a medium of high ion product values, substantial amounts of $N_2H_5^+$ and OH^- exist in the pure neutral solvent. For example, if the ionic product is 10^{-5}, the pure neutral solvent is $10^{-2.5}$ M or 3.10^{-3} M in $N_2H_5^-$ and in OH^-. A buffering effect is observed and the measurement of the pH of the pure neutral solvent is possible. The same reasons are valid for the determination of the acidity functions in such media (vide infra).

TABLE VII

ION PRODUCT OF N_2H_4–H_2O MIXTURES

Weight of N_2H_4 (%)	pK_i (ionic strength = 0.1)[108]	pK_i (ionic strength = 0)[109]
5	5.4	—
6	—	5.8
10	5.0	—
19	—	5.5
25	4.6	—
31	—	5.5
43	5.4	—
50	—	5.9
55	—	6.4
64	5.9	—
85	8.1	—
93.5	11.0	—

Some acidity constants of weak acids have been determined in water–hydrazine mixtures[99,103,108] (see Table VIII). Owing to the high basicity of N_2H_4, a weak acid in water appears not to be so weak in water–hydrazine. The fact that weak acids become stronger is more evident for acids which correspond to uncharged bases (amines). In the case of acids which correspond to charged bases, this effect is compensated for by the diminution of the dielectric constant of the medium by hydrazine addition, which tends to render the acids weaker.

Acid–base titrations do not have a good precision in water–hydrazine mixtures of high ion product because of the nonquantitative nature of the reactions. The theoretical precision of 1% for the titration of a strong acid by a strong base may be reached in the hydrate $N_2H_4 \cdot H_2O$ (64% N_2H_4 in weight) or at a higher concentration of N_2H_4.

B. Formation of Complexes

The only complexes systematically studied are those of Fe^{2+}.[110] One can predict from the strong solvation of Fe^{2+} by hydrazine[105] that the Fe^{2+} complexes are not as stable in hydrazine as in water.

The behavior of the 2,2′-dipyridyl (Dp) and the 1,10-phenanthroline complexes are similar and very close to those of the 1,10-phenanthroline-5-sulfonate (phen-SO_3^-) complexes. They contain 3 moles of organic ligand for 1 mole of Fe^{2+}, and are stable up to about 64% N_2H_4 by weight. For higher concentrations in hydrazine, formation of mixed complexes with organic ligand and OH^- occurs.

With ligands such as quinolines, the decrease in the stability of the com-

TABLE VIII

pK_a of Some Acids in Water–Hydrazine Mixtures

Composition of N_2H_4 (weight %)	pK_i	Positively charged acids						Other acids	
		Ammonia	Glycol	Ethylamine	Diethylamine	n-Butylamine	Guanidine	HBO_2	HCO_3^-
Water	14.0	9.4	9.9	11.0	11.3	11.1	13.7	9.1	10.2
5	5.4	—	1.3	2.6	2.8	2.5	4.9	0.9	2.0
10	5.0	Strong	0.6	2.0	1.9	1.6	4.1	1.2	1.5
25	4.6	—	Strong	1.1	1.0	0.8	3.3	1.5	1.6
50	5.4	—	—	0.7	—	—	2.9	2.1	2.4
64	5.9	—	—	—	—	—	2.8	2.4	—
85	8.1	—	—	—	—	—	2.6	—	—
93.5	11.0	—	—	—	—	—	—	—	—
95	—	—	—	—	—	—	2.2	—	—

TABLE IX

STABILITY CONSTANTS OF SOME COMPLEXES OF Fe^{2+}

N_2H_4 (%)	pK_{OH^-}	pK_{phen}	pK_{D_p}	$pK_{phen\ SO_3^-}$	pK_{Q^-}	$pK_{QSO_3^{2-}}$
0	—	20.2	17.9	—	15.1	15.0
5	0.9	11.3	9.0	11.7	6.5	6.2
10	1.0	10.5	8.2	11.0	6.5	5.7
25	1.9	10.1	7.9	9.9	6.9	6.2
50	4.1	9.15	6.8	9.1	8.6	8.0
64	6.3	8.8	—	8.4	10.7	9.3
85	—	—	—	—	13.0	—
90	—	—	—	—	13.7	—

plexes due to the solvation of iron by hydrazine is compensated for by the decrease in the dielectric constant with hydrazine addition. The formula for the complexes contains two ligands per atom of iron for either 8-hydroxyquinoline (Q^-) or 8-hydroxyquinoline 5-sulfonate (QSO_3^{2-}). The results are reported in Table IX.

C. Redox Reactions

The half-wave potentials of several redox couples have been measured by polarography.[111] They depend on the composition of the mixtures as given in Table X. For a given mixture, they vary with the acidity of the solution.

TABLE X

$E_{1/2}(V)$ AT $pH = 0$ FOR CATIONS $(10^{-4}\ M)^a$

N_2H_4 (%):	5	10	25	50	64	80	95	100
Cr^{3+}/Cr^{2+}	−0.53	−0.54	−0.55	−0.65	−0.59	−0.51	−0.54	—
Co^{2+}/Co	−0.44	−0.47	−0.50	−0.57	−0.59	−0.59	−0.57	−0.53
$Zn^{2+}/Zn(Hg)$	−0.42	−0.43	−0.43	−0.46	−0.47	−0.51	−0.51	−0.46
V(IV)/V(III)	−0.42	−0.42	−0.42	−0.44	−0.38	−0.47	—	—
As(III)/As	−0.42	−0.42	−0.42	−0.48	−0.51	−0.53	−0.50	—
Ni^{2+}/Ni	−0.22	−0.23	−0.29	−0.35	−0.41	−0.41	−0.41	−0.35
Sb(III)/Sb	0.00	−0.00	−0.05	−0.14	−0.13	−0.15	−0.15	—
$Cd^{2+}/Cd(Hg)$	0.00	+0.02	−0.03	−0.07	−0.11	−0.09	−0.08	−0.05
$Cu^+/Cu(Hg)$	+0.32	+0.28	+0.24	+0.17	+0.16	+0.14	+0.14	—
$Pb^{2+}/Pb(Hg)$	+0.25	+0.25	+0.26	+0.25	+0.22	0.24	+0.26	+0.29
$Ti^+/Tl(Hg)$	+0.25	+0.28	+0.31	+0.34	+0.33	0.33	+0.33	—

^{a} Reference electrode: normal hydrogen electrode in each mixture. From Furlani[106] and Bauer.[111]

TABLE XI

$E_{1\,2}(V)$ AT DIFFERENT pH VALUES[a]

	$[H^+] = 1$	$[H^+] = 10^{-1}$	$[OH^-] = 10^{-2}$	$[OH^-] = 10^{-1}$
Cr_3/Cr^{2+}	-0.55	-0.60	---	---
Co^{2+}/Co	-0.50	-0.51	-0.52	-0.53
Zn^{2+}/Zn	-0.43	-0.46	-0.52	-0.59
$V(IV)/V(III)$	-0.42	-0.54	—	—
$As(III)/As$	-0.42	-0.42	—	—
Ni^{2+}/Ni	-0.29	-0.30	-0.31	-0.32
$Sb(III)/Sb$	-0.07	-0.14	-0.22	-0.41
Cd^{2+}/Cd	$+0.03$	-0.04	-0.06	—
Cu^+/Cu	$+0.24$	$+0.23$	$+0.23$	$+0.22$
Pb^{2+}/Pb	$+0.26$	$+0.24$	$+0.16$	—
Tl^+/Tl	$+0.35$	$+0.34$	$+0.34$	$+0.34$

[a] Solvent at $25°_0$ N_2H_4.[111] Reference electrode: normal hydrogen electrode. Cation concentrations: 10^{-4} M.

These variations are especially important for zinc, vanadium, and antimony (Table XI).

From the data in Table XI, it appears that the order of the various cations on the potential scale is maintained by addition of hydrazine to water except for Ni^{2+} and Co^{2+} which become less oxidizing than Tl^+, Cd^{2+}, and even for Co^{2+} which less oxidizes than Zn^{2+}. The proton H^+ does not oxidize lead and thallium, and above 25% N_2H_4, lead becomes more reducing than thallium and copper more reducing than lead.

VII. SOLVATION BY HYDRAZINE; COMPARISON WITH OTHER SOLVENTS

A. Anhydrous Hydrazine

The solvation energies of several inorganic ions have been calculated by Izmailov.[112] Cations seem to be more solvated than anions (Table XII). As cations give more stable complexes with hydrazine in aqueous solution, their solvation energy increases.

In H_2O–N_2H_4 mixtures, a preferential solvation of the cations as well as the anions by hydrazine occurs. A modern way to express the difference of solvation energy in two solvents is to use the transfer activity coefficient γ_t,[113] the definition of which results from the following relation.

$$\log {}_t(A)_{water\ s} = \frac{(\Delta G_t^0)_{water \to s}}{2.3\ RT}$$

TABLE XII

SOLVATION ENERGIES[a]

	N_2H_4	H_2O
I^-	59.5	1.0
Cs^+	61.5	1.5
Rb^+	70.0	2.5
Br^-	71.0	-2.0
Cl^-	73.5	1.0
K^+	75.0	2.0
Na^+	92.0	2.0
Li^+	113.5	3.5
Ag^+	128.5	-18.5
H^+	274.5	-18.0
Ca^{2+}	363.0	8
Cd^{2+}	446.5	-18.5
Zn^{2+}	508.5	-16.5

[a] In kcal/mole. From
Izmailov.[112]

ΔG_t^0 is the change of free enthalpy occurring during the transfer of a molecule A from water to a solvent S. This transfer activity coefficient permits the calculation of the activity of A in a solvent S. If we know the activity of A in water, a_W, the two solutions of A (in water and in S) being in an equilibrium state, it follows that

$$a_W = \gamma_{t\,W\to S}\,a_S$$

In dilute solution, a_W and a_S are equal in first approximation to the concentrations. γ_t is greater than unity if the molecule A is more solvated by water than by the solvent and the contrary holds if $\gamma_t < 1$.

The definition presents no difficulty for molecular solutes (NH_3, H_2, etc.). It is quite different for ions because of the impossibility of transferring an ion without the counterion. So only the mean transfer activity coefficient $\gamma_{t\pm}$ is theoretically measurable. To overcome this difficulty, several assumptions on the solvation of selected species have been proposed. Discussions on their validity have been published elsewhere.[113] In the case of hydrazine as a solvent, the transfer activity coefficient has been determined using two assumptions: that of Pleskov, based on the assumed negligible solvation of the rubidium cation Rb^+, and that of Izmailov, based on the extrapolation to infinite radius of the difference of free enthalpy $G_M^0 - G_H^0$ for a series of metals of varying radii. The results are summarized in Table XIII.[104,106,114,115]

TABLE XIII

COMPARISON OF log $\gamma_{t\,H_2O\rightarrow N_2H_4}$

log γ_t	Pleskov's assumption and results[104,114]	Pleskov's assumption and Furlani results[106]	Izmailov's assumption[115]
Li^+	1.5	—	2.6
Na^+	0	—	1.4
K^+	0	—	1.4
Rb^+	0	—	1.8
Ca^{2+}	0	—	6
H^+	−13.2	—	−14
Ag^+	−16	—	—
$Hg(I)$	−7.1	—	—
Cu^+	−20.3	—	—
Pb^{2+}	−3.8	−4.2	—
Zn^{2+}	−5.0	−6.0	—
Cd^{2+}	−5.0	−6.2	—
Ni^{2+}	—	−17.3	—
Co^{2+}	—	−19.5	—
Cl^-	—	—	0.7
Br^-	—	—	1.4
I^-	—	—	0.7

Recently the cobaltocene/cobalticinium redox potential has been used as a reference.[116] Consequently, Strehlow's assumption based on the same change of solvation of cobaltocene and cobalticinium ion by transfer from one solvent to another, could be utilized in hydrazine. Unfortunately, the results cannot be used because of the lack of data concerning the potential of this couple versus common reference electrodes in N_2H_4.

The donor number, DN, of hydrazine is 44,[90] determined from the NMR chemical shift of ^{23}Na. It is a measure of the energy of the reaction of $SbCl_5$ on N_2H_4 and reflects the basicity of the solvent. Slightly higher than the donor number of pyridine (DN = 33) and lower than the donor number of amines (ethylenediamine: 55, isopropylamine: 57.5, tert-butylamine: 57.5, ammonia: 59), the value of the donor numbers give a good measure of the relative properties of these solvents.

B. Acidity Functions

The variations of the basicity of the water–hydrazine mixtures have been determined by means of acidity (or Hammett) functions.

Classic Hammett acidity functions have been determined[99,103] using pH

TABLE XIV

ACIDITY FUNCTIONS OF $H_2O-N_2H_4$ MIXTURES

N_2H_4 (weight %)	H^{-103}	H_{GC}^{116}	H^{-109}
5	11.18	—	—
10	11.55	—	12.0
15	11.93	—	—
16.49	—	13.05	—
20	12.29	—	12.5
23.88	—	14.13	—
25	12.72	—	12.8
30	13.15	—	13.2
30.77	—	14.93	—
35	13.56	—	—
37.21	—	15.70	—
40	14.03	—	13.8
43.24	—	16.28	—
45	14.52	—	—
48.9	—	16.78	—
50	14.99	—	14.5
54.33	—	17.20	—
55	15.43	—	—
59.25	—	17.60	—
60	15.93	—	15
64.00	—	17.90	—

indicators. A more modern way, which avoids the stepwise use of indicators involves the determination of electrochemical methods. One measures the difference between a reference electrode, the potential of which does not depend on the solvent, and an electrode which indicates the activity of the proton. Two examples have been proposed. The first uses the hydrogen electrode in the $H_2O-N_2H_4$ mixture and the saturated calomel electrode in water with the assumption that a liquid junction potential does not exist between the two media.[109] The second uses the glass electrode cobaltocene–cobalticinium couple with the assumption that the potential of this couple is not dependent on the solvent.[116] The results are summarized in Table XIV.

VIII. EXPERIMENTAL DATA

It is necessary to protect the solvent in an inert atmosphere to avoid oxidation by oxygen, absorption of CO_2, and in the case of anhydrous hydrazine, to avoid water absorption.

A. Anhydrous Hydrazine

Most of the experimental methods employed for the study of reactions in solutions can be used with anhydrous hydrazine, namely, spectroscopic, conductometric, and electrochemical methods.

Conductometric measurements should be made with a bright platinum electrode to avoid the spontaneous decomposition of hydrazine in contact with the smooth platinum. The limiting ionic conductances of Na^+, K^+, Cl^-, and ClO_4^- ions have been determined by transference studies.[117] Their values are comparable to those of the same ions in water; $N_2H_5^+$ is much more conductive in hydrazine than in water.

For electrochemical methods, the use of a platinized platinum electrode is still valid. A good reference electrode in anhydrous hydrazine[102,106] is based on the following cell: $Cd(Hg)/CdSO_4\downarrow/K_2SO_4(sat.)$ in N_2H_4. Its potential is 0.205 V vs. the normal hydrogen electrode in anhydrous hydrazine. An analogous electrode has been made using zinc instead of cadmium.[118]

The hydrogen electrode indicates proton activity and may be used for pH measurements and titrations. A glass electrode filled with anhydrous hydrazine and using the $Cd(Hg)/CdSO_4$ reference electrode inside the bulk of the glass membrane gives satisfactory results.[119] Its response in acidic media is the same as that of the hydrogen electrode. In a basic media, there is a difference between the glass electrode and the hydrogen electrode potentials. Nevertheless, titrations of acids by various quaternary ammonium hydroxides are possible. Titration of alkaline hydroxides requires the use of an hydrogen electrode. The strongest acid is $N_2H_5^+$ which is commercially available or may be prepared in solution by addition of HCl or H_2SO_4 to N_2H_4. The strongest base in hydrazine is $N_2H_3^-$. It can be prepared by heating $NaNH_2$ in N_2H_4 at 60°C.[30] For the recording of voltammetric curves, several working electrodes such as mercury and platinum have been used.[105,106]

Glass or Pyrex vessels can be used for N_2H_4 storage. The resistance to anhydrous hydrazine of several alloys has been tested[120]; these can be arranged in order of decreasing susceptibility: 4130 steel > 410 stainless steel > 718 Inconel > Ti-6 Al-4% V > 6061 Ti6 Al.

Fires caused by hydrazine can be extinguished by bromotrifluoromethane.[121] Before systematic use of this solvent it may be useful to consult Back and Thomas,[41] in which safety precautions are summarized.

B. H_2O–N_2H_4 Mixtures

Not as unstable as anhydrous hydrazine, the H_2O–N_2H_4 mixtures are easier to manipulate. Nevertheless they require about the same precautions as anhydrous hydrazine. Here we will report only the noticeable differences.

The reference electrodes using zinc or cadmium systems cannot be employed because of the slow precipitation of zinc or cadmium hydroxides. The following hydrogen electrode has been used as a reference in $N_2H_4-H_2O$: Pt, $H_2/N_2H_5{}^+Cl^-(10^{-1}\ N)$.

An ordinary aqueous glass electrode gives satisfactory results in acidic media and presents a deviation in alkaline media. Nevertheless, the aqueous glass electrode becomes slowly (3 or 4 weeks) unsatisfactory because small quantities of hydrazine which diffuse inside the filling liquid reduce the Ag/AgCl internal reference electrode.[108]

The only possible electrode for the recording of voltammetric curves is the dropping mercury electrode. At this electrode the electroactivity range is about 1.2 V.

On a mercury surface the oxidation of hydrazine limits the oxidation range, and the reduction of water limits the reduction one.[111] At a platinum electrode, we do not benefit from the overpotential for water reduction and the electroactivity range becomes so narrow (100–200 mV) that measurements are no longer possible.[108]

REFERENCES

1. L. F. Audrieth and B. A. Ogg, "The Chemistry of Hydrazine." Wiley, New York, 1951.
2. R. Ohme and A. Zubek, Z. Chem. **8**, 41 (1968).
3. K. Tachiki and R. Endo, Japan Kokai, 73/37,397; C. A. **79**, 68267d (1973).
4. F. Weiss, J. P. Schirmann, and H. Mathais, German Patent 2,143,516 (1972); C. A. **77**, 154577v (1972).
5. R. Powell, "Hydrazine Manufacturing Processes." Noyes Development Corporation, Park Ridge, New Jersey, 1968.
6. S. G. McGriff and W. A. McRae, French Patent 1,513,854 (1968); C. A. **70**, 92681u (1969).
7. P. L. Speeding, Chem. Eng. Prog., Symp. Ser. **67**, 20 (1971); C. A. **75**, 89605v (1971).
8. I. Heroin, Rev. Fiz. Chim., Ser. A **10**, 265 (1973); C. A. **79**, 151520b (1973).
9. A: Sakumoto, Nippon Kagaku Kaishi **9**, 1591 (1972); C. A. **78**, 65101p (1973).
10. B. G. Dzantiev, I. A. Savushkin, G. V. Nichipor, and V. T. Kazazyan, Nucl. Sci. Abstr. **26**, 14888 (1972); C. A. **77**, 55291r (1972).
11. G. Gaussens, French Patent 95,916 (1900); C. A. **79**, 80953g (1973).
12. G. A. Mazzochin, F. Magno, and G. Bontempelli, J. Electroanal. Chem. **45**, 471 (1973).
13. V. L. Syaduk and E. B. Eremin, Zh. Fiz. Khim. **47**, 1538 (1973); C. A. **79**, 97440k (1973).
14. V. L. Syaduk and E. N. Eremin, Zh. Fiz. Khim. **46**, 2894 (1972); C. A. **78**, 62741m (1973).
15. K. H. Linke and R. Taubert, Chem. Ber. **03**, 2008 (1970).
16. C. A. Lobry de Bruyn, Recl. Trav. Chim. Pays-Bas **13**, 433 (1894); **18**, 297 (1899).
17. G. R. Gedye, T. E. Allibone, and E. K. Rideal, J. Chem. Soc. p. 1158 (1932).
18. M. S. Spencer, J. Inorg. Nucl. Chem. **31**, 2611 (1969).
19. C. Bighi, G. Saglietto, and A. Betti, Ann. Univ. Ferrara, Sez. 5 **2**, 163 (1967); C. A. **68**, 84136j (1968).
20. C. Bighi, A. Betti, and G. Saglietto, Bull. Soc. Chim. Fr. p. 4637 (1967).
21. C. Bighi, A. Betti, and G. Saglietto, Ann. Chim. (Rome) **57**, 1142 (1967).
22. J. L. Spigarelli, Diss. Abstr. Int. B **31**, 4547 (1971).

23. E. Santacesaria and L. Giuffre, *Riv. Combust.* **23**, 438 (1969).
24. H. P. Meissner and R. F. Baddour, U.S. Patent 3,458,283 (1969).
25. J. G. Rigsby, U.S. Patent 3,740,436 (1973); *C. A.* **79**, 80954h (1973).
26. V. Talasek, *Collect. Czech. Chem. Commun.* **33**, 731 (1968).
27. W. A. Schroeder, *in* " Methods in Enzymology" (C. H. W. Hirs and S. N. Timasheff, eds.), Vol. 25, p. 138. Academic Press, New York, 1972; *C. A.* **77**, 11031v (1972).
28. R. A. Penneman and L. F. Audrieth, *J. Am. Chem. Soc.* **71**, 1644 (1949).
29. N. I. Ampelogova and A. G. Puzikov, USSR Patent 279,155 (1970); *C. A.* **74**, 71425f (1971).
30. E. Santacesaria, *Chim. Ind. (Milan)* **51**, 283 (1969); *C. A.* **50**, 690g (1956).
31. C. D. Good and D. R. Poole, U.S. Patent 3,598,546 (1971); *C. A.* **75**, P89695z (1971).
32. W. Swiecicki, *Med. Pr.* **24**, 71 (1973); *C. A.* **79**, 101177z (1973).
33. S. H. Lee and H. Aleyassine, *Arch. Environ. Health* **21**, 615 (1970).
34. C. Biancifiori, *J. Natl. Cancer Inst.* **44**, 943 (1970).
35. S. S. Mirvish, *Int. J. Cancer* **4**, 318 (1969).
36. E. Krüger, *Biochem. Pharmacol.* **19**, 1825 (1970).
37. D. Lopez-Mendoza and S. Villa-Trevino, *Lab. Invest.* **25**, 68 (1971).
38. R. P. Peterson, *Br. J. Ind. Med.* **27**, 141 (1970).
39. M. Ronne, Thesis, University of Paris, Paris (1966).
40. H. P. Silverman and G. Giarusso, U.S. Patent 3,676,188 (1972); *C. A.* **77**, 156049y (1972).
41. K. C. Back and A. A. Thomas, *Annu. Rev. Pharmacol.* **10**, 395 (1970).
42. C. A. Lobry de Bruyn, *Recl. Trav. Chim. Pays-Bas* **15**, 174 (1896).
43. L. D. Barrick, G. H. Drake, and H. L. Lochte, *J. Am. Chem. Soc.* **58**, 160 (1936).
44. D. M. Mason, O. W. Wilcox, and B. H. Sage, *J. Phys. Chem.* **56**, 1008 (1952).
45. J. W. Brühl, *Ber. Dtsch. Chem. Ges.* **30**, 158 (1872).
46. J. W. Brühl, *Z. Physik. Chem.* **23**, 564 (1897).
47. H. Ulich and W. Nespital, *Z. Physik. Chem., Abt. B* **16**, 221 (1932).
48. P. Pascal, "Traité de chimie minérale," Vol. X, p. 542. Masson, Paris, 1958.
49. A. L. Tsykalo, V. K. Sasenkov, V. I. Selevanyuk, and A. P. Yakushev, *Zh. Prikl. Khim.* **47**, 441 (1974); *C. A.* **81**, 42312y (1974).
50. X. X. Krivich, *C. R. Acad. Sci. URSS* **23**, 36 (1939).
51. Y. Hamada, A. Y. Hirakawa, K. Tamagake, and M. Tsuboi, *J. Mol. Spectrosc.* **35**, 420 (1970).
52. Y. Kamishina, *J. Chem. Phys.* **60**, 4626 (1974).
53. P. A. Giguiere and V. Schomaken, *J. Am. Chem. Soc.* **65**, 2025 (1943).
54. W. G. Penney and G. B. Sutherland, *Trans. Faraday Soc.* **30**, 898 (1934).
55. W. G. Penney and G. B. Sutherland, *J. Chem. Phys.* **2**, 492 (1934).
56. W. Fresenius and J. Karweil, *J. Phys. Chem.* p. 44 (1939).
57. J. O. Jarvie and A. Rauk, *Can. J. Chem.* **52**, 2785 (1974).
58. J. O. Jarvie, A. Rauk, and C. Edminston, *Can. J. Chem.* **52**, 2778 (1974).
59. N. Bodor, M. J. S. Dewar, W. B. Jennings, and S. D. Worley, *Tetrahedron* **26**, 4109 (1970).
60. A. L. Tsykalo, V. K. Ososkov, and W. K. Selevanyuk, *Zh. Fiz. Khim.* **48**, 1609 (1974); *C. A.* **81**, 177808t (1974).
61. M. Sieprawsi and R. Cohen-Adad, *Bull. Soc. Chim. Fr.* p. 2630 (1973).
62. L. V. Litvinova, V. V. Kushchenko, and K. D. Mischchenko, *Zh. Obshch. Khim.* **42**, 1815 (1972).
63. E. Meyer and H. G. Wagner, *Z. Phys. Chem.* **89**, 329 (1974).
64. G. Charlot, "Analyse quantitative minérale," pp. 8, 42, and 614. Masson, Paris, 1966.
65. F. A. Cotton and G. Wilkinson, "Advanced Inorganic Chemistry," p. 351. Wiley (Interscience), New York, 1972.

66. S. Syamsunder and T. K. S. Murthy, *Indian J. Chem.* **11**, 669 (1973).
67. S. R. Cooper and J. B. Morris, *Anal. Chem.* **24**, 1360 (1952).
68. Y. K. Gupta and S. K. Mishra, *J. Chem. Soc. A* **18**, 2918 (1970).
69. R. C. Paul, R. K. Chauman, and R. Parkash, *Talanta* **21**, 663 (1974).
70. G. Charlot, "Les Reactions Chimiques en Solution." Masson, ed., Paris, 1969.
71. L. V. Nikitiṅa and M. T. Kozlovskii, *Izv. Akad. Nauk Kaz. SSR, Ser. Khim.* **17**, 12 (1967);
    C. A. **68**, 72713m (1968).
72. A. A. Grinberg and E. M. Marshak, *Zh. Neorg. Khim.* **13**, 1213 (1968).
73. D. Moso and J. Szabo, *Energ. Atomtech.* **27**, 132 (1974).
74. H. Yamaguchi, H. Kume, and I. Manabe, U.S. Patent 3,808,138 (1974); *C. A.* **81**, 41280g
    (1974).
75. E. H. Sprague, U.S. Patent 3,823,225 (1975).
76. T. Yamazaki, Japan Kokai, 74/20,089 (1974). *C. A.* **81**, 59923y (1974).
77. C. E. Marks, British Patent 1,351,441 (1974).
78. Y. Yamashita and M. Sata, U.S. Patent 3,785,988 (1974).
79. G. Lecointre, Thèse 3ème cycle, University of Paris, Paris (1974).
80. J. Heitbaum and W. Vielstich, *Electrochim. Acta* **18**, 501 (1973).
81. T. Nakata, M. Yamashita, T. Shimada, and M. Kubokawa, *Denki Kagaku* **41**, 119 (1973).
82. N. V. Korovin and A. G. Kicheev, *Elecktrokhimiya* **6**, 1330 (1970).
83. T. Kodera, H. Kita, and M. Honda, *Electrochim. Acta* **17**, 1361 (1972).
84. G. V. Vitvitskaya, T. N. Glazatova, and V. S. Daniel-Bek, *Katal. Reakst. Zhidk. Faze, Tr.
    Vses. Konf., 2nd, 1966* p. 411 (1967); *C. A.* **68**, 92446e (1968).
85. I. G. Shmachkova, V. F. Stenin, and N. V. Koronin, *Ref. Zh., Khim.* Abstr. No. 19B (1973);
    C. A. **78**, 66207g (1973).
86. H. Kohmueller, *Siemens Forsh.- Entwicklungsber.* **2**, 153 (1973); *C. A.* **79**, 99775x (1973).
87. M. H. Milles and P. M. Kellet, *J. Electrochem. Soc.* **117**, 60 (1970).
88. P. E. Iversen and H. Lund, *Anal. Chem.* **41**, 1322 (1969).
89. M. Schmeisser and F. Huber, *Z. Anorg. Allg. Chem.* **367**, 62 (1969).
90. M. Herlem and A. I. Popov, *J. Am. Chem. Soc.* **94**, 1431 (1972).
91. P. Walden and H. Hilbert, *Z. Phys. Chem. (Leipzig)* **165**, 241 (1933).
92. T. W. B. Welsh and H. J. Broderson, *J. Am. Chem. Soc.* **37**, 816 (1915).
93. Zh. G. Sakk and V. Y. Rosolovskii, *Zh. Neorg. Khim.* **17**, 1783 (1972).
94. V. Y. Rosolovskii and Zh. G. Sakk, *Zh. Neorg. Khim.* **15**, 2262 (1970).
95. N. A. Gokcen and E. T. Chang, *U.S. Gov. Res. & Dev. Rep.* **69**, 154 (1969).
96. E. T. Chang, N. A. Gokcen, and T. M. Poston, *J. Spacecr. Rockets* **6**, 1177 (1969).
97. E. T. Chang, N. A. Gokcen, and T. M. Poston, *J. Phys. Chem.* **72**, 638 (1968).
98. E. T. Chang, N. A. Gokcen, and T. M. Post, *J. Chem. Eng. Data* **16**, 404 (1971).
99. L. J. Vieland and R. P. Seward, *J. Phys. Chem.* **59**, 466 (1955).
100. V. A. Pleskov, *Acta Physicochim. URSS* **13**, 662 (1940).
101. V. Y. Rosolovskii, N. V. Krivstsov, and K. V. Tifova, *Zh. Neorg. Khim.* **13**, 681 (1968).
102. J. C. Goudeau, M. Broussely, F. Souil, and M. L. Bernard, *Electrochim. Acta* **17**, 2025
     (1972).
103. N. C. Deno, *J. Am. Chem. Soc.* **74**, 2039 (1952).
104. V. A. Pleskov, *Usp. Khim.* **16**, 254 (1947).
105. C. Furlani, *Gazz. Ital.* **87**, 371 (1957).
106. C. Furlani, *Ann. Chim.* **45**, 264 (1955).
107. G. Charlot, "Oxidation-Reduction Potentials." Pergamon, Oxford, 1958.
108. D. Bauer, *Bull. Soc. Chim. Fr.* **11**, 3302 (1965).
109. J. L. Brisset, R. Gaboriaud, and R. Schaal, *J. Chim. Phys. Phys.-Chim. Biol.* **67**, 1726 (1970).
110. D. Bauer, *Bull. Soc. Chim. Fr.* **8**, 2631 (1966).

111. D. Bauer, *Bull. Soc. Chim. Fr.* **3**, 944 (1967).
112. N. A. Izmailov, *Dokl. Akad. Nauk SSSR* **149**, 1103 (1963).
113. D. Bauer and M. Breant, "Electroanalytical Chemistry" (A. J. Bard, ed.) Vol. 8, p. 280, Marcel Dekker, New York, 1975.
114. V. A. Pleskov, *Zh. Fiz. Khim.* **14**, 1477 (1940).
115. N. A. Izmailov, *Dokl. Akad. Nauk SSSR* **149**, 1103 (1963).
116. J. Janata and R. D. Holtby-Brown, *J. Electroanal. Chem.* **44**, 137 (1973).
117. G. Baca, *Diss. Abstr. Int. B* **30**, 3581 (1970).
118. H. Ulich and K. Biastoch, *Z. Phys. Chem., Abt. A* **178**, 136 (1936).
119. J. C. Goudeau, M. Broussely, and F. Souil, *C. R. Hebd. Seances Acad. Sci., Ser. C* **269**, 200 (1969).
120. W. P. Gilbreath and M. J. Adamson, *NASA Tech. Note* **D-7604** (1974).
121. D. S. Burgess, T. A. Christos, J. C. Cooper, T. A. Kubala, and G. H. Martindill, *Sci. Tech. Aerosp. Rep.* **7**, 2620 (1969).

Author Index

A

Abraham, M. H., 77(42), *117*
Abrahamson, E. A. A., Jr., 237(205), *249*
Abegg, R., 190(90), 233(90), 234(90, 195), *247, 249*
Adams, R. N., 124(34), *141*
Adamson, M. J., 271(120), *275*
Ahrland, S., 6(2, 3, 74), *17*, 23(31), 24(3, 32a), 26(33), 31(33), 32(2, 47), *33*, 34(2, 32a, 49), *35*, 37(51), *41, 44*, 46(32a, 33, 49, 58, 59, 64), 47(49, 58), *48*, 49(32a, 49, 71), *50* (58), 51(59, 74), *52*, 53(74), *54*, 56(79, 80), 57(85, 88), 58(88), *60, 61, 62*
Aitken, H. W., 97, 100(82a), *118*
Akimoto, N., 170, *177*
Akisnov, V. K., 169(78, 79), *177*
Aknes, G. A., 89(62), *93, 117*
Alderman, R. L., *86*
Alessi, J. T., 169, *177*
Alexander, R., 12(11), 18(19), *60*, 69(20), 71 (29), 73(20, 36), *75*, 76(20), *116, 117*, 236 (200), *249*
Alexandrova, A. M., 164, *177*
Aleyassine, H., 254(33), *273*
Alfenaar, M., 70, *116*, 159(29), 161(29), *176*
Allen, G., 202(141), 214(141), *248*
Allerhand, A., 91, *118*
Allibone, T. E., 253(17), *272*

Allred, A. L., 90(70), 96, *118*
Ambrose, B. A., 183, *245*
Amis, E. S., 64, *116*
Ampelogova, N. I., 254(29), *273*
Anderson, E. X., 199(127), *247*
Anderson, J., *86*
Andon, R. J. L., 183(33), *245*
Andrussow, K., *30*
Angenot, P., 183(35), *245*
Angerstein, H., 190(94), *247*
Anisimova, K. M., 241(222), *250*
Arndt, F., 181(3, 4), *245*
Arnek, R., *38*, 46(60), *54, 61*
Arnett, E. M., 18(25), *25*, 29(36), *61*, 64(1, 7, 9), 77, 79, 80, 82, 84, *86*, 87, 89(7), 91, *94*, 96, 99(1), 100, 101, *108*, 111(9), 112, 114, *116, 117*
Arvia, A. J., *30*
Aten, A. C., 124(20, 21), *141*
Audrieth, L. F., 181(13, 14), 194(14), 195 (14), 199, 232, 240(13), 242, *245, 249*, 252, 254(28), 263(1), *272, 273*
Aue, D. H., 105, 106, *108, 119*

B

Baca, G., 271(117), *275*
Bacarella, A. L., 164(59), *177*
Back, K. C., 254, 271, *273*
Badawi, F. A. K., 210(157), *248*

277

U

Ueoka, S., 225(180), *249*
Ulich, H., 254(47), 255(47), 271(118), *273, 275*

V

Van Allan, J. A., 169(77), *177*
van Audenhaege, A., *93, 95*
van der Heijde, H. B., 191(102, 104), 211 (159), 213(159), 221(159), 223(159), *247, 248*
Vanderwerf, C. A., 186(71), 243(71), *246*
Van Der Werf, C. A., 242(224), 243(224, 227), *250*
Vanderzee, C. E., *35, 42, 52*
Vanier, N. R., 29(38), *61*
Van Raalte, D., 103(93, 95), *118*
Varga, K., 181(22), *245*
Varsányi, M., 161(31), *176*
Vaughn, J. W., 8(8), *60*
Verastegin, J., 72(34), 73(34), *117*
Vetchinkin, S. I., 103(92), *118*
Vetter, K. J., 150(18), *176*
Vieland, L. J., 260(99), 261(99), 262(99), 264 (99), 269(103), *274*
Vielstich, W., 258, *274*
Villa-Trevino, S., 254(37), *273*
Vitvitskaya, G. V., 258, *274*
Vogel, G. C., 91(76), *118*
Vogel, W., *30*
von Hevesy, G., 184, *246*
Vrestal, J., 181(15), 189(15), 190(15), 191(15), 206, 232, 233(15), 240(214), *245, 249*
Vystrčil, A., 130(66, 67), *142*

W

Waentig, P., 241, *249*
Wagenknecht, J. H., 124(13), 128(13), *141*
Waghorne, W. E., 12(11), *60*, 69(20), 73(20), 75, 76(20), *116*, 207(151), 208(151), 236 (200), *248, 249*
Wagner, H. G., 256(63), *273*
Wait, E., *17*
Walden, P., 181(14), 194(14), 195(14), 199, *245*, 259(91), 260, 261(91), *274*
Walker, J. L., Jr., 148(8), 155(8), *176*
Walton, M. S., 192(113), *247*

Walz, H., 181(16), 241(16), *245*
Wang, I., *86*
Warburg, E., 164(63), *177*
Watson, P., 70, *116*
Watts, D. W., 67(17), 75, *116*
Wawzonek, S., 123(10), 124(13, 15, 18, 19, 28, 31, 32), 128(13, 63), *141, 142*
Wayland, B., *93, 99*
Wearring, D., 124(19), *141*
Webb, H. M., 106(113, 114), *108, 119*
Weinbernred, N. L., 139(101), *143*
Weiss, F., 253(4), *272*
Wel, L. Y., 104(100), 105(100), *118*
Wells, A. F., 49(72), *62*
Welsh, T. W. B., 259(92), 260, *274*
Wendt, H., 72(33), *117*
West, R., *93*
Weston, R. E., *86*
Whang, K. J., 175(115), *178*
Whatley, L. S., *93*
Whisman, M. L., 123(5), *140*
Wibaut, J. P., 183, 243(228), *245, 250*
Wieneke, A. A., 162(40, 41), *177*
Wilchek, M., 191(106), 217(106), 219(106), *247*
Wilcox, O. W., 254(44), 255(44), *273*
Wilcox, F., Sr., 166, 168, *177*
Wiley, G. R., 90(72), 96, 99(72), 100(72), *118*
Wilkinson, G., 256(65), 257(65), *273*
Willeboordse, F., 236, *249*
Williams, A. A., 56(80), 57(85), *58, 62*
Williams, R. J. P., 45, *61*
Williams, T., *44*
Wilski, H., 181(12, 17), 182(17), 183, 192(12), 199(12, 17), 212(12), 216, 243, 244(17), *245*
Wilson, H. N., 181(8), *245*
Wintersteiner, O., 184(45), 232(45), *246*
Wisotsky, M., *86*
Wolf, J. F., 84(54), *86*, 87(54), *117*
Wood, J. L., 242(225), *250*
Woodgate, S. D., 105(107), *108, 118*
Wooley, E. M., 67(19), *116*
Worley, S. D., *273*
Wright, E. H. M., 181(24), *245*
Wright, G., 57(87), *62*
Wu, C., *86*
Wu, Y.-C., 18(24), *25, 61*
Wuhrmann, H. R., 152, 154(23), 156(20, 23), *176*

Subject Index